W9-AFB-963

ANNUAL REVIEW OF NURSING RESEARCH

VOLUME 29, 2011

SERIES EDITOR

Christine E. Kasper, PhD, RN, FAAN
Department of Veterans Affairs
Office of Nursing Services, Washington, DC
and
Professor, Graduate School of Nursing
Uniformed Services University of the Health Sciences,
Washington, DC

VOLUME EDITORS

Ginette A. Pepper, PhD, RN, FAAN
Professor and Helen Bamberger
Colby Presidential Endowed Chair in
Gerontological Nursing
Associate Dean for Research and
PhD Program
Director, Hartford Center of
Geriatric Nursing
University of Utah, College of Nursing,
Salt Lake City, UT

Kenneth Wysocki, PhD, RN, FNP, FAANP
Genomics Fellow
University of Pittsburgh
School of Nursing
Pittsburgh, PA

Annual Review of Nursing Research

Genetics

VOLUME 29, 2011

Series Editor

CHRISTINE E. KASPER, PhD, RN, FAAN

Volume Editors

GINETTE A. PEPPER, PhD, RN, FAAN
KENNETH WYSOCKI, PhD, RN, FNP, FAANP

Springer Publishing Company, LLC
11 West 42nd Street
New York, NY 10036
www.springerpub.com

Acquisitions Editor: Allan Graubard
Composition: Absolute Service, Inc.

ISBN: 978-0-8261-5754-6
E-book ISBN: 978-0-8261-5755-3
ISSN: 0739-6686
Online ISSN: 1944-4028

12 13 14/ 5 4 3 2 1

Printed in the United States of America by Hamilton Printing.

Contents

About the Volume Editors

Ginette A. Pepper, PhD, RN, FAAN, is the Helen Lowe Bamberger Presidential Endowed Chair in Gerontological Nursing and Associate Dean for Research and the PhD Program at the University of Utah College of Nursing in Salt Lake City, Utah. Her research relates to adverse drug effects on the functional status of older adults. As the nation's first Geriatric Nurse Practitioner, she pioneered prescriptive authority for advanced practice nurses. A nurse pharmacologist who received her PhD in nursing and pharmacology from the University of Colorado, she currently is co-investigator and consultant on studies using eccentric exercise in fall prevention in frail elders, two meta-analysis studies of psychotropic drugs and falls and medication adherence, and the role of interprofessional communication in preventing adverse drug events in hospitalized older adults. Ginny directs the University of Utah Hartford Center of Geriatric Nursing Excellence and is active in the University of Utah Interdiciplinary Personalized Medicine Initiative.

Kenneth J. Wysocki, PhD, RN, FNP-BC, FAANP, is currently a Genetics Fellow at the University of Pittsburgh School of Nursing working on genomic asthma research with the Asthma Institute at the University of Pittsburgh Medical Center, and the Center for Human Genomics and Personalized Medicine at Wake Forest School of Medicine. As a family nurse practitioner since 1995, he has served as a clinician and educator in Arizona and New Zealand. Committed to health policy excellence and NP advancement, he has served in numerous state, national, and international organizations, lobbied health care leaders, and coordinated NP conferences in Arizona and New Zealand. He is a recipient of the 2002 AANP Arizona state award of excellence, the 2007 National Institute of Nursing Research Summer Genetics Fellowship, and a 2009 NINR National Research Service Award Fellowship. He served as review editor for *Smart Ideas* and currently is a review editor for *Health Monitor*.

Contributors

Rebecca Anderson, BSN, RN
PhD Candidate, University of Utah College of Nursing
Assistant Director, Genetic Science in Society (GeneSIS Center)
University of Utah, Salt Lake City, UT

Kelley Baumgartel, BSN, RN
Doctoral Student, University of Pittsburgh, Pittsburgh, PA

Mandy Bell, BSN, RN
Doctoral Student, University of Pittsburgh, Pittsburgh, PA

Jeffrey R. Botkin, MD, MPH
Professor, Department of Pediatrics
Director, Genetic Science in Society (GeneSIS Center)
Associate Vice President for Research
University of Utah, Salt Lake City, UT

Kathleen A. Calzone, PhD, RN, APNG, FAAN
Senior Nurse Specialist, Research, National Institutes of Health, National Cancer Institute, Center for Cancer Research, Genetics Branch, Bethesda, MD

Yvette P. Conley, PhD
Associate Professor of Nursing and Human Genetics, University of Pittsburgh, Pittsburgh, PA

Susan G. Dorsey, PhD, RN, FAAN
Associate Professor, University of Maryland School of Nursing, Department of Organizational Systems & Adult Health, Baltimore, MD

Martha Driessnack, PhD, PNP-BC, RN
Assistant Professor, College of Nursing, The University of Iowa, Iowa City, IA

Matthew J Gallek, PhD, RN
Assistant Professor, College of Nursing, The University of Arizona, Tucson, AZ

Agatha M. Gallo, PhD, APN/CPNP, RN, FAAN
Professor, College of Nursing, The University of Illinois at Chicago, Chicago, IL

Karen E. Greco, PhD, RN, ANP-BC, FAAN
Nurse Specialist, Research, (Contractor), National Institutes of Health, National Cancer Institute, Center for Cancer Research, Genetics Branch, Bethesda

Margaret M. Heitkemper, PhD, RN
Professor, Department of Biobehavioral Nursing and Health Systems, University of Washington, Seattle, WA

Monica E. Jarrett, PhD, RN
Professor, Department of Biobehavioral Nursing and Health Systems, University of Washington, Seattle, WA

Jean Jenkins, PhD, RN, FAAN
Senior Clinical Advisor, National Institutes of Health, National Human Genome Research Institute, Bethesda, MD

Sang-Eun Jun, PhD, RN
Lecturer, College of Nursing, Keimyung University, Daegu, South Korea

Christine E. Kasper, PhD, RN, FAAN
Department of Veterans Affairs, Office of Nursing Services, Washington, DC
Professor, School of Nursing, Uniformed Services University of the Health Sciences, Washington, DC

Ruth Kohen, MD
Assistant Professor, Department of Psychiatry & Behavioral Sciences, University of Washington, Seattle, WA

Emma L. Kurnat-Thoma, PhD, MS, RN
Director, Research Services, URAC, Washington, DC

Gwen Latendresse, PhD, CNM
Assistant Professor, College of Nursing, University of Utah, Salt Lake City, UT

Pei-Chao Lin, MSN, RN
Instructor, Department of Nursing, Yun-Ing Junior College of Health Care & Management, Kaohsiung City, Taiwan
College of Nursing, Kaohsiung Medical University, Kaoshiung, Taiwan

Suzanne M. Mahon, DNSc, RN
Division of Hematology/Oncology, Department of Internal Medicine, Saint Louis, University, St. Louis, MO

Ginette A. Pepper, PhD, RN, FAAN
Professor and Helen Bamberger Colby Presidential Endowed Chair in Gerontological Nursing
Associate Dean for Research and PhD Program,
Director, Hartford Center of Geriatric Nursing
University of Utah, College of Nursing, Salt Lake City, UT

Cynthia L. Renn, PhD, RN
Assistant Professor, University of Maryland School of Nursing, Department of Organizational Systems & Adult Health, Baltimore, MD

Leslie Ritter, RN, PhD, FAAN
Professor and the William M. Feinberg, MD, Endowed Chair in Stroke Research, College of Nursing, University of Arizona, Tucson, AZ

Erin Rothwell, PhD, CTRS
Assistant Professor (Research), College of Nursing
University of Utah, Salt Lake City, UT

Tonya A. Schneidereith, PhD, CRNP, PNP-BC, CPNP-AC
Assistant Professor of Nursing, Stevenson University, Stevenson, MD

Diane Seibert, PhD, CRNP, FAANP
Associate Professor, Graduate School of Nursing, Program Director, Family Nurse Practitioner Program, Uniformed Services University of the Health Sciences

Chantel Snyder, BSN, RN
Doctoral Student, University of Pittsburgh, Pittsburgh, PA

Kathleen J. H. Sparbel, PhD, FNP-BC
Clinical Assistant Professor, University of Illinois at Chicago, College of Nursing, Chicago, IL

Theresa Timcheck, BSN, RN
Doctoral Student, University of Pittsburgh, Pittsburgh, PA

Audrey Tluczek, PhD, RN
Associate Professor, University of Wisconsin-Madison, School of Nursing, Madison, WI

Susan Tinley, PhD, RN, CGC
Associate Professor of Nursing, Creighton University, Omaha, NE

Patricia C. Underwood, PhD, RN, FNP-BC
Brigham and Women's Hospital, Harvard Medical School, Boston, MA

Shu-Fen Wung, PhD, RN, ACNP, FAAN
Associate Professor, College of Nursing, The University of Arizona, Tucson, AZ

Kenneth Wysocki, PhD, RN, FNP-BC, FAANP
Genomics Fellow, University of Pittsburgh School of Nursing, Pittsburgh, PA

Jamie Zelazny, MPH, RN
Doctoral Student, University of Pittsburgh, Pittsburgh, PA

Preface

This 29th volume in the *Annual Review of Nursing Research* (ARNR) series delves into the continually expanding area of genetics and genomics. Nursing as a profession is changing the landscape of health care and staying cutting edge with genomic research, dissemination, and counseling. Drs. Ginette Pepper and Kenneth Wysocki, well-known scholars and researchers in the field of genetics, have served as the volume editors. Content for the chapters was carefully chosen and edited them into this review of nursing research in genetics and genomics. As series editor, it is my hope that these topically based chapters will be used not only by those conducting research studies, but also as texts and supplements to nursing curricula for both the undergraduate and graduate students.

While it is impossible to provide a detailed and comprehensive review of the vast outpouring of genetics research related to nursing practice and research, this issue thoughtfully and clearly presents those areas of study, which lend themselves for translation to clinical setting or expansion into other related areas of research. We were mindful that this volume would only be able to sample a small portion of nurses engaged in genomics research, education and policy development around the world. Thus, we were able to capture a genomic spectrum of what nurses are engaged to date and contribute to the depth and breath of genomics in nursing.

This volume is comprised of four sections. The current status of genomic molecular science is the focus of Part I and includes four chapters. Yvette Conley and her team address genomic research designs in Chapter 1. In Chapter 2, Emma Kurnat-Thoma reviews the use of genetic information to create personalized medicine for individual patients. Kenneth Wysocki and Leslie Ritter discuss what is known as the "diseasome" in Chapter 3, which refers to all the disorders and diseases of an organism, viewed as a whole, with special reference to genetic features. Susan G. Dorsey and Cynthia Renn examine the use of various animal models in genetics research in Chapter 4.

In Part II, the focus is on ethical, legal, and social issues in genomics led by an in-depth review by Erin Rothwell, Rebecca Anderson, and Jeff Botkin on newborn screening in Chapter 5. Karen Greco and Suzanne Mahon tackle the emerging issue of direct marketing to consumers for cancer testing and its legal and ethical implications in Chapter 6. In Chapter 7, Martha Driessnack and Agatha Gallo review the ethical issues to be considered in genetic testing in child and adolescent diseases. In Chapter 8, Kathleen Calzone and Jean Jenkins tackle building the nursing capacity in genetics and genomics by structuring competencies for all nurses. This is followed by a second discussion of genetic competencies, this time at the graduate level, by Karen Greco, Susan Tinley, and Diane Seibert in Chapter 9.

Part III includes six chapters, which explore the genetics of diseases and symptoms. Christine E. Kasper will discuss some of the important research involved in maintenance of strength and mobility in relation to striated muscle in Chapter 10. Matthew Gallack and Leslie Ritter explore central nervous system genetic disorders in Chapter 11. In Chapter 12, Shu-Fen Wung and Pei-Chao Lin discuss cardiac disorders resulting from genetic factors. The distinguished scholar, Margaret Heitkemper, with Ruth Kohen, Sang-Eun Jun, and Monica E. Jarrett, present the important and emerging area of gastrointestinal genomics in Chapter 13. Patricia Underwood reviews the genomic implications of diabetes to nursing and personalized care in Chapter 14. The final chapter in this section by Kathleen Sparbel and Audrey Tluczek reviews the research involved in cystic fibrosis genetics.

Part IV discusses genomics across the lifespan. In Chapter 16, Gwen Latendresse presents the genomics of pregnancy. And in the final chapter, Tonya Schneidereith discusses pediatric aspects of sickle cell anemia. Together, these thoughtful scholarly discussions highlight what is known, what needs to be studied, and areas for future action. It is expected that this issue will lead the way for setting the foundation for the important research in the many and varied areas of nursing research which incorporate genetics and which needs to be accomplished in the near future. It will also guide the way for students who are drawn to establish their research careers in these areas of study.

As one moves through the various topic areas of genetics research in nursing it is advised to carefully consider the depth of knowledge of genetic science required to apply rapidly emerging findings to the patient population at the various levels of nursing practice, as well as the knowledge required to conduct useful and novel research. In 1997, the National Institute of Nursing Research (NINR) began the national discussion of the need for nursing research in genetics. The various genetics organizations, such as the International Society of Nursing in Genetics (ISONG), the National Coalition of Health Professional Education in Genetics (NCHPEG), and the National Health Service (NHS) of the United Kingdom, are to be commended for their diligence in constructing competencies for both undergraduate and graduate students of nursing. By clarifying areas of knowledge necessary to practice, they have provided a place in academic nursing curricula for genetics. However, clinical knowledge of the science of genetics is insufficient to engage in nursing research in the scientific realm of genetics. It is hoped that PhD students will be nurtured in genetics research where there are strong faculty with programs of study in the field; while, DNP students incorporate genetics in their advanced clinical practice and build the evidence of genetic practice. Interdisciplinary training with genetics departments or research clusters can be invaluable to support the work of the student or faculty member. Successful examples of these collaborations can be found currently at the Universities of Utah and Pittsburgh. In addition, the NINR Graduate Partnership Program (GPP) at the National Institutes of Health (NIH) and the NINR Summer Genetics Institute (SGI) program have also been an excellent source of training for nurse scientists in genetics.

Whole genome sequencing (WGS) to personalize treatment is operational, albeit on a small scale. As the cost to determine WGS for a patient has exponentially decreased, genomic based medicine and health care have been implemented in various large hospital systems and particularly in Oncology. I had the privilege to attend the 2011 meeting on "Personal Genomes" at Cold Spring Harbor Meetings & Courses WGS reveals that each patient has his own disease. Observable signs and symptoms form the basis of our current classification system and these are no more than clusters of information that guide us in our attempt to remediate loss of function and pain. Actual detailed sequencing of similar disease classifications from patients reveal that each has a unique aberrant sequence. And, there is growing knowledge of imprinting and epigenetics that play a role in gene expression, which impacts future generations with potential new targets for pharmacogenomics therapies. Clearly, we do not understand evolutionary biology at all (Childs, Wiener, & Valle, 2005). It can be argued that genetic dysfunction is the normal state as there are evolutionary loss-of-function variants in "healthy" human genomes. So, from a nursing point of view, the question should be, "why does this patient have this disease at this time?" Nursing has always viewed each individual patient as a unique entity; now advances in genetic technologies support this view.

Christine E. Kasper, PhD, RN, FAAN
Series Editor

REFERENCE

Childs, B., Wiener, C., & Valle, D. (2005). A science of the individual: Implications for a medical school curriculum [Review]. *Annual Review of Genomics and Human Genetics, 6,* 313–330. doi: 10.1146/annurev.genom.6.080604.162345

STATE OF GENOMIC MOLECULAR SCIENCE

CHAPTER 1

Molecular Genomic Research Designs

Kelley Baumgartel, Jamie Zelazny, Theresa Timcheck, Chantel Snyder, Mandy Bell, and Yvette P. Conley

ABSTRACT

Genetic and genomic research approaches have the capability to expand our understanding of the complex pathophysiology of disease susceptibility, susceptibility to complications related to disease, trajectory of recovery from acquired injuries and infections, patient response to interventions and therapeutics, as well as informing diagnoses and prognoses. Nurse scientists are actively involved in all of these fields of inquiry, and the goal of this chapter is to assist with incorporation of genetic and genomic trajectories into their research and facilitate the design and execution of these studies. New studies that are going to embark on recruitment, phenotyping, and sample collection will benefit from forethought about research design to ensure that it addresses the research questions or hypotheses being tested. Studies that will use existing data or samples will also benefit from forethought about research design for the same reason but will also address the fact that some designs may not be feasible with the available data or samples. This chapter discusses candidate gene association, genome-wide association, candidate gene expression, global gene expression, and epigenetic/epigenomic study designs. Information provided includes rationale for selecting an appropriate study design, important methodology considerations for each design, key technologies available to accomplish each type of study, and online resources available to assist in executing each type of study design.

http://dx.doi.org/10.1891/0739-6686.29.1

In the last decade, we have progressed from a rough draft of the human genome sequence to availability of an abundance of publicly available databases and high throughput data collection technologies to facilitate genetic and genomic study design. Genetic (focus on one gene at a time) and genomic (focus on entire genome as well as gene–gene interactions) research continues to hold great promise for understanding a wealth of human conditions, providing objective data for diagnosis and prognosis, informing therapeutics, and providing the cornerstone for evidence based practice for genomic health care (Green, Guyer, & National Human Genome Research Institute [NHGRI], 2011; Lander, 2011). The research programs of many nurse scientists are ripe for incorporating a genetic/genomic research component or movement of existing genetic or genomic research in a new direction.

The goal of this article is to bring together key information about designing studies with a molecular genetic or genomic focus coupled with dynamic resources offered to the reader to expand their understanding and ensure access to state of the science information. It is not meant to be an exhaustive resource, but one that sets the stage for contemplation of embarking on such research designs and key issues to ponder during study design phase. This article is written for the researcher who has a basic understanding of genetics and is contemplating adding a genetic or genomic component to their research or designing the next step in their genetic or genomic program of research. Readers are encouraged to visit an extremely useful resource, the National Human Genome Research Institute's talking glossary at http://www.genome.gov/glossary/ for clarification of unfamiliar terms and expansion of knowledge about genetic terminology. Technology to collect genetic and genomic data changes rapidly, therefore proper study design and selection of appropriate methodology to accomplish a study also change rapidly. This article incorporates a large number of online resources that are continuously updated in an attempt to keep the article as up to date as possible. Readers are encouraged to visit these online resources when designing their study to ensure that their study design is state of the science.

DNA POLYMORPHISM-BASED ASSOCIATION STUDIES

The overall objective of a polymorphism-based association study is to examine the relationship between DNA variation and a phenotype (e.g., diabetes, fatigue). A *polymorphism* is defined as a DNA variation that is present in at least one percent of the population (NHGRI, n.d.b). One advantage of this approach compared to other genetic/genomic approaches is the use of DNA. DNA is a very stable template for experiments, allowing for use of previously collected samples. Such a retrospective approach could save time and money that would be needed to prospectively recruit participants and collect samples; however, attention must be given to subject consent to assure that informed consent was obtained for future genetic/genomic evaluation related to the phenotype of interest. Another advantage is that this approach does not require that subjects be related, which is a requirement for linkage analysis—an approach not discussed in this article. It should be noted that although related individuals are not required, newer software has been developed to allow for the analyses of related individuals within the context of an association study. Two very appealing additional advantages of polymorphism-based studies are the fact that polymorphisms do not change over time, and the DNA template that is used can be extracted from any tissue. The sample for DNA extraction and collection of polymorphism data only need to be collected once, yet that polymorphism data can be evaluated within the context of a phenotype that changes over time. Although blood and saliva are the most frequently used cell/tissue type for DNA extraction, any cells/tissues that have a nucleus can serve as samples for polymorphism-based studies. Because DNA polymorphisms do not change and are not tissue specific, investigators need not worry about collection of DNA samples over time or from what tissue DNA extraction occurs. These advantages are not carried over to other genomic approaches detailed in this chapter.

Candidate Gene Association Studies

Rationale for Taking a Candidate Gene Association Approach

Candidate gene association studies investigate polymorphisms representing a specific gene(s) to determine if it is associated with a phenotype of interest. With this hypothesis-driven approach, the investigator preselects the candidate gene(s) to be evaluated. This approach is only appropriate when a priori assumptions about the gene(s) that may be involved in the phenotype of interest can be justified.

Genome-wide association studies (GWAS), discussed in the next section, have large sample size requirements (e.g., 1,000 cases/1,000 controls), and one relative advantage of the candidate gene approach is that it often requires half that number or less. This reduced sample size requirement compared to a GWAS is because of the focused evaluation of a candidate gene(s), which reduces multiple testing concerns. The candidate gene association approach is also ideal when studying rarer phenotypes because attainment of a large sample may not be feasible for a condition with a low population frequency.

Subject and Sample Considerations

Clearly defined inclusion/exclusion criteria, which include a detailed definition of the phenotype, are essential to the candidate gene association approach. Structured inclusion/exclusion criteria help to ensure that individuals with/without the phenotype of interest are similar in all aspects except for the condition being investigated. Moreover, phenotypic assessment of controls should be as comprehensive as the phenotypic assessment of cases. Ultimately, carefully crafted criteria and thorough phenotypic assessments help reduce the impact of confounding variables.

Population stratification represents another potential source of confounding in candidate gene association studies using a case-control design. The case-control design compares allele, genotype, or haplotype frequencies between the groups. Because these frequencies can be extremely disparate for different ancestries, it is important to control for ancestry to avoid spurious results/conclusions (e.g., concluding that there is an association between a phenotype/allele when in reality the association is fueled by ancestral differences in allelic frequencies). The risk for population stratification can be mitigated. Subgroup analysis represents one option, but it relies on self-report to categorically measure race/ethnicity. An option that controls for population stratification statistically is the use of ancestral informative markers (AIM), which are polymorphisms in the DNA that allow one to calculate an admixture proportion for an individual. The application of these proportions are used for analysis rather than the traditionally used, though unreliable, method of self-reported race/ethnicity. In a recent study, only 30 AIMs were needed to estimate European admixture in a group of African American women (Ruiz-Narváez, Rosenberg, Wise, Reich, & Palmer, 2011). Although different AIMs may be needed to estimate other admixture proportions, this example demonstrates that population stratification can be successfully controlled through the analysis of genetic markers.

Another aspect of the candidate gene association study that should be considered is sample size requirements. Quanto (http://hydra.usc.edu/gxe/) is a freely downloadable computer program that can assist with sample size and/or power calculations for candidate gene association studies. User defined criteria can be manipulated according to the polymorphisms that have been selected for evaluation and according to study design specifications.

Candidate Gene Selection

Candidate gene selection is often based on biologic plausibility. This plausibility can be based on biological pathways implicated in the condition, biomarker data implicating a gene–gene product in the phenotype of interest, pharmacologic treatments for the condition that may indicate a target gene(s), or data from animal models (Hattersley & McCarthy, 2005). Bio-informatics databases, such as the Gene Ontology (http://www.geneontology.org/), may also aid in the identification of genes whose products may impact the phenotype of interest (The Gene Ontology, 1999–2011). Moreover, consideration should be given to several genes on which to focus, ranging from a single gene to genes within a candidate biological pathway. Because more biologically global conclusions can be drawn, the study of a biologic pathway has the advantage of being more informative than the singular gene approach in most situations (Jorgensen et al., 2009).

Polymorphism Selection

Once selection of the candidate gene(s) is finalized, polymorphisms must be selected to evaluate candidate gene variability, and these are the genetic data used for analyses. The candidate gene association approach includes the evaluation of single nucleotide polymorphisms (SNPs), repeat polymorphisms, insertion/deletion polymorphisms (INDEL), and copy number variants (CNV).

Resources for Polymorphism Selection. The SNP is the most common type of polymorphism and is a nucleotide (also known as a base) in the DNA where the nucleotide present (e.g., A, T, C, G) varies in the population (Genetics Home Reference, 2011). The scientific literature and various online databases provide excellent resources for SNP identification and selection (See Table 1.1). A simple literature search combining the candidate gene(s) with the keyword "functional polymorphism" will help to identify SNPs known to alter the function of the candidate gene(s). Because functional polymorphisms modify the function of a gene regardless of phenotype, the literature search should not be limited to just the phenotype of interest. In addition to the literature, investigators also commonly use the database of Single Nucleotide Polymorphisms (dbSNP) (http://www.ncbi.nlm.nih.gov/projects/SNP/) and the International HapMap Project (http://hapmap.ncbi.nlm.nih.gov/) to identify/select SNPs and tagging SNPs, respectively.

HapMap is accessed for the selection of tagging SNPs (tSNP), which represent the current gold standard for the evaluation of genetic variation in the candidate gene association study. The goal of HapMap is to develop a haplotype map of the human genome and to describe common patterns of genetic variation in humans (International HapMap Project, 2006). Essentially, HapMap is based on the premise that DNA is inherited in chunks/blocks (haploblock). Within these haploblocks, certain variants are inherited together. If the genotype of one variant within that block of DNA is known, the genotype of a second variant within the same block can be determined because they are inherited together. Thus, HapMap assists the user in selecting SNPs that tag a certain haploblock of DNA (tagging SNPs or tSNPs). Ultimately, use of tSNPs allows one to fully evaluate the genetic variability of the candidate genes with the least number of SNPs (International HapMap Project, n.d.).

Repeat polymorphisms are characterized by repeating units of DNA bases. The number of times these DNA units repeat is variable in the population (Passarge, 2007). Although repeat polymorphisms are less frequent in the

TABLE 1.1

Online Genome Databases and Resources

Name and Address	Description
Database of Short Genetic Variations (also known as SNP database) http://www.ncbi.nlm.nih.gov/snp/?term	This database houses documented SNPs, microsatellites, and small-scale INDELs. It provides population-specific allele frequencies; genotype data, genome location, and information on function (e.g., change in an amino acid).
International HapMap Project http://hapmap.ncbi.nlm.nih.gov/	This database is used to identify and select tagging SNPs. User-defined criteria under the configure tab include population selection, R^2 cutoff values, and mean allele frequency cutoff. SNPs identified in the literature or dbSNP can also be included in the tagger SNP configuration.
Database of Genomic Structural Variation http://www.ncbi.nlm.nih.gov/dbvar/	This database houses information on documented structural variants including CNVs. User-defined limits include criteria such as study design, method type (e.g., SNP genotyping, FISH), project ID, and variant type.
The Copy Number Variation (CNV) Project http://www.sanger.ac.uk/research/areas/humangenetics/cnv/	This database provides CNV data from two projects (Global CNV assessment and High-resolution CNV discovery).
Genetics Home Reference http://ghr.nlm.nih.gov/	This website by the National Library of Medicine contains information concerning genetic conditions, genes, and chromosomes.
Talking Glossary of Genetic Terms http://www.genome.gov/glossary/index.cfm	This glossary provides definitions, illustrations, and animations of commonly used genetic/genomic terms.
The Gene Ontology Project http://www.geneontology.org/	This database can be used to identify genes whose products may impact a phenotype of interest. The domains covered include cellular component, molecular function, and biological process.

TABLE 1.1
Online Genome Databases and Resources (Continued)

Name and Address	Description
Catalog of Published Genome-Wide Association Studies http://www.genome.gov/gwastudies/	A database containing all published GWA studies attempting to genotype at least 100,000 SNPs in the initial stage.
Genome-Wide Association Studies Data Repository http://gwas.nih.gov/02dr2.html	Website for the NIH Genome-Wide Association Study Portal
The Genes, Environment, and Health Initiative http://www.genesandenvironment.nih.gov	Website for Genes, Environment and Health Initiative (GEI)
Database of Genotypes and Phenotypes http://www.ncbi.nlm.nih.gov/sites/entrez?db=gap	Database containing results of studies investigating genotype–phenotype interaction. Currently houses NIH GWAS repository.
Center for Inherited Disease Research http://www.cidr.jhmi.edu/	Provides genotyping and statistical genetic services to investigators approved for access through competitive peer review process
Understanding the Basics of Microarrays http://www.ncbi.nlm.nih.gov/About/primer/microarrays.html	This publication from the National Center for Biotechnology Information (NCBI) provides an overview of DNA microarrays explaining gene expression, the technology underlying microarrays, the purpose and importance of microarrays, and the basics of microarray experiments.
Gene Expression Omnibus http://www.ncbi.nlm.nih.gov/geo/	GEO: The Gene Expression Omnibus. GEO serves as public repository and online resource for storage and retrieval of gene expression data. GEO currently maintains microarray and serial analysis of gene expression (SAGE) data on over 100 organisms
European Bioinformatics Institute http://www.ebi.ac.uk/	The European Bioinformatics Institute (EBI) is a nonprofit organization that focuses on research and services in bioinformatics. EBI's website enables access to gene expression databases (Array Express Archive and Gene Expression Atlas) and microarray analysis tools (Expression Profiler, Next Generation and Bioconductor).

(*Continued*)

TABLE 1.1
Online Genome Databases and Resources (*Continued*)

Name and Address	Description
Serial Analysis of Gene Expression Portal http://www.sagenet.org	Sagenet provides a detailed description of serial analysis of gene expression (SAGE). This website also provides SAGE applications, publications, and resources.
Histone Database http://www.research.nhgri.nih.gov/histones/	NHGRI histone database. Histone sequence information, including posttranslational modifications
Antibody Validation Database http://compbio.med.harvard.edu/antibodies/about	Collect and to share experimental results on antibodies that would otherwise remain in individual laboratories, thus aiding researchers in selection and validation of antibodies.
Chromatin Structure and Function http://www.chromatin.us/chrom.html	Information on chromatin biology, histones, and epigenetics (hosted by Jim Bone)
Database for DNA Methylation and Environmental Epigenetic Effects http://www.methdb.de/	Human DNA methylation Database. DNA methylation data readily available to public. Future development includes environmental impact on methylation
CpG Island Searcher http://www.uscnorris.com/cpgislands2/cpg.aspx	CpG island searcher. CpG Island sequence search algorithm allows for selection of percentage methylation and length of (ISLAND?) and gaps between islands.
Catalogue of Parent of Origin Effects http://igc.otago.ac.nz/home.html	Imprinted Gene Catalogue. Catalogue of parent of origin effects. Can search by taxon, chromosome, gene name, or key word
Database of Noncoding RNAs http://www.noncode.org/NONCODERv3/	Knowledge database dedicated to ncRNA—information on class, name, location, related publications, mechanism through which it exerts its function. Includes all traditional ncRNAs, but excludes tRNAs and rRNAs.
MicroRNA Database http://www.mirbase.org/	MicroRNA data resource Searchable database of > 16,000 published miRNA sequences and annotation; includes location and sequence of mature miRNA. Can search by name, keyword, reference, and/or annotation.

TABLE 1.1
Online Genome Databases and Resources (Continued)

Name and Address	Description
The Epigenome Network of Excellence http://www.epigenome-noe.net/WWW/index.php	Epigenome Network of Excellence. Website of European interdisciplinary epigenetics research network; includes protocols, an antibody database, and reference information on epigenetics.
Human Epigenome Project http://www.epigenome.org/	The Human Epigenome Project Research Consortium. Collaborative effort to catalogue and interpret genome-wide methylation patterns of all human genes and major tissues

genome than SNPs, they are often more informative as they usually have more alleles in the population than SNPs, which typically only have two. The short tandem repeat (STR) is typically composed of a repeating unit of two to four DNA bases (e.g., CAG CAG CAG), whereas the variable number tandem repeat (VNTR) is composed of a larger repeating unit (Passarge, 2007) usually greater than five bases. For the evaluation of STRs and VNTRs, the literature continues to be the best source for identification and characterization.

An INDEL polymorphism occurs when a base(s) is added or subtracted from a place in the DNA. It is the presence or absence of the INDEL that is variable in the population (Nussbaum, McInnes, & Huntington, 2007). Like SNPs, the dbSNP can be freely accessed to identify small-scale INDELs.

The CNV occurs when the number of copies of a particular genomic sequence/segment is variable in the population (NHGRI, n.d.a). CNVs can be identified through scientific literature and online databases. The Database of Genomic Structural Variation (dbVar) (http://www.ncbi.nlm.nih.gov/dbvar) and The Copy Number Variation Project by the Wellcome Trust Sanger Institute (http://www.sanger.ac.uk/humgen/cnv/) are two online resources that may assist in CNV identification.

Genotype Data Collection Technologies

Multiple options are available for SNP genotyping, including the polymerase chain reaction-restriction fragment length polymorphism (PCR-RFLP) technique, real-time PCR allelic discrimination (e.g., TaqMan), multiplexing via mass spectrometry, and bead chip technology. Selection of the genotyping technique is guided by the number of samples and polymorphisms to be genotyped and available resources. PCR-RFLP, which is used to genotype SNPs based on differences in fragment lengths, is suitable when the number of SNPs and samples to be genotyped is relatively small. Real-time PCR allelic discrimination (http://www.appliedbiosystems.com; http://www.roche-applied-science.com), which genotypes SNPs based on allele-specific fluorescence intensity signals, is suitable for a medium number of SNPs and sample size. Because PCR-RFLP and real-time PCR allelic discrimination can only genotype one SNP at a time, the use of high throughput technologies have become the gold standard for SNP genotype collection when the number of SNPs to be evaluated approaches 24. The iPLEX Gold-SNP Genotyping assay (http://www.sequenom.com), which genotypes SNPs based on differences in molecular mass, allows for the analysis of up to 36 SNPs per assay (Sequenom, 2010) in larger sample sizes. Not only can an investigator analyze multiple SNPs simultaneously but time, assay to assay variability, and costs are reduced. The GoldenGate Genotyping Assay (http://www.illumina.com) is another high throughput bead-based technology that can be used when the number of SNPs and samples to be analyzed is too large for other technologies.

There are several genotyping technologies also available for repeat polymorphisms, INDELs, and CNVs. PCR amplification followed by fragment sizing can be used for genotyping repeat polymorphisms. As with SNPs, real-time PCR allelic discrimination can be used to genotype small INDELs. Finally, TaqMan Copy Number Assays (http://appliedbiosystems.com) or cytogenetic techniques (e.g., Fluorescence In Situ Hybridization) can be used for genotyping candidate CNVs.

Genome-Wide Association Studies

Rationale for Taking a GWAS Approach

A GWAS genotypes thousands to millions of polymorphisms across the genome for individuals who are phenotypically well-characterized (DiStefano & Taverna, 2011). If genetic variability is significantly different between cases and controls, those variations may be associated with susceptibility to or protection from the phenotype of interest and can provide direction as to which region of the genome these differences might be located. Ongoing efforts of the Human Genome Project and the International HapMap Project have made this approach possible through the generation of large databases that reference and map both sequence and variability.

The major advantage of a GWAS approach is that the biology of the phenotype of interest does not need to be completely understood prior to implementing this approach and the SNPs or genes of interest do not need to be defined a priori. Instead of selecting genes and polymorphisms a priori, polymorphisms that cover haploblocks across the entire genome are used for genotype data collection; and nonparametric-based analyses determine what genes/regions of the genome are relevant to the phenotype of interest (Hakonarson & Grant, 2011). The data derived from GWAS will provide direction regarding which areas of the genome warrant additional study.

There are several limitations to GWAS. The variant identified may not be what is accounting for the association, but is rather "tagging along" with the actual causal variant(s). This obstacle is also present for candidate gene association studies, particularly those that use a tSNP approach. Therefore, it may be necessary to follow up with more focused genotype data collection, including denser polymorphism evaluations and/or sequencing of that specific region of the genome to identify the exact allele accounting for the association (www.genome.gov/20019523). A major limitation for the GWAS approach and perhaps a reason why many investigators are unable to pursue this approach is the need for thousands of subjects who are phenotypically well characterized and for which DNA is available. The need for large samples sizes for GWAS is because of the inherent issue of multiple testing that accompanies the evaluation of thousands to millions of different genetic variables. Additionally, the need for very large sample sizes coupled with the cost of commercial genome-wide scanning techniques makes this approach very costly. GWAS approaches are also not optimal to assess rare polymorphisms as the data collection approaches for the GWAS are more focused on optimizing informativeness of the data (Ku, Loy, Pawitan, & Chia, 2010).

Subject and Sample Considerations

The cross-sectional case-control study design is the most frequently used approach for a GWAS. Study subjects should be selected based on a well-defined and heritable phenotype. Cases are defined as individuals who meet criteria for a phenotype of interest. Controls are individuals who have never met criteria for the phenotype and ideally have passed through the age or period of risk for the phenotype (Hakonarson & Grant, 2011). Like candidate gene associations studies, ancestry must be considered to avoid issues related to population substructure, and this is why some investigators have conducted these types of studies with homogeneous populations (Psychiatric GWAS Consortium Coordinating Committee, 2009). Case and control groups should be matched on ancestry as much as possible to avoid false-positives. Despite this consideration, an advantage of GWAS is that whole genome data can provide adequate data to identify stratification, and inflation of test statistics because of population substructure can be addressed (Hakonarson & Grant, 2011).

Obtaining a sufficiently large sample size is essential to ensure sufficient statistical power for a GWAS approach. Approximately 1,000 cases and a similar number of controls are required to detect 1–5 variants associated with a given trait. A larger sample is needed to uncover additional variants that may have diminishing contributions to the disease (Hakonarson & Grant, 2011).

Informed Consent Issues. Although informed consent is of paramount importance with any research study, researchers who are considering a GWAS should be cognizant of issues related to conducting such as study and the National Institutes of Health (NIH) policy on data sharing for GWAS. In January 2008, the NIH adjusted its policy mandating the sharing of GWAS data obtained in NIH-funded or conducted studies. The details of this policy can be found at http://gwas.nih.gov/. Most NIH-funded GWAS are required to include language in the consent document that addresses public sharing of de-identified genotype and phenotype data. Researchers who are planning to study existing samples must ensure that the original consent signed by the subjects is consistent with conducting a GWAS.

Genotype Data Collection Technologies

There are currently two commonly used vendors that provide technology for collection of GWAS data, Affymetrix and Illumina. The companies use different technological approaches, which are both widely used in the research community. The Affymetrix Genome-Wide SNP Array 6.0 features 1.8 million genetic markers, including 906,600 SNPs and more than 946,000 probes for the detection of CNVs. This platform also includes a high-resolution reference map and a copy number polymorphism (CNP) algorithm (see http://www.affymetrix.com for additional information). The Illumina Omni Microarrays provide a multiple BeadChip option that will soon include nearly 5 million markers per sample, including both common and rare variants identified by the 1,000 Genomes project. Omni microarrays assess structural variation, including CNVs and copy neutral variants (inversions and translocations) that may also be significant contributors to disease (see http://www.illumina.com for additional information).

Resources of Interest for GWAS. The Center for Inherited Disease Research (CIDR) at Johns Hopkins University (http://www.cidr.jhmi.edu/requirements/applications.html) is funded by NIH Institutes and provides genotyping and statistical genetic services to investigators who have received access after a competitive peer review process. Interested investigators are required to submit an application for projects supported by the NIH. In order to maximize access to resources, the application process to CIDR should ideally take place before or at the time of grant application, though this is not a requirement.

The repository for GWAS data is currently the database of Genotypes and Phenotypes (dbGaP; http://www.ncbi.nlm.nih.gov/entrez/query.fcgi?db=gap). This database was developed to archive the results of studies that have investigated the genotype–phenotype interaction and serves as a useful resource in reviewing the work that has already been completed and aids in planning future research. The dbGaP provides the opportunity for in silico research. Researchers have the option of two levels of access (open and closed) to dbGaP: Open-access data are aggregate data that are publicly available, whereas closed level access requires an application and approval process that includes de-identified subject specific data. The genotype data and their linked phenotype data are invaluable resources, and researchers are encouraged to investigate this database as it pertains to their phenotypes of interest prior to designing a study.

GENE EXPRESSION STUDIES

Gene expression studies evaluate the activity of a gene using the level of messenger RNA (mRNA) from a gene(s) and determine if that level is associated with the phenotype of interest. DNA contains a code to generate mRNA through a process called transcription. The amount of mRNA produced from a gene, if at all, depends on many factors including tissue type, local cell environment, and point in the cell cycle.

A gene expression study is different from a polymorphism-based study because an expression study evaluates mRNA levels that can change over time, uses less stable mRNA instead of DNA, and mRNA levels can be dramatically different based on what tissue is used for analysis because gene expression is tissue-specific. Gene expression studies therefore should address whether multiple samples over time are needed for evaluation (similar to other types of biomarkers that change over time), RNA stabilization, and what cell/tissue type is most appropriate to evaluate for the phenotype of interest. For these reasons, many stored samples may not be appropriate for this approach.

Candidate Gene Expression Studies

Rationale for Taking a Candidate Gene Expression Approach

Candidate gene expression studies investigate mRNA levels for a specific gene(s) to determine if it is associated with a phenotype of interest. Similar to a candidate gene association approach, this is a hypothesis-driven approach where the investigator a priori selects the candidate gene(s) to be evaluated. This approach is only appropriate if the investigator has ample justification for investigating a specific gene(s).

Subject and Sample Considerations

Gene expression studies often involve relative comparisons of mRNA levels between two groups (these groups can be different types of tissues, groups that vary for a particular exposure, or groups that vary by the presence or absence of a phenotype of interest). Clearly defined inclusion/exclusion criteria are necessary because of the relative comparison nature of this approach.

RNA Stabilization. Stabilization of RNA is essential to obtain accurate gene expression profiles of biological samples. Immediately after sample collection, RNA degradation and other transcriptional changes begin to occur. These alterations may result in false up or down regulation of gene expression levels. RNA stabilization preserves a representative gene expression profile for later analysis (e.g., quantitative RT-PCR and microarray analysis). RNA stabilization methods vary based on the type of biological sample. Five of the most common RNA stabilization methods are: (a) *PAXgene Blood RNA System* (http://www.preanalytix.com), which uses a single tube (pre-filled with RNA stabilization reagent) for blood collection, RNA stabilization, sample transport and storage, and purification of total RNA (PreAnalytiX, 2010); (b) *LeukoLOCK System* (http://www.lifetechnologies.com), which filters and isolates leukocytes from whole blood. RNA later solution is then used to stabilize the RNA of the leukocytes. A notable advantage to the LeukoLOCK System is the ability to remove a large proportion of reticulocyte-derived globin mRNA. Depletion of the globin mRNA allows for the detection of thousands of additional genes on microarray (Life Technologies Corporation, 2010a); (c) *RNAlater* (http://www.ambion.com; http:www.qiagen.com) stabilizes RNA in various fresh samples including animal tissue, tissue culture cells, leukocytes, yeast, and bacteria. After collection, the sample is submerged in the RNAlater stabilization solution. This solution permeates and stabilizes the sample eliminating the need for immediate processing or freezing of samples (Life Technologies Corporation, 2010b; Qiagen, 2006); (d) *Oragene RNA for Expression Analysis Self Collection Kit* (http://www.dnagenotek.com) allows for the non-invasive collection of RNA from saliva. Donors are instructed to expectorate into a vial, cap the container, and shake vigorously to release a stabilization solution from the cap. Oragene RNA samples can remain stable for months at room temperature (DNA Genotek, 2011); and (e) *Snap Freeze* quick-freezes solid tissues with liquid nitrogen, and dry ice can be used to preserve RNA; however, disruptions during freezing and thawing can lead to RNA degradation. Because of potential RNA degradation and difficulty of obtaining and working with liquid nitrogen and dry ice, RNAlater described previously may be a more viable option for solid tissue RNA stabilization.

Candidate Gene Selection

Candidate gene selection must be justified and rationale for selection is similar to selection of candidate genes in the candidate gene association section. The same bioinformatics databases mentioned in that section are also applicable to aiding in the selection of candidate genes for an expression study; and as with a candidate gene association study, a candidate gene expression study should consider focusing on a group of genes in a biological pathway versus the value of focusing on a single gene.

Expression Data Collection Technologies

Selection of a data collection technology for a candidate gene expression study should take into account the number of genes/loci and the number of samples to be evaluated. The most frequently used technologies for a candidate gene approach include Northern blotting, quantitative real-time PCR (qRT-PCR), and multiplex platforms that support 3–36 genes/loci per reaction.

Northern blotting requires electrophoresis of RNA, transfer to a membrane and hybridizing the membrane with a probe specific for detection of the mRNA of interest. The advantages of blotting are that most laboratories will have the equipment to conduct this type of data collection, and assessment of RNA size is possible. Disadvantages of blotting are that RNA degradation is common, it requires more RNA as a template for the experiment compared to other methods, it is laborious, and is not optimal for quantification of mRNA levels.

Currently, one of the most popular techniques for assessing the level of mRNA for a gene/locus is qRT-PCR. QRT-PCR requires conversion of RNA to a more stable template called cDNA (complementary DNA), PCR amplification and probe hybridization for the gene/locus of interest. The probe is fluorescently labeled and liberation of this fluorescent label is quantified, reflecting the amount of starting mRNA template in the sample. One crucial step in conducting qRT-PCR is normalization of the data generated. Normalization of the data allows for sample-to-sample comparisons that have been corrected for noise, such as what's introduced when sample dispensing between samples isn't uniform. This is often done using qRT-PCR data collected simultaneously for an endogenous control, which usually represents a stably expressed gene (often referred to as a "housekeeping gene") and allows for normalization of data across samples (Guénin et al., 2009). Thought needs to be given to selection of an appropriate endogenous control given that different tissues will have different stably expressed genes (Guénin et al., 2009). If in doubt, endogenous control panels are available for assessment prior to conducting qRT-PCR. Advantages of qRT-PCR include high sensitivity and reduced RNA template requirements, high throughput capabilities, quantification of starting mRNA template is possible with use of proper exogenous reference controls, and for many genes/loci/pathways off the shelf-optimized assays are available (http://www.appliedbiosystems.com; http://www.roche-applied-science.com).

Multiplex gene expression assays are available when the number of genes/loci to be evaluated is in the range of approximately 3–36. One example is the QuantiGene Plex 2.0 assay (for more information see http://www.panomics.com/index.php?id=product_6) that uses Luminex technology to collect the data and the assay can be customized.

Global (Genome-Wide) Gene Expression Studies

Rationale for Taking a Global Gene Expression Approach

Whole genome expression (also known as global gene expression or gene expression profiling) offers a comprehensive view of gene activity within a biological sample by examining mRNA levels for all known genes across the genome. In this way, whole genome expression provides functional information regarding "when and where a protein is expressed, when it is degraded, and with which other proteins it may interact" (Altman & Raychaudhuri, 2001, p. 340). Because of the dynamic nature of expression, gene expression profiles are often relatively compared under multiple conditions (such as comparing different tissue types, comparing normal versus abnormal tissues, comparing tissues before and after an exposure) or over a period of time (Altman & Raychaudhuri, 2001; Arcellana-Panlilio & Robbins, 2002). The use of global gene expression profiling is extremely advantageous when little to nothing is known about the genes influencing a condition, a similar advantage held by the GWAS approach. Thus, whole genome expression can identify novel candidate hypotheses through a nonparametric analysis of genome-wide expression data.

Subject and Sample Considerations

Sample Selection. Although this is an approach similar to GWAS, with evaluation of thousands of genes in a nonparametric manner, sample size requirements for global gene expression are usually smaller, requiring approximately 10 subjects per variable. Matching of subjects for key variables known to influence the phenotype under investigation can reduce the number of variables that need to be accounted for in the analyses. A sample size calculator for global gene expression experiments can be found at http://bioinformatics.mdanderson.org/MicroarraySampleSize/. Additionally, as with candidate gene expression studies, mRNA stabilization of the collected samples is crucial.

Gene Expression Data Collection Technologies

Microarrays. Microarrays are used to examine the expression profile of a single sample (often referred to as single dye array) or to compare expression levels between two different samples/conditions (often referred to as two dye array). The microarray itself is a solid surface covered with an "ordered arrangement of unique nucleic acid fragments derived from individual genes"

(Arcellana-Panilio & Robbins, 2002, p. G397). Fluorescently labeled template hybridizes to these nucleic acid fragments (referred to as probes) on the solid surface through complementary pairing. The intensity of the florescence at each spot on the microarray corresponds to the amount of sample binding to a particular nucleic acid fragment and thus, the gene expression level. If the microarray reveals any interesting findings, qRT-PCR should be carried out for validation purposes. For a visual representation of microarray methodology, visit this web address: http://www.bio.davidson.edu/courses/genomics/chip/chip.html.

Microarrays have revolutionized gene expression analyses, as this technology is able to simultaneously survey thousands of genes in a short period. However, the ability to detect novel genes is limited to the hybridization probes represented on the microarray. Off-the-shelf probe sets that contain reference sequences can be used, or custom probe sets are designed based on specific genes of interest or pathways. Additionally, microarrays require specialized lab equipment and are very useful when analyzing a small sample size but become costly as sample size increases. Two popular microarrays platforms include Affymetrix's GeneChip and Illumina's BeadChip.

Affymetrix's GeneChip platform (for more information see http://www.affymetrix.com) uses traditional solid support microarray technology. Affymetrix's latest product, the GeneChip Human Gene 1.0 ST Array, is able to interrogate 28,869 genes and covers over 700,000 distinct probes. A greater number of samples can be processed simultaneously (with this same probe set) using the Human Gene 1.1 Array Strip (4 samples/strip) and the Human Gene 1.1 Array Plate (16, 24, or 96 samples/plate). Affymetrix also provides whole transcript expression analysis technology for mice and rats.

Instead of using a solid support platform, the Illumina BeadChip platform (for more information see http://www.illumina.com) employs silica beads (each covered with thousands of copies of a specific oligonucleotide) self-assembled in microwells of fiber optic bundles or planar silica slides. Illumina's most recent whole genome expression array, the HumanHT-12 v4 BeadChip, provides high throughput processing of 12 samples and covers over 47,000 probes. Illumina also offers whole genome expression BeadChip technology for mice and rats.

Normalization of gene expression data is also important with microarray data collection. Unlike qRT-PCR where an appropriate endogenous control needs to be selected and included in the data collection, microarrays already include a range of endogenous controls for which data is simultaneously collected and from which the investigator can select to use for normalization of the data.

Sequence-based technologies that use next-generation sequencing (NGS; high throughput sequencing) are also available for collection of genome-wide gene expression data. An example of such a technology is the RNA-Seq method (for more information see http://www.illumina.com). This method requires conversion to cDNA, ligation of the cDNA fragments, creation of a library, sequencing of the template, and collection of frequency data for a transcript. An advantage of this approach over microarrays is that it does not require primers or probes, therefore novel transcripts that would not be detectable with a microarray can be identified.

Serial Analysis of Gene Expression. Serial analysis of gene expression (SAGE) provides comprehensive quantitative gene expression data. SAGE technology is based on three main principles: (a) a short sequence tag (9–17 bases) contains sufficient information to distinctively identify a transcript, (b) sequence tags can be linked together to form one long molecule that can be cloned and sequenced to allow efficient analysis of transcripts, and (c) the number of times a particular tag is observed corresponds to the expression level of the transcript (Sagenet, 2003; Velculescu, Zhang, Vogelstein, & Kinzler, 1995). One of the main advantages of SAGE, similar to RNA-Seq, is the ability to detect novel genes as it does not require prior sequence information or hybridization probes for each transcript like microarrays (Velculescu et al., 1997). Another advantage of SAGE is that it uses common laboratory equipment and techniques. Any laboratory that performs PCR and manual sequencing could also execute SAGE. Nonetheless, because of cloning and sequencing, SAGE can be expensive, time consuming, and labor intensive.

EPIGENETIC STUDIES

An epigenetic mechanism is a biochemical alteration to the DNA molecule that does not change the sequence of the DNA but does influence gene expression. *Epigenetics* is often defined as the "study of mitotically and/or meiotically heritable changes in gene function that cannot be explained by changes in DNA sequence" (Russo, Martienssen, & Riggs, 1996, p. 1).

The epigenetic/epigenomic approach shares many advantages and disadvantages with DNA polymorphism-based approaches and gene expression-based approaches. Like DNA polymorphism-based approaches, the epigenetic/epigenomic approach uses DNA as its template for data collection. Because both DNA sequence and its chemical modifications are stable, stored samples are more likely to be appropriate for this approach than gene expression approaches. Similar to a gene expression-based approach, epigenetic/epigenomic alterations can change over time and can differ dramatically between cell/tissue types. Although template stability is not an issue, the investigator should give great consideration to whether multiple samples over time are needed for evaluation and what cell/tissue type is most appropriate to evaluate for the given phenotype of interest. For these reasons, similar to gene expression studies, many stored samples may not be appropriate for this approach.

Chromatin remodeling, noncoding RNAs, histone modifications, and DNA methylation are all epigenetic/epigenomic alterations that impact gene expression. Chromatin remodeling is an enzymatic process that results in altered chromatin and nucleosome composition. This transformed structure provides regulatory proteins access to the DNA molecule. Noncoding RNAs are not translated into protein but have considerable involvement in gene expression through interactions with DNA/mRNA. Although chromatin remodeling and noncoding RNAs are important to gene regulation, this article will focus primarily on the commonly examined epigenetic mechanisms for which the most technology for data collection is available: histone modifications and methylation.

Rationale for Taking an Epigenetic/Epigenomic Approach

The decision to take an epigenetic (candidate gene) or an epigenomic (genome-wide) approach is based on wanting to evaluate the mechanism for gene regulation. There are many environmental factors that impact the severity and frequency of epigenetic/epigenomic alterations and subsequent gene expression; therefore, this approach is often used to examine multifactorial diseases that have an environmental component associated with it. Epigenetic approaches to examine transcriptional regulation have contributed to a more comprehensive understanding of complex conditions that demonstrate aberrant gene expression, including: cancer (Wilop et al., 2011), mental health (Read, Bentall, & Fosse, 2009), and cardiovascular disorders (Ordovás & Smith, 2010). Furthermore, the investigation of diseases for which DNA mutations have not been revealed may benefit from an epigenetic approach.

Subject and Sample Considerations

The epigenome is subject to frequent alterations; therefore, longitudinal sample collection is recommended if evaluating time-sensitive trends. Subject size recommendations for an epigenetic study follow similar guidelines to a gene expression study and vary on whether the investigator will examine the entire genome (hypothesis generating/larger sample size) or a candidate gene profile (hypothesis driven/smaller sample size). Like the other approaches described, an epigenetic study does not require that subjects be related. The advantages and disadvantages of conducting a genome-wide versus candidate gene epigenetic study are similar to those described in previous sections.

The epigenome is largely determined by cell type, and this is especially true for methylation patterns; therefore, tissue source is extremely important to consider for this type of approach. For example, the methylation profile of a skin cell is very different than the methylation profile of a liver cell because different genes are expressed in each cell type, and methylation is a driving force behind tissue specific gene expression. Similar to a gene expression study, an epigenetic design requires the samples for epigenetic analyses be from a tissue that appropriately addresses the phenotype of interest. Tissue-specific sample collection will capture epigenetic patterns that impact gene expression, which are potentially contributing to the disease. Unlike a gene expression study that examines RNA, this design requires DNA, which is advantageous for the investigator who has access to previously collected samples, assuming they were collected from an appropriate tissue for the phenotype under investigation.

Epigenetic and Epigenomic Data Collection Technologies

This section will focus on the two epigenetic mechanisms most frequently studied: (a) histone modification and (b) methylation. Posttranslational histone modifications include alteration of the histone tail through biochemical changes that ultimately impact gene activity. Genome-wide histone modifications can be captured with chromatin immunoprecipitation technology (ChIP) and quantified with a microarray (ChIP-chip). Methylation refers to the addition of a methyl group to a cytosine, often at CpG islands, which are regions of the genome that are rich in CG base sequences. Hypermethylation of a gene typically leads to gene suppression, whereas hypomethylation results in gene expression. Genome-wide methylation intensities can also be measured with affinity-based immunoprecipitation (MeDIP) and quantified with a microarray (MeDIP-chip, Infinium platform). Methylation of candidate genes can also be measured with restriction enzymes that recognize only demethylated CpG regions (HELP assay) or pyrosequencing. Next Generation Sequencing approaches are also becoming increasingly popular, more cost-effective, and provide global sequencing for histone modification (ChIP-seq) and methylation (MeDIP-seq), often integrating these with other epigenetic mechanisms. This section will describe each method and provide the reader with technologies and recommendations to aide in the design and implementation of an epigenetic study.

Histone Modification Analysis

Histone modification signals can be captured with chromatin immunoprecipitation (ChIP), which provides modification position approximation on the genome (Collas, 2010). The ChIP-chip technique combines this ChIP technology with a microarray (chip) to quantify the sum of binding sites on the genome (Aparicio, Geisberg, & Struhl, 2004). The ChIP-seq technique (see Next Generation Sequencing) has become a popular technique compared to ChIP-chip. Unlike ChIP-seq, ChIP-chip requires more amplification, multiplexing is not possible

(Park, 2009), and the results have a lower resolution that are limited to the coverage provided by the selected microarray (Evertts, Zee, & Garcia, 2010). Nimble Gen offers a whole-genome ChIP-chip tiling array that allows the investigator to choose between ordering the entire genome set or individual arrays within a set (http://www.nimblegen.com/products/chip/wgt/index.html). Single gene ChIP technologies are available that target antibodies against specific histone modifications. Mass spectrometry also allows the measurement of mass-to-charge ratio of peptides (Evertts et al., 2010) and allows for changes in modification to be quantified during chromatin assembly (Deal & Henikoff, 2010).

When performing any microarray experiment, it is important to address concerns that may compromise the integrity of the experiment, including image acquisition, background subtraction, standard normalization, and the need to control for biases from dye (Buck & Lieb, 2004). Additionally, the reproducibility of the histone-modification results depends on the quality and specificity of antibodies used. Antibodies may exhibit appropriate specificity but are ineffective when subjected to ChIP reagents (Egelhoffer et al., 2010). The Center for Biomedical Informatics at Harvard Medical School has developed an online repository that allows investigators to search for antibodies subjected to validation tests (http://compbio.med.harvard.edu/antibodies/about). It is important to note that this validation data should be used as a guide and investigators are encouraged to validate their own findings.

Bisulfite-Conversion Based Methylation Analyses

Bisulfite-conversion of unmethylated cytosines to uracils remains the gold standard to evaluate methylation (Huang, Huang, & Wang, 2010). Bisulfite-conversion based microarrays use probes that hybridize targets to methylated and unmethylated regions and release a fluorescent intensity that denotes methylation status (Huang et al., 2010). Recent research indicates that tissue-specific methylation occurs in CpG island shores rather than previously targeted CpG islands (Irizarry et al., 2009); therefore, CpG islands alone are not sufficient to reveal differentially methylated regions, and methylome evaluation should also include CpG shores (Gupta, Nagarajan, & Wajapeyee, 2010). Like other nonsequencing-based methods, the results of this platform are "susceptible to certain polymorphisms that were not known or considered at the time the array was designed" (Rakyan, Down, Balding, & Beck, 2011, p. 532). Illumina offers the Infinium HumanMethylation450K that provides a whole-genome analysis of methylation intensities of more than 450,000 sites, including CpG islands, shores, and other CpG sites outside of islands (for more information see http://www.illumina.com/products/methylation_450_beadchip_kits.ilmn). Candidate gene methylation assessment can be accomplished through technologies such as the EpiTYPER (for more information see http://www.sequenom.com)

that uses bisulfite converted DNA as a template for PCR and after modification and cleavage of the PCR product, mass spectrometry is performed to quantify methylated and non-methylated DNA.

Bisulphite-based sequencing (BS-seq) uses bisulphite converted DNA as a template, PCR amplification occurs, and sequencing of the resulting fragments provide a global view of methylation with minimal bias toward CpG dense regions. This approach provides the highest level of coverage and resolution but is not capable of distinguishing between methylated and hydroxymethylated cytosine bases. BS-seq can be used for both a genome-wide or candidate gene approach. Pyrosequencing examines the methylation intensity of specific sites or genes of interest. Illumina offers a single-site resolution methylation assay that uses bisulfite conversion and pyrosequencing to produce high-resolution results (http://www.illumina.com/technology/veracode_methylation_assay.ilmn).

Affinity-Based Methylation Analyses

Genome-wide affinity-based microarrays use enzyme recognition sites within CpG sites that enrich the methylated fraction of the genome. The MeDIP-chip technique (methylated DNA immunoprecipitation-chromatin immunoprecipitation) immunoprecipitates the methylated portion of genomic DNA with an antibody and is followed by quantification of methylation with a microarray. This technique yields a restricted resolution that is limited by the type of array used. MeDIP-chip should be validated with quantitative PCR, although referencing is not required because bisulfite conversion does not occur. ArrayStar offers MeDIP-chip services that include quality assessments for both methods (http://www.arraystar.com/Microarray/service_main.asp?id=181).

Restriction Endonuclease-Based Methylation Analysis

Restriction endonucleases have been adapted to discriminate methylated from unmethylated regions in the DNA (Edwards et al., 2010). This approach uses restriction enzymes that recognize only unmethylated sites and are therefore unable to cut methylated portions of DNA. This method, combined with high throughput sequencing is limited by the availability of restriction enzyme sites in the target DNA (Gupta et al., 2010). Additionally, this technique requires large amounts of DNA (Biotage, 2007). Advantages for this approach include a simplified data analysis, straightforward protocol, and it does not require bisulfite conversion. The use of restriction enzymes to analyze methylation can be used for either candidate-gene or genome-wide studies (Gupta et al.) and has been used as a method of methylation mapping analysis (Edwards et al., 2010).

Data Quality assessments are important to incorporate into an epigenetic study. Quantile and LOESS normalization is recommended, which assumes a similar total strength (source). Additionally, bisulfite-based experiments, especially pyrosequencing because PCR is highly variable, should include verification in independent samples to distinguish methylation from incomplete bisulfite conversion (Laird, 2010). Incomplete conversion of methylated cytosines remains a major weakness of bisulfite-conversion based analysis techniques. Fully methylated and fully unmethylated controls should be provided by commercial vendors that allow the investigator to evaluate bisulphite-conversion efficiency.

Next Generation Sequencing (NGS) for Histone Modification Analysis
DNA sequencing from epigenetic events may provide a first step toward quantification of epigenetic mechanisms. Similar to ChIP-chip, ChIP-seq uses antibodies to enrich for histone modifications but is instead followed by high throughput sequencing that measures gene expression levels (Evertts et al., 2010). This technique determines the genome-wide patterns of modified chromatin, including histone methylation, acetylation status, and binding regions for proteins (Werner, 2010). Unlike ChIP-chip, ChIP-Seq offers higher resolution with fewer artifacts, greater coverage, and requires less DNA. Illumina offers a ChIP-seq assay that provides a wide range of binding sites with varying strength (http://www.illumina.com/technology/chip_seq_assay.ilmn).

Next Generation Sequencing (NGS) for Methylation Analysis
MeDIP-seq (Methylated DNA Immunoprecipitation-Sequencing) is a high throughput sequencing technique of methylated DNA fragments that is aligned to a referenced genome. This technique is comparatively easier to analyze and interpret (Gupta et al., 2010); however, this method is best used to study hypermethylation of CpG-rich areas because methylated CpG-rich sequences are more efficiently enriched than methylated CpG-poor sequences (Bibkova & Fan, 2009).

CONCLUSIONS

Nurse scientists should give much thought to how a genetic or genomic study could positively impact and move forward their program of research. When designing a genetic or genomic research study, it is paramount that one decides if they will take a polymorphism-based, gene expression-based or epigenetic-based approach and then within the context of that study whether they will take a genetic or a genomic approach. This article, although not providing an exhaustive review of available technologies, demonstrates various technologies available for commonly used approaches, each with advantages and disadvantages. Availability of databases housing information to facilitate study design, data collection, interpretation of findings, and dissemination of data have greatly improved over the past decade. Investigators are encouraged to visit and use in silico resources when designing a research study to ensure they are conducting novel investigations and using up to date information.

ACKNOWLEDGMENT

The authors would like to acknowledge support available through the National Institutes of Health, National Institute of Nursing Research award "Targeted Research and Academic Training Program for Nurses in Genomics" (T32 NR009759).

REFERENCES

Altman, R. B., & Raychaudhuri, S. (2001). Whole-genome expression analysis: Challenges beyond clustering. *Current Opinion in Structural Biology*, *11*(3), 340–347. http://dx.doi.org/10.1016/S0959-440X(00)00212-8

Aparicio, O., Geisberg, J. V., & Struhl, K. (2004). Chromatin immunoprecipitation for determining the association of proteins with specific genomic sequences in vivo. *Current Protocols in Cell Biology*, Chapter 17: Unit 17.7. http://dx.doi.org/10.1002/0471143030.cb1707s23

Arcellana-Panlilio, M., & Robbins, S. M. (2002). Cutting-edge technology. I. Global gene expression profiling using DNA microarrays. *American Journal of Physiology. Gastrointestinal and Liver Physiology*, *282*(3), G397–G402. http://dx.doi.org/10.1152/ajpgi.00519.2001

Bibkova, M., & Fan, J. (2009). Genome-wide DNA methylation profiling. *Wiley Interdisciplinary Reviews: Systems Biology and Medicine*, *2*(2), 210–223. http://dx.doi.org/10.1002/wsbm.35

Biotage (2007). CpG methylation analysis by pyrosequencing: Benchmarks and application. Retrieved from http://www.pyrosequencing.com/graphics/7424.pdf

Buck, M. J., & Lieb, J. D. (2004). ChIP-chip: Considerations for the design, analysis, and application of genome-wide chromatin immunoprecipitation experiments. *Genomics*, *83*(3), 349–360. http://dx.doi.org/10.1016/j.ygeno.2003.11.004

Childs, B., Wiener, C., & Valle, D. (2005). A science of the individual: Implications for a medical school curriculum. [Review]. *Annual Review of Genomics and Human Genetics, 6*, 313-330. doi: 10.1146/annurev.genom.6.080604.162345

Collas, P. (2010). The current state of chromatin immunoprecipitation. *Molecular Biotechnology*, *45*(1), 87–100. http://dx.doi.org/10.1007/s12033-009-9239-8

Deal, R., & Henikoff, S. (2010). Capturing the dynamic epigenome. *Genome Biology, 11*(10), 218. http://dx.doi.org/10.1186/gb-2010-11-10-218

DiStefano, J. K., & Taverna, D. M. (2011). Technological issues and experimental design of gene association studies. In J. K. K. DiStefano (Vol. ed.), *Disease Gene Identification: Vol. 700, Part 1. Methods and Protocols (Methods in Molecular Biology)* (pp. 3–16). New York, NY: Humana Press. http://dx.doi.org/10.1007/978-1-61737-954-3_1

DNA Genotek. (2011). Oragene RNA. Retrieved from http://www.dnagenotek.com/DNA_Genotek_Product_RNA_Overview.html

Edwards, J. R., O'Donnell, A. H., Rollins, R. A., Peckham, H. E., Lee, C., Milekic, M. H.,. . . Bestor, T. H. (2010). Chromatin and sequence features that define the fine and gross structure of genomic methylation patterns. *Genome Research*, *20*(7), 972–980. http://dx.doi.org/10.1101/gr.101535.109

Egelhoffer, T. A., Minoda, A., Klugman, S., Lee, K., Kolasinska-Zwierz, P., Alekseyenko, A. A., . . . Lieb, J. D. (2010). An assessment of histone-modification antibody quality. *Nature Structural & Molecular Biology*, *18*(1), 91–93. http://dx.doi.org/10.1038/nsmb.1972

Evertts, A. G., Zee, B. M., & Garcia, B. A. (2010). Modern approaches for investigating epigenetic signaling pathways. *Journal of Applied Physiology*, *109*(3), 927–933. http://dx.doi.org/10.1152/japplphysiol.00007.2010

Genetics Home Reference. (2011). What are single nucleotide polymorphisms (SNPs)? Retrieved from http://ghr.nlm.nih.gov/handbook/genomicresearch/snp

Green, E. D., Guyer, M. S., & National Human Genome Research Institute. (2011). Charting a course for genomic medicine from base pairs to bedside. *Nature*, *470*(7333), 204–213. http://dx.doi.org/10.1038/nature09764

Guénin, S., Mauriat, M., Pelloux, J., Van Wuytswinkel, O., Bellini, C., & Gutierrez, L. (2009). Normalization of qRT-PCR data: The necessity of adopting a systematic, experimental conditions-specific, validation of references. *Journal of Experimental Botany*, *60*(2), 487–493. http://dx.doi.org/10.1093/jxb/ern305

Gupta, R., Nagarajan, A., & Wajapeyee, N. (2010). Advances in genome-wide DNA methylation analysis. *Biotechniques*, *49*(4), iii–xi. http://dx.doi.org/10.2144/000113493

Hakonarson, H., & Grant, S. F. (2011). Planning a genome-wide association study: Points to consider. *Annals of Medicine*, *43*(6), 451–460. http://dx.doi.org/10.3109/07853890.2011.573803

Hattersley, A. T., & McCarthy, M. I. (2005). What makes a good genetic association study? *The Lancet*, *366*(9493), 1315–1323. http://dx.doi.org/10.1016/S0140-6736(05)67531-9

Huang, Y. W., Huang, T. H., Wang, L. S. (2010). Profiling DNA methylomes from microarray to genome-scale sequencing. *Technology in Cancer Research & Treatment*, *9*(2), 139–147.

International HapMap Project. (n.d.). What is the HapMap? Retrieved from http://hapmap.ncbi.nlm.nih.gov/whatishapmap.html.en

International HapMap Project. (2006). About the International HapMap Project. Retrieved from http://hapmap.ncbi.nlm.nih.gov/abouthapmap.html

Irizarry, R. A., Ladd-Acosta, C., Wen, B., Wu, Z., Montano, C., Onyango, P., . . . Feinberg, A. P. (2009). The human colon cancer methylome shows similar hypo- and hypermethylation at conserved tissue-specific CpG island shores. *Nature Genetics*, *41*(2),178–186. http://dx.doi.org/10.1038/ng.298

Jorgensen, T. J., Ruczinski, I., Kessing, B., Smith, M. W., Shugart, Y. Y., & Alberg, A. J. (2009). Hypothesis-driven candidate gene association studies: Practical design and analytical considerations. *American Journal of Epidemiology*, *170*(8), 986–993. http://dx.doi.org/10.1093/aje/kwp242

Ku, C. S., Loy, E. Y., Pawitan, Y., & Chia, K. S. (2010). The pursuit of genome-wide association studies: Where are we now? *Journal of Human Genetics*, *55*(4), 195–206. http://dx.doi.org/10.1038/jhg.2010.19

Laird, P. W. (2010). Principles and challenges of genome-wide DNA methylation analysis. *Nature Reviews Genetics*, *11*(3), 191–203. http://dx.doi.org/10.1038/nrg2732

Lander, E. S. (2011). Initial impact of the sequencing of the human genome. *Nature*, *470*(7333), 187–197. http://dx.doi.org/10.1038/nature09792

Life Technologies Corporation. (2010a). LeukoLOCK total RNA isolation system. Retrieved from http://www.ambion.com/techlib/prot/fm_1923.pdf

Life Technologies Corporation. (2010b). RNA later tissue collection: RNA stabilization solution. Retrieved from http://www.ambion.com/techlib/prot/bp_7020.pdf

National Human Genome Research Institute [NHGRI]. (n.d.a). Talking glossary of genetic terms: Copy Number Variation (CNV). Retrieved from http://www.genome.gov/glossary/index.cfm?id=40

National Human Genome Research Institute [NHGRI]. (n.d.b). Talking glossary of genetic terms: Polymorphism. Retrieved from http://www.genome.gov/glossary/index.cfm?id=160

Nussbaum, R. L., McInnes, R. R., & Huntington, F. W. (2007). *Thompson & Thompson Genetics in Medicine* (7th ed.). Philadelphia, PA: Saunders Elsevier.

Ordovás, J. M., Smith, C. E. (2010). Epigenetics and cardiovascular disease. *Nature Reviews Cardiology*, *7*(9), 510–519. http://dx.doi.org/10.1038/nrcardio.2010.104

Park, P. (2009). ChIP-seq: Advantages and challenges of a maturing technology. *Nature Reviews*, *10*(10), 669–680. http://dx.doi.org/10.1038/nrg2641

Passarge, E. (2007). *Color Atlas of Genetics*. (3rd ed.). New York, NY: Thieme.

PreAnalytiX. (2010). PAXgene Blood RNA: The better the source, the more to explore. Retrieved from http://www.qiagen.com/literature/render.aspx?id=200337

Psychiatric GWAS Consortium Coordinating Committee. (2009). Genomewide Association Studies: History, rationale, and prospects for psychiatric disorders. *American Journal of Psychiatry*, *166*(5), 540–556. http://dx.doi.org/10.1176/appi.ajp.2008.08091354

Qiagen. (2006). RNA*later* handbook. Retrieved from www.qiagen.com/literature/render.aspx?id=403

Rakyan, V. K., Down, T. A., Balding, D. J., & Beck, S. (2011). Epigenome-wide association studies for common human diseases. *Nature Reviews Genetics*, *12*(8), 529–541. http://dx.doi.org/10.1038/nrg3000

Read, J., Bentall, R. P., & Fosse, R. (2009). Time to abandon the bio-bio-bio model of psychosis: Exploring the epigenetic and psychological mechanisms by which adverse life events lead to psychotic symptoms. *Epidemiologia E Psichiatria Sociale*, *18*(4), 299–310. http://dx.doi.org/10.1017/S1121189X00000257

Ruiz-Naváez, E. A., Rosenberg, L., Wise, L. A., Reich, D., & Palmer, J. R. (2011). Validation of a small set of ancestral informative markers for control of population admixture in African Americans. *American Journal of Epidemiology*, *173*(5), 587–592. http://dx.doi.org/10.1093/aje/kwq401

Russo, V. A., Martienssen, R. A., & Riggs, A. D. (1996). *Epigenetic Mechanisms of Gene Regulation* (Vol. 32). Plainview, NY: Cold Spring Harbor Laboratory Press.

Sagenet (2003). Description of SAGE. Retrieved from http://www.sagenet.org/findings/index.html

Sequenom. (2010). *MassARRAY® iPLEX® gold-SNP genotyping: From target discovery to HTP validating* (version 2) [Brochure]. San Diego, CA: Author.

The Gene Ontology. (1999–2011). An Introduction to the Gene Ontology. Retrieved from http://www.geneontology.org/GO.doc.shtml

The University of Texas MD Anderson Cancer Center: Department of Bioinformatics and Computational Biology (2003–2010). Sample size for microarray experiments. Retrieved from http://bioinformatics.mdanderson.org/MicroarraySampleSize/

Velculescu, V. E., Zhang, L., Vogelstein, B., & Kinzler, K. W. (1995). Serial analysis of gene expression. *Science*, *270*(5235), 484–487. http://dx.doi.org/10.1017/S1121189X00000257

Velculescu, V. E., Zhang, L., Zhou, W., Vogelstein, J., Basrai, M. A., Bassett, D. E., Jr., . . . Kinzler, K. W. (1997). Characterization of the yeast transcriptome. *Cell*, *88*(2), 243–251. http://dx.doi.org/10.1016/S0092-8674(00)81845-0

Werner, T. (2010). Next generation sequencing in functional genomics. *Briefings in Bioinformatics*, *11*(5), 499–511. http://dx.doi.org/10.1093/bib/bbq018

Wilop, S., Fernandez, A. F., Jost, E., Herman, J. G., Brümmendorf, T. H., Esteller, M., & Galm, O. (2011). Array-based DNA methylation profiling in acute myeloid leukaemia. *British Journal of Haematology*, *155*(1), 65–72. http://dx.doi.org/10.1111/j.1365-2141.2011.08801.x

CHAPTER 2

Genetics and Genomics

The Scientific Drivers of Personalized Medicine

Emma L. Kurnat-Thoma

ABSTRACT

Scientific advances in genetics and genomics will be incorporated into health care soon. The tailoring of treatment to an individual's genetic make up has been termed Personalized Medicine. These advances are promising and are receiving significant attention; however, many nurses are caught in the gap between technologic advances and clinical diffusion and uptake. Aiming to reduce this gap, this chapter provides an overview of the science driving Personalized Medicine, outlines areas of research and clinical translation where nurses may expect to see its fruits, and briefly identifies obstacles preventing its full realization. Four scientific elements of Personalized Medicine are described: (1) discovery of novel biology that guides clinical translation mechanisms, (2) genetic risk assessment, (3) molecular diagnostic technology, and (4) pharmacogenetics and pharmacogenomics. Successful design and implementation of Personalized Medicine will hinge on the roles of nurses conducting or participating in collaborative initiatives that are furthering genetic/genomic applications within these contexts.

http://dx.doi.org/10.1891/0739-6686.29.27

INTRODUCTION

Formally completed in 2003, the Human Genome Project provided a reference map of human genome DNA sequence (International Human Genome Sequencing Consortium, 2001, 2004). Similarly, the International Haplotype Map Project was completed in 2005 and provided a map of variation in human DNA and how this variation was arranged in populations (International HapMap Consortium, 2005). Fruits of these research efforts are revolutionizing biology, medicine, and ultimately, health care delivery (Collins, 2010; Collins, Green, Guttmacher, & Guyer, 2003; Collins & McKusick, 2001; Lander, 2011). Clinical translation of these landmark scientific achievements will contribute to Personalized Medicine where an individual's DNA can be used to finely tailor health care practices. Health advances include personalized risk assessments that predict disease development years before clinical appearance, use of precisely tuned molecular screening tests and clinical diagnostics, and safer and more effective pharmaceutical agents prescribed to match individuals' genetic constitution. Further developing and fueling these innovative applications is the continued discovery of novel biology in high-tech laboratory settings across the country and throughout the world.

Scientific achievement beyond the Human Genome Project and the HapMap is being fueled by unprecedented technological advancement in next generation sequencing (Metzker, 2010). For example, cost of sequencing an individual's genome in 2002 was 1 billion dollars, but through Illumina, Inc., Individual Genome Sequencing Service, it can now be commercially performed for $7,500 (Illumina, Inc., 2011). Furthermore, it is estimated that this price will drop to less than $1,000 per genome by 2012 or 2013—a magnitude and rate of cost reduction that surpasses the gains made by the microprocessor industry. Because the human genome is approximately 6 billion data points, the sheer volume of scientific data is overwhelming, and data flowing from studies examining the associations between genetic sequences and disease has been likened to "drinking water from a fire hose" (Hunter & Kraft, 2007). A PubMed search using the terms "genetics" and "genomics" comparing 1991–2001 and 2001–2011 demonstrates this rapid increase: 662,516 genetics and 9,212 genomics results for 1991–2001, compared to 1,244,899 genetics and 60,710 genomics for the past decade (http://www.ncbi.nlm.nih.gov/sites/entrez/ queried on September 16, 2011).

Intense mainstream media coverage of scientific progress in genetics and genomics has resulted in public expectation that these advances are already incorporated into clinical practice. Capitalizing on this environment, several companies have launched direct-to-consumer marketing campaigns for genomic profiling tests. Some testing companies provide test results and interpretations that are not supported by scientific data (Government Accountability Office, 2010; Janssens et al., 2008). Others may include information that is difficult for the consumer to interpret in the absence of knowledgeable practitioner guidance (Gollust, Wilfond, & Hull, 2003; Kutz, 2006; McGuire, Evans, Caulfield, & Burke, 2010). Current examples include services offered by *23andMe*, *DeCODE*, and *Navigenics*, where for $99–$1,100 a consumer can submit a sample of their DNA and obtain personalized analyses of their genetic code. Information contained in these companies' reports includes a person's lifetime risk for development of health conditions such as osteoporosis, cancer, diabetes, and heart disease.

Previous scholarly works in nursing have extensively discussed genetics and genomics and associated nursing implications presented by these fields, but none outline Personalized Medicine and the molecular science driving it. In this chapter, I outline four conceptual Personalized Medicine scientific elements, describe their health care implications and identify some translation obstacles. It can be concluded that although the biologic advances fueling Personalized Medicine offer sound proof of technologic possibility, it will likely be one or two decades until there is enough application evidence for nurses to use the entirety of an individual's genome to guide personalized patient health care. During this long incubation period, health care patterns and practices will begin to incorporate elements of genetics. The field of nursing should continue to anticipate and contribute to these developments and play an integral role in their translation and application.

CLINICAL APPLICATION OF GENETICS AND GENOMICS: PERSONALIZED MEDICINE

Personalized Medicine, in lay definition, is "using information about a person's genetic makeup to tailor strategies for the *detection*, *treatment*, or *prevention of disease*" (Collins, 2005). This new paradigm of health care rests on four key elements, the first driving the latter three, which contain direct clinical relevance to the nursing community (Figure 2.1): (1) discovery of novel biologic processes, which serve as the foundation for all clinical translation; (2) personalized risk assessments; (3) enhanced diagnostic accuracy through molecular profiling; and (4) pharmacogenetics and pharmacogenomics. Refer to Table 2.1 for a concept glossary and Table 2.2 for relevant reference genetics and genomics websites.

The Foundation of Personalized Medicine—New Biology, New Drugs

Novel biologic knowledge arising from the Human Genome Project and HapMap has great implications for advances in future health care practices. Validated Genome-Wide Association Study (GWAS) associations identify 1,319 loci for more than 221 health and disease traits and are publicly searchable in an online catalogue maintained by the National Human Genome Research Institute at www.genome.gov/26525384 (Hindorff et al., 2009; Manolio, Brooks, & Collins,

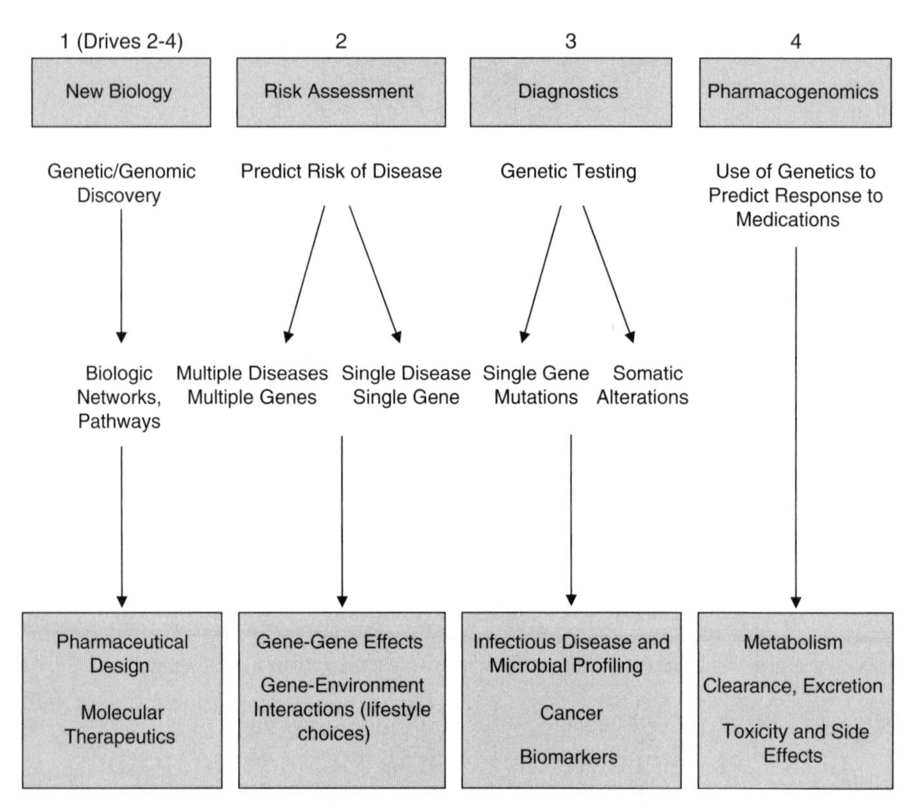

FIGURE 2.1 The four elements of personalized medicine.

2008; National Human Genome Research Institute, 2011a). This knowledge translates to understandings of biologic networks not previously known as involved with common diseases (Figure 2.2). Through these new pathway connections, opportunities for improved clinical measurement and therapeutic manipulation will arise. A relevant example is illustrated by the case of age-related macular degeneration (Exhibit 2.1), one of the first conditions to be analyzed with the GWAS approach.

Although gene/disease associations like those for age-related macular degeneration arose from the Human Genome Project and HapMap, several additional ongoing scientific initiatives are expected to yield similar biomedical impact. These include ENCylopedia Of DNA Elements (ENCODE), 1000 Genomes Project, Human Microbiome Project, and the Cancer Genome Atlas. Because nurses play integral roles in generation of new knowledge and clinical realization of novel technologies, understanding significance of these efforts can help drive their clinical translation and application.

TABLE 2.1
Personalized Medicine Working Glossary

Term/Acronym	Definition
Allele	Varying (or variant) forms of a gene at a specific location in the genome.
Analytic validity	Accuracy and reliability, a genetic/genomic test detects a particular genetic characteristic.
Biomarkers	Biological molecules in blood or other body fluids whose parameters can be associated with disease presence and severity. Can be detected and measured by many different methods such as laboratory assays and imaging technology.
CNV	Copy number variation where combinations of two to three nucleotides are continually repeated in non-coding portions of the genome (i.e., CACACACACACACA).
Clinical utility	Degree with which a genetic or genomic test will provide benefit to patients after accounting for potential harms.
Clinical validity	Ability of a genetic/genomic test to detect or predict a clinical condition or outcome.
Deletion	A mutation caused by the removal of DNA from the chromosome. Deletions can be of any length, from one base pair to a large chromosomal segment (millions of base pairs).
Enhancer	Short stretch of regulatory DNA sequence that signals where transcription factors should bind. Enhancers modulate rate of transcription and can be found great distances away from the gene it regulates.
Epigenetics	Study of heritable differences in gene function that occurs without changes in DNA sequence (i.e., methylation patterns).
Exon	Region of a gene encoding for a particular portion of the complete protein.
Gene	The functional and physical unit of heredity passed from parent to offspring. Genes contain information for making a specific protein.
Genetic testing	Generic term for an array of techniques that analyze DNA, RNA, or proteins for general health or medical identification purposes. Currently, more than 2,000 tests are clinically available.
Genome	The entirety of an individual's genetic code, approximately 6 billion nucleotides comprising approximately 20,500 genes.

(*Continued*)

TABLE 2.1
Personalized Medicine Working Glossary (Continued)

Term/Acronym	Definition
Genomics	Scientific study of a genome—including any or all combinations of genes, their functions, and their interactions with each other and the surrounding environment.
Genotype	An individual's two alleles at specific loci.
Haplotype	Combinations of SNP alleles located close to one another on a chromosome. If close together, haplotypes can be inherited as units or blocks.
HapMap	The haplotype map. A map of all inherited genetic variation (haplotypes) in the human genome.
Heterozygous	Having two different forms of a particular gene (AB).
Homozygous	Having two identical forms of a particular gene (AA).
Indel	An insertion/deletion polymorphism where AA, AB, BB yield: insertion/insertion, insertion/deletion, deletion/deletion.
Insertion	A mutation caused by the insertion of DNA from the chromosome. Insertions can also be of variable length (one to many base pairs).
Intron	Noncoding sequences of DNA that are removed from the RNA transcript prior to exportation from the nucleus.
Locus	The physical location of a gene or gene segment on a chromosome.
Loci	The plural of locus.
Methylation	Chemical reactions that place a methyl group (three hydrogen atoms and one carbon atom) on the DNA nucleotide cytosine (C); presence of methylation silences genetic expression.
MicroRNA	Subtype of RNA that binds to messenger RNA strands in order to block protein translation at the ribosomes.
Multifactorial Dx	Diseases caused by interactions of numerous genes with environmental factors. Examples include obesity, diabetes, heart disease, and cancer.
Mutation	Permanent and structural alteration in DNA. Most cause little, if any, harm. If in a critical location, such as the DNA repair genes in BRCA 1 and 2, can cause severe disease such as early onset cancer.
Negative predictive value	Probability that patients with a negative genetic/genomic test result will not get a specific disease or condition.

TABLE 2.1

Personalized Medicine Working Glossary (Continued)

Term/Acronym	Definition
Phenotype	A patient's observable clinical and physiologic characteristics as a result of inherited genotype interacting with their environment.
Polymorphism	The existence of multiple genotypes in a population, at one locus. Variants are not caused by mutations in DNA because they occur at a frequency greater than can occur by evolutionary (slow) means. Polymorphisms may take several forms, including SNPs, CNVs, and insertion/deletion (indel's).
Positive predictive value	Probability that patients with a positive genetic/genomic test result will get a specific disease or condition.
Promoter	Short stretch of regulatory DNA sequence that signals where transcription should start in a gene (for the RNA polymerase).
Sensitivity (clinical)	Percent of patients with positive genetic/genomic test result that are correctly identified as having the defined clinical trait.
Silencer	Short stretch of regulatory DNA sequence that signals where chromatin should become condensed. This blocks other enzymes from accessing the DNA strands to prevent transcription.
Specificity (clinical)	Percent of patients with a negative genetic/genomic test result that are correctly identified as *not* having the defined clinical trait.
SNPs	Single nucleotide polymorphism(s). The difference of a single-base pair at a specific position in the genome between two different individuals in a population. Most are inconsequential; but if in a coding region, may cause changes in gene efficiency and/or function.
Variant	Another word for polymorphism. There are different types of variants such as SNPs, CNVs, insertion/deletion (indels), and RFLPs.

Source: Centers for Disease Control and Prevention, 2011; National Cancer Institute, 2011; National Center for Advancing Translational Sciences, Office of Rare Diseases Research, 2011; National Center for Biotechnology Information, GeneReviews, 2011; National Human Genome Research Institute, 2011c.

TABLE 2.2
Genetics and Genomics Scientific and Clinical Translation Resources

Resource Title	Reference Website Locations
General Information	
American Society of Human Genetics	http://www.ashg.org/
The Cancer Genome Atlas	http://cancergenome.nih.gov
Centers for Disease Control and Prevention	http://www.cdc.gov
ClinicalTrials.gov	http://www.clinicaltrials.gov/
Database of Genotype and Phenotype	http://www.ncbi.nlm.nih.gov/sites/entrez?Db=gap
U.S. Food and Drug Administration	http://www.fda.gov/
Genomic Careers Resource for Students	http://www.genome.gov/27538514
Gene Tests	http://www.ncbi.nlm.nih.gov/sites/GeneTests/
Published Genome-Wide Association Study Catalog	http://www.genome.gov/gwastudies/
HapMap Project and Data	http://hapmap.ncbi.nlm.nih.gov/
Human Genome Epidemiology Network	http://www.cdc.gov/genomics/hugenet/
Human Microbiome Project	https://commonfund.nih.gov/hmp/
Illumina, Individual Genome Sequencing	http://www.everygenome.com/
International Society of Nurses in Genetics	http://www.isong.org/
National Coalition for Health Professional Education in Genetics	http://www.nchpeg.org/
National Cancer Institute's Early Detection Research Network	http://edrn.nci.nih.gov/
National Institutes of Health	http://www.nih.gov
National Human Genome Research Institute	http://www.genome.gov/
National Office of Public Health Genomics	http://www.cdc.gov/genomics/
Pharmacogenomics Knowledge Base	http://www.pharmgkb.org/
1000 Genomes Project	http://www.1000genomes.org/

TABLE 2.2
Genetics and Genomics Scientific and Clinical Translation Resources (Continued)

Resource Title	Reference Website Locations
Clinical Translation	
Evaluation of Genomic Applications in Practice and Prevention	http://www.egappreviews.org/
Genomic Applications in Practice and Prevention Network	http://www.gappnet.org/gappnet/
Human Genome Epidemiology Network	http://www.cdc.gov/genomics/hugenet/default.htm
Professional Society Recommendations[a]	
American Society for Human Genetics	http://www.ashg.org/pdf/dtc_statement.pdf
American College of Medical Genetics	http://www.acmg.net/AM/Template.cfm?Section=Policy_Statements&Template=/CM/HTMLDisplay.cfm&ContentID54157
International Society of Nurses in Genetics	http://www.isong.org/ISONG_PS_direct_consumer_marketing_genetic_tests.php
National Society of Genetic Counselors	http://www.nsgc.org/Advocacy/PositionStatements/tabid/107/Default.aspx#DTC

[a]Direct-to-consumer testing and complex disease management.

Encode

ENCODE is a public research consortium for identification of all parts of the human genome (within each gene) that regulate normal physiologic and biologic functions (National Human Genome Research Institute, 2011b). Findings from ENCODE have shown that regions of the human genome previously thought to have no obvious or direct protein coding function are extremely important in gene regulation; expansive areas of intronic "junk" DNA actually regulate protein production in distant genes and chromosomes (ENCODE Project Consortium et al., 2007). Demonstrating enormous complexity, new structural categories are realized through ENCODE; instead of clear translation boundaries between exons and introns, there are alternative start and stop sites that have temporal and spatial uniqueness. When completed, ENCODE will demonstrate how an individual's biologic make-up may be linked to a person's responses in health, illness, and injury states (ENCODE Project Consortium et al., 2011).

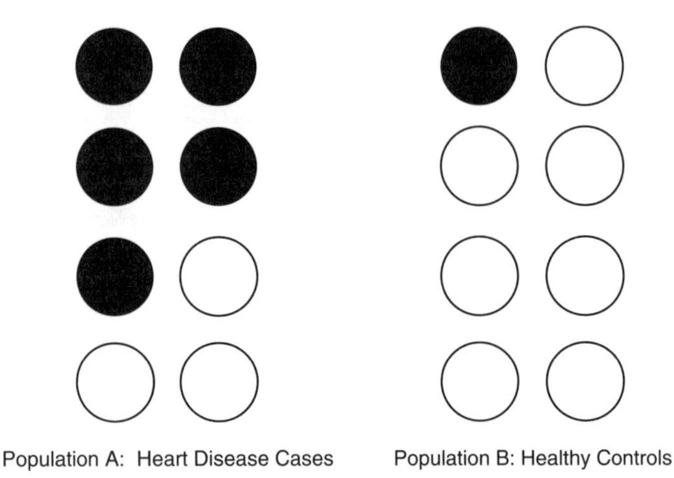

FIGURE 2.2 Basic foundation of genome-wide association studies.

Shown in Figure 2.2 is the basic model for genome-wide association studies (GWAS), where DNA from a population of patients with a disease is compared to healthy individuals to detect genetic differences. Patient's DNA is placed on a chip and millions of Single Nucleotide Polymorphisms (SNPs) can be tested at once.

In this figure, each circle represents a patient in two different populations; population A represents a cohort of heart disease cases and population B represents a group of healthy controls. Darkened circles indicate patients whose SNP sequence combinations contains a higher risk of heart disease development than patients with blank circles whose SNP sequence combinations carries a normal or low-risk of heart disease development. By genotyping individuals in both cohorts and comparing their SNP sequences, genetic patterns are identified that may be associating more frequently with heart disease, as demonstrated in this figure. If these patterns are located within a specific gene that was previously unknown, new molecular pathways and therapeutic targets could possibly arise as a result of the GWAS discovery. Thus, the GWAS approach permits identification of genes that cannot be predicted by selecting candidates from known disease pathways, allowing for discovery of novel genes in complex conditions.

It is important to remember that increased frequency of high-risk genotypes in a population of diseased patients does not infer causation. In addition to validating any positive signals obtained from GWAS research, scientists must then conduct functional laboratory experiments to determine if the gene identified is actually involved in the molecular cause of the disease.

1000 Genomes Project

Through HapMap and GWAS efforts, common genetic variants have been identified that contribute to complex chronic diseases. But the effects of those identified variants account for very little, in terms of risk, of developing a particular trait or phenotype. For example, obesity GWAS research identifies two common variants in the FTO and MC4R (melanocortin-4 receptor) genes carried by approximately 20% of those of European descent. These variants exert physiologic effects

EXHIBIT 2.1 Age-Related Macular Degeneration: From HapMap to Clinical Trial in Less Than 5 Years

Age-related macular degeneration (AMD) is a complex and multifactorial disease, is the leading cause of blindness with 1.75 million affected Americans, and is projected to affect 1 in 4 individuals older than 75 years old (Klein, Peto, Bird, & Vannewkirk, 2004; Ting, Lee, & MacDonald, 2009). In 2004–2005, researchers using HapMap identified a genetic variant responsible for 50% increased risk of developing AMD in the complement factor H gene (Edwards, et al., 2005; Haines, et al., 2005; Klein, et al., 2005). Complement factor H is part of the inflammatory complement cascade and helped lend credibility to concurrent investigations on inflammatory pathways as causative mechanisms of AMD, such as those involving C-reactive protein (Seddon, Gensler, Milton, Klein, & Rifai, 2004; Seddon, George, Rosner, & Rifai, 2005). These research efforts permitted development of prospective assessments of the role of the contributory SNP in patients; findings validated the role of genetics and inflammation in AMD (Schaumberg, et al., 2006). Follow-up investigation ascertained interactive effects of the complement factor H variant with modifiable environmental risk factors, including body mass index, smoking, regular aspirin intake, and dietary habits (Schaumberg, Hankinson, Guo, Rimm, & Hunter, 2007). This culminated in the formation of a large, randomized, controlled clinical trial enrolling 4,757 patients from 11 medical centers (Klein, et al., 2008). The multidisciplinary collaborative effort, "The Age Related Eye Disease Study," ascertained pharmacogenetic effects of treatment with zinc and antioxidant supplements according to SNP genotypes, one of which is in the complement factor H gene.

on body mass index (BMI) but account for only 2% of adult variability of the trait (Frayling et al., 2007; Loos et al., 2008). Rare and difficult to find genetic variants may explain more of the variability in BMI, but are not currently known. Sequencing the genomes of more individuals will allow for identification of rare variants with greater and more sizeable genetic effects. In September 2008, an international consortium launched the "1000 Genomes Project" to generate an extensive catalogue of human genetic variation capable of increasing discoveries of new biology from a population perspective. Initial data stemming from the project identifies extraordinarily complex patterns of heredity (Sudmant et al., 2010).

Human Microbiome Project

Understanding how bacterial colonies interact in our bodies is increasingly important (Hsiao & Fraser-Liggett, 2009). Bacteria and the colonies with which they live interact elegantly and intricately in contexts of greater host microenvironments. Recently, presence of specific types of bacteria in gut flora were found to predict occurrence of Type 1 Diabetes in experimental mouse models (Wen et al., 2008). Although symbiotic bacteria outnumber human body cells by 10:1, the species living on and within us have yet to be cataloged and information on how they communicate with each other and their human hosts is unknown (Turnbaugh et al., 2007). Using genomic sequencing, analyses of microbes from intestinal epithelial tissue in animals and individuals with obesity, intestinal disease, and cancer identified unique microbial profiles (Chu et al., 2004; Eckburg & Relman, 2007; Turnbaugh et al., 2006). To ascertain the extent that microbial communities participate in health and disease, the National Institutes of Health (NIH) Roadmap for Medical Research established the Human Microbiome Project (HMP, 2007). A global consortium, the HMP will provide genome sequence for floral communities across various human epithelial tissues. Initial data identify great diversity within and across individuals dependent on tissue sampling site (Grice et al., 2008). Funded grants stemming from the consortium will characterize microbiomes across variable disease states, tissue sites, and patient subpopulations (NIH HMP Working Group et al., 2009).

Cancer Genome Atlas

In 2007, the National Cancer Institute and the National Human Genome Research Institute began a large-scale genome sequencing of human tumor cells from cancer cohorts to provide a comprehensive catalogue of malignancy abnormalities (Collins & Barker, 2007; Cancer Genome Atlas, 2011). Characterization of glioblastoma, a common brain malignancy in adults noted for poor survival outcomes, yielded molecular profiles that may be used to reclassify and target clinical treatments (Cancer Genome Atlas Research Network, 2008). Continued progress with other common tumor types, such as ovarian cancer, are revolutionizing how molecular derangements in malignancies are understood, and are expanding diagnostic, prognostic, and treatment options (Cancer Genome Atlas Research Network, 2011).

Genetic Risk Assessment

Human genetic variation holds great promise in predicting the development of costly chronic and preventable diseases. For example, carriers of Single Nucleotide Polymorphism (SNP) risk alleles in the Transcription Factor 7-like 2 (TCF7L2) gene have significantly increased lifetime risk of developing diabetes (Helgason et al., 2007; Prokunina-Olsson et al., 2009). Using the information contained in a person's genetic constitution for these purposes is increasingly attractive given that recent genome-wide association studies (GWAS) delineate numerous novel and quantifiable genetic risks for common health conditions including heart disease, depression, colorectal cancer, and osteoporosis (Figure 2.2; McPherson et al., 2007; Scott et al., 2007; Tomlinson et al., 2007; Wellcome Trust Case Control Consortium, 2007).

Of particular interest is whether SNPs identified in GWAS reports can be used in preventive medicine to predict illnesses. Sponsored by the Agency for Healthcare Research and Quality (AHRQ), the U.S. Preventive Services Task Force (USPSTF) evaluates health screening mechanisms for efficacy, cost, and accuracy, and includes well-known examples of screening standards (Atkins, Fink, & Slutsky, 2005; Harris et al., 2001; U.S. Preventive Services Task Force, 2002a, 2002b, 2003a, 2003b, 2004, 2005). Presently, there is not enough evidence to guide a USPSTF review of using genomic SNP sequence for preventive screening (Burke & Psaty, 2007; Gwinn & Khoury, 2006; Khoury, Yang, Gwinn, Little, & Dana Flanders, 2004). In order to properly inform a USPSTF investigation, novel genetic variant associations must be consistently replicated across various settings (Burke & Psaty, 2007). However, many gene–gene and gene–environment interactive effects are unknown, attributable risk over one's lifetime is small to modest for many SNPs, and population SNP allele frequencies are not yet fully established (Janssens et al., 2007; Khoury, Little, Gwinn, & Ioannidis, 2007). Further contributing to delays in clinical translation are inadequate experimental designs, biased or erroneous interpretation of scientific data, and public dissemination of results from less rigorous research (Little et al., 2002; Moonesinghe, Khoury, & Janssens, 2007). To address concerns regarding the quality and completeness of genetic association reporting, a framework was recently identified for the evaluation of risk prediction models that include genetic variants (Janssens et al., 2011).

Knowledge gaps aside, the most heated debates involving SNP genetic testing concern provision of maximal benefit while limiting harm to patients (Khoury, Gwinn, Burke, Bowen, & Zimmern, 2007). Possible adverse unintended consequences of SNP genomic profiling include increased anxiety stemming from a test that does not impact health outcomes and excessive financial cost for little clinical advantage (Burke & Zimmern, 2004). Equally important is need for scientific data demonstrating how patients will use personalized genetic risk information once they receive it (McBride & Brody, 2007; Thompson, 2007).

As mentioned previously, several companies market and sell genomic "risk" profiles directly to consumers (Burke & Press, 2006; Janssens et al., 2008). Proposed advantages of direct-to-consumer genomic profiling are increased availability, privacy, convenience, and enhanced market translation of molecular research advances (Goddard et al., 2007). Test results are interpreted and communicated to customers via a report that is mailed to them or accessed via the Internet. Detailed test interpretations and lifestyle recommendations may also be provided. Regulatory protection against false health claims for consumers using these services is a key deficiency as genomic technologies flood health care markets (Katsanis, Javitt, & Hudson, 2008; Secretary's Advisory Committee on Genetics Health and Society, 2008).

To enhance accurate clinical translation of SNP associations, numerous resources are available to guide how biologic information can be used (Table 2.2). The Centers for Disease Control has developed and maintained *HuGENet* (Human Genome Epidemiology Network), a research database similar to *The Cochrane Collaboration Reviews*, where results of population-based gene–environment associations can be searched and obtained (Higgins et al., 2007; Lin et al., 2006; Seminara et al., 2007). Launched in 2004, the Evaluation of Genomic Applications in Practice and Prevention (EGAPP) is an independent and multidisciplinary panel that critically evaluates evidence supporting use of genomic tests in clinical practice through assessment of analytic validity, clinical validity, and clinical utility (Teutsch et al., 2009). Genetic test recommendations are available for various conditions, including venous thromboembolism, breast cancer, and Lynch syndrome (or Hereditary Nonpolyposis Colorectal Cancer; HNPCC). Building on the foundations of EGAPP, the Genomic Applications in Practice and Prevention Network (GAPPNet) brings together more collaborative stakeholders in order to better outline and disseminate current knowledge; develop an evidence-based recommendation process for review of newly released genetics/genomics technologies; translate research into real-world dissemination; and develop comprehensive education, outreach, and surveillance programs (Khoury et al., 2009). Professional organizations are also assuming positions of leadership by clarifying the clinician's role when managing patient inquiries about genomic profiling. For example, the American Society for Human Genetics issued a position statement outlining scope of clinical services that can be safely and legitimately provided to consumers (Hudson, Javitt, Burke, & Byers, 2007).

Diagnostics

Nurses can expect that technology-driven increases in diagnostic capacity will yield dramatic advancements for Personalized Medicine. Efficient diagnosis is presently a benchmark standard for health care quality and patient safety, and improved specificity and sensitivity for clinical decision-making is a key feature of many programs of research (Bissonnette & Bergeron, 2006; Dietel & Sers, 2006; Institute of Medicine, 2001; Snyderman & Langheier, 2006). The ability to generate and analyze enormous amounts of unique biologic data is fueling three fields of clinical application: infectious disease, cancer, and biomarker discovery.

Infectious Disease

In prescribing antimicrobial agents, the inability to reach a definitive clinical diagnosis is a significant challenge. Diagnostic certainty in treating infectious disease decreases patient inflammatory responses, transmission risks, and harmful exposure to clinicians (Bissonnette & Bergeron, 2006; Diekema et al., 2004; McGowan & Tenover, 2004; Raoult, Fournier, & Drancourt, 2004; Tenover, 2006; Alliance for the Prudent Use of Antibiotics, 2005). However, culturing of patient specimens typically requires 1–2 days for results, and until they are obtained, incorrect use of broad-spectrum prescriptive agents yields high pharmaceutical costs and the formation of antibiotic-resistant bacterial strains.

Genomic science is revealing why it is so challenging to effectively identify organisms resulting in a patient's clinical deterioration. Microbial genome sequence comparison shows evidence of horizontal gene transfer, resulting in enormous differences across and between singular pathogenic strains, shattering standard dogma of bacterial classification and clinical treatment (Fraser & Rappuoli, 2005). Applying these findings to clinical settings, it is hoped that ultra-fast sequencing of microbial genomes from infectious specimens can provide agent identification in 1–2 hours versus the traditional culturing standard of 24–48 hours (Bissonnette & Bergeron, 2006). Although rapid on-site diagnostics are the ultimate goal of ultra-fast microbial sequencing, it will require much more scientific effort before clinicians can fully harness these tools. For example, clinically relevant and predictive information must first be able to be extracted from expansive bioinformatic datasets of microbial genome sequence (Relman, 2011).

Cancer

As a professional discipline, oncology currently experiences the most clinical progress in working towards Personalized Medicine where molecular tumor profiles can be used to correlate targeted treatment regiments. Classic examples include breast cancers expressing human epidermal growth factor receptor-2 protein (HER2/neu) and chronic myelogenous leukemia with positive Philadelphia chromosome status. Molecular understanding of both conditions has yielded extraordinarily successful response rates to drug therapy with the monoclonal antibody Herceptin, and tyrosine kinase inhibitor Gleevec. These therapeutic agents target specific and clinically measurable genetic changes (Fischer, Streit, Hart, & Ullrich, 2003). Further expansion of molecular cancer profiling beyond these examples is being driven by the Cancer Genome Atlas Project (see previous discussion) and high-throughput screening technologies (Ludwig & Weinstein, 2005; Srivastava, 2006).

As with other areas of genomic scientific advancement, translation into clinical treatment and screening mechanisms is proving challenging. Presently, a formidable challenge preventing Personalized Medicine from being realized in cancer care is reconciling the unexpected difference between meaningful clinical diagnostic standards with personalized genomic tumor profiling (Ludwig & Weinstein, 2005). For example, genetic and genomic molecular profiles challenge current oncologic survival and treatment curves, necessitating redefined scoring criteria for the tumor, nodes, and metastasis (TNM) staging system, the widely used clinical grading system for cancer diagnosis stratification (Lam, Shvarts, Leppert, Figlin, & Belldegrun, 2005; Leong, 2006; Nguyen & Schrump, 2006; Piccaluga et al., 2008). Also contributing to the translation gap is the need for improved clinical research standards when validating positive scientific findings or performing cross-comparison analyses across different study populations. Spanning both research and clinical settings—differences in tumor specimen collection, laboratory processing and testing procedures, and clinical phenotype documentation are proving to be a significant source of confounding variability when trying to validate high-throughput genomic screening signals (Compton, 2007; Srivastava, 2006).

Biomarker Discovery

Nurses are widely familiar with biomarkers, or singular proteins and molecules in body fluids associated with diseases, as a way to facilitate personalized approaches to patient health. The linking of biomarkers to clinical outcome measurements presently represents a fruitful area of biologic research, and efforts can be classified into genetic, proteomic, antigen, and autoantibody classes (Srivastava, 2006). Powerful molecular approaches, such as microarray platforms, are fueling biomarker discovery because of technological ability to simultaneously evaluate thousands of molecules within and across numerous samples and patients (He, 2006). In the coming years, biomarkers will contribute greatly to Personalized Medicine by providing increasingly precise mechanisms of disease detection, prognostication, and therapeutic monitoring (Rai, 2007). Presently, the technology-driven increases in data volume and precision have yet to replace less accurate and widely used screening tests, such as the prostate-specific antigen, and ovarian cancer antigen-125 tests. A chief reason for biomarker adoption delay is the vast amount of time and work that is required to appropriately screen and validate initial findings before reliable reproducibility is obtained (Ransohoff, 2004). Development chronology occurs over many years with five distinct basic and clinical research development phases: (1) preclinical discovery, (2) assay analytical validation, (3) case/control cohort investigation, (4) longitudinal observation, and lastly, (5) prospective case/control investigation (Pepe et al., 2001). Stage progression occurs in a "funnel" format where initial screens of many molecules are examined and eliminated in order to meet rigorous specificity and sensitivity criteria while also demonstrating potential for predictive capacity (Srivastava, 2006). Poor reproducibility early on prohibits further application and progression to clinical translation.

Pharmacogenomics

Pharmacogenomics evolved from pharmacogenetics, a field established 50 years ago when it was demonstrated that a person's genetic inheritance could effect drug metabolism (Alving, Carson, Flanagan, & Ickes, 1956; Evans & Relling, 2004; Meyer, 2004). Building on the pharmacogenetic paradigm of one-gene and one-drug, pharmacogenomics studies how numerous genes interact with each other and the environment (Evans & McLeod, 2003; Goldstein, Tate, & Sisodiya, 2003; Weinshilboum, 2003; Weinshilboum & Wang, 2006). Presently, pharmacogenomic science understands and predicts adverse reactions to medical therapeutics based on an individual's DNA sequence (Ginsburg, Konstance, Allsbrook, & Schulman, 2005; Roses, 2004). Pharmacogenomics is projected to have broad clinical utility in multifactorial diseases in order to reduce adverse drug reactions (Phillips, Veenstra, Oren, Lee, & Sadee, 2001).

A recent illustrative case study for Personalized Medicine is warfarin (Coumadin), a frequently prescribed anticoagulant. Warfarin demonstrates costly challenges such as limited therapeutic windows, large dosage differences across individuals, risk for serious bleeding sequelae, inability to estimate patient drug responses from medical and physical criteria, and need for frequent international normalized ratio (INR) monitoring via phlebotomy (Daly & King, 2003). Traditional dosing factors incorporated include diet, age, gender, dietary intake, body weight, and use of other medications (Wadelius et al., 2007). More recently, variation in genes encoding the hepatic microsomal enzyme cytochrome P450 2C9 (CYP2C9) and the vitamin K epoxide reductase complex 1 (VKORC1) demonstrated influence respectively on warfarin pharmacokinetics and pharmacodynamics (Higashi et al., 2002; Rieder et al., 2005). Single nucleotide polymorphisms (SNPs) in CYP2C9 and VKORC1 genes explained 35%–50% of dosage difference variability among patients—leading to development of a clinical genetic test for alleles resulting in altered dose requirements (Aquilante et al., 2006; Gage et al., 2004; Geisen et al., 2005). On August 16, 2007, the Federal Drug Administration (FDA) updated the warfarin label to support combined CYP2C9 and VKORC1 genetic testing, with up to 35% of individuals benefiting from lower starting doses (FDA Press Conference on Warfarin Transcript, 2007). On January 22, 2010, this package insert was upgraded to a formal recommendation to conduct genetic testing prior to dosing warfarin and included dosing ranges for CYP2C9 and VKORC1 allelic variation combinations (FDA Package Insert Label, 2010).

Although warfarin is a mainstream drug for which the FDA provides genomic recommendations, it is not yet successful from a combination of uptake, utility, and cost perspectives. Recent warfarin genetic testing studies show that CYP2C9/VKORC1 is predictive of INR level but not bleeding or thrombotic sequelae, and that cost effectiveness remains a key concern (Eckman, Rosand, Greenberg, & Gage, 2009; Mahajan, Meyer, Wall, & Price, 2011; Millican et al., 2007; Rieder et al., 2005). Thus, Centers for Medicare and Medicaid Services (CMS) could not conclude from available evidence that warfarin pharmacogenetic testing was better than existing care coverage, and denied financial reimbursement for CYP2C9/VKORC1 testing on May 4, 2009 (Jensen et al., 2009). Opening the door for decision reversal if clinical trials demonstrate significant and cost-effective clinical utility, CMS issued the decision "pursuant to Coverage with Evidence Development (CED)." However, for CMS approval and widespread clinical uptake in the United States to occur, use of the CYP2C9/VKORC1 genetic test (currently priced at $300–$500) in current clinical trials will need to demonstrate ability to affordably improve current standards of care through prevention of adverse bleeding and thrombotic events for CMS coverage to be issued. Additional genetic/genomic and clinical research challenges for CYP2C9/VKORC1 testing, in addition to pharmacogenomic policy implications, are presented in Exhibit 2.2.

CONCLUDING REMARKS

This chapter outlined how personalization of health care practices will be affected through continued waves of scientific knowledge in genetics and genomics. Personalized medicine, a term used to denote how an individual's DNA guides tailored health care interventions, is being fueled by advancements in four concept domains: (1) discovery of novel biology, (2) genetic risk assessment, (3) molecular diagnostics, and (4) pharmacogenetics and pharmacogenomics. Despite the great potential for improved health across the four domains, scientific initiatives outlined in this article will fall short of desired outcomes without sufficient clinical application evidence concerning its implementation. Thus, the field of nursing should anticipate and contribute to these needs and developments, and play an integral role in their clinical translation and application. For example, successful design and implementation of Personalized Medicine will likely hinge on the roles of nurses conducting or participating in collaborative research, education, and clinical practice initiatives that are furthering genetics/genomics content applications. Because nurses "hold the keys" to accurate patient observation, client education, and health care practice dissemination, they are the foremost health discipline capable of bringing full investment in scientific advancement

EXHIBIT 2.2 The Pharmacogenomic and Policy Implications of CYP2C9 and VKORC1 Genetic Testing

Despite strong scientific genetic evidence of efficacy, warfarin classically demonstrates the systematic, structural, and clinical evidentiary challenges awaiting pharmacogenomics in personalized medicine (Phillips & Van Bebber, 2006). Because tests must demonstrate high predictive value for adverse events, metabolically accurate dosing ranges and treatment effects will need to be established according to allele prevalence in populations (Smits et al., 2005). Study designs specific to population genetics are not reflected in current standards for FDA drug evaluation and approval, where criteria are dependent on 20th century infrastructure and measurements (Califf, 2004; Haga, Thummel, & Burke, 2006; Woodcock, Witter, & Dionne, 2007). Moving forward, a key concern is that randomized clinical trials cannot account for genomic variability across study participants or clearly delineate patient groups per genetic responses to pharmaceutical agents. Common approaches to address the issue involve patient stratification according to race or ethnicity, but question of appropriateness is necessary as genetics research identifies serious flaws in these categories (Bamshad, Wooding, Salisbury, & Stephens, 2004; Doyle, 2006; Foster & Sharp, 2002; Lee, 2007). To better account for biotechnologic advancement, many proposals call for overhauled federal regulatory networks allowing improved focus on pharmacogenomic clinical research including synchronized payment systems, genetic/genomic education for patients and health care providers, and clinical care focusing on an individual's dynamic probability for disease development (Califf, 2004). What may achieve this most quickly is integration across government networks to form clinically accurate, large-scale, prospective observational cohorts (Roses, 2007).

A barrier to large-scale clinical genetic research efforts is reluctance in wholly embracing research efforts, stemming from public fears of genetic discrimination. To better address these concerns, the Genetic Information Nondiscrimination Act was passed in 2008. This legislation guarantees basic protection for individuals receiving genetic testing, so that they are protected against employer and health insurance discrimination practices (Slaughter, 2008). Limitations of the legislation are that it does not protect individuals from discrimination in the military or for long-term care coverage determinations, in addition to considerations that the legislative language is not specific enough regarding what is considered a "genetic test" (Baruch & Hudson, 2008; McGuire & Majumder, 2009).

to fruition—yielding advanced health for patients, families, and communities. As these technologic efforts expand and accelerate, it is hoped that professional nurses of all backgrounds can increase their practical involvement in the progress towards, and realization of, Personalized Medicine.

ACKNOWLEDGMENTS

This chapter partially fulfilled doctoral requirements at University of Utah's College of Nursing and was financially supported by the National Institutes of Health (NIH) pre-doctoral Intramural Research Training Award through the National Institute of Nursing Research (NINR). Warm appreciation for mentored faculty support is extended to Dr. Lawrence Brody, Chief of the Genome Technology Branch at the National Human Genome Research Institute in Bethesda, MD. The views and written content expressed in this chapter are solely that of the author and are not affiliated with URAC.

REFERENCES

Alliance for the Prudent Use of Antibiotics. (2005). Executive summary: Select findings, conclusions, and policy recommendations. *Clinical Infectious Diseases*, *41*(Suppl. 4), S224–S227.

Alving, A. S., Carson, P. E., Flanagan, C. L., & Ickes, C. E. (1956). Enzymatic deficiency in primaquine-sensitive erythrocytes. *Science*, *124*(3220), 484–485.

Aquilante, C. L., Langaee, T. Y., Lopez, L. M., Yarandi, H. N., Tromberg, J. S., Mohuczy, D., . . . Johnson, J. A. (2006). Influence of coagulation factor, vitamin K epoxide reductase complex subunit 1, and cytochrome P450 2C9 gene polymorphisms on warfarin dose requirements. *Clinical Pharmacology and Therapeutics*, *79*(4), 291–302.

Atkins, D., Fink, K., & Slutsky, J. (2005). Better information for better health care: The Evidence-based Practice Center program and the Agency for Healthcare Research and Quality. *Annals of Internal Medicine*, *142*(12 Pt 2), 1035–1041.

Bamshad, M., Wooding, S., Salisbury, B. A., & Stephens, J. C. (2004). Deconstructing the relationship between genetics and race. *Nature Reviews Genetics*, *5*(8), 598–609.

Baruch, S., & Hudson, K. (2008). Civilian and military genetics: Nondiscrimination policy in a post-GINA world. *American Journal of Human Genetics*, *83*(4), 435–444. http://dx.doi.org/10.1016/j.ajhg.2008.09.003

Bissonnette, L., & Bergeron, M. G. (2006). Next revolution in the molecular theranostics of infectious diseases: Microfabricated systems for personalized medicine. *Expert Review of Molecular Diagnostics*, *6*(3), 433–450.

Burke, W., & Press, N. (2006). Genetics as a tool to improve cancer outcomes: Ethics and policy. *Nature Reviews. Cancer*, *6*(6), 476–482.

Burke, W., & Psaty, B. M. (2007). Personalized medicine in the era of genomics. *JAMA: The Journal of the American Medical Association*, *298*(14), 1682–1684. http://dx.doi.org/10.1001/jama.298.14.1682

Burke, W., & Zimmern, R. L. (2004). Ensuring the appropriate use of genetic tests. *Nature Reviews. Genetics*, *5*(12), 955–959.

Califf, R. M. (2004). Defining the balance of risk and benefit in the era of genomics and proteomics. *Health Affairs (Millwood)*, *23*(1), 77–87.

Cancer Genome Atlas Research Network. (2008). Comprehensive genomic characterization defines human glioblastoma genes and core pathways. *Nature*, *455*(7216), 1061–1068. http://dx.doi.org/10.1038/nature07385

Cancer Genome Atlas Research Network. (2011). Integrated genomic analyses of ovarian carcinoma. *Nature*, *474*(7353), 609–615.

Centers for Disease Control and Prevention. (2011). *Pediatric genetics. Glossary of terms*. Retrieved from http://www.cdc.gov/ncbddd/pediatricgenetics/glossary.html

Chu, F. F., Esworthy, R. S., Chu, P. G., Longmate, J. A., Huycke, M. M., Wilczynski, S., & Doroshow, J. H. (2004). Bacteria-induced intestinal cancer in mice with disrupted Gpx1 and Gpx2 genes. *Cancer Research*, *64*(3), 962–968.

Collins, F. S. (2005, July 17). Personalized medicine: A new approach to staying well. *The Boston Globe*. Retrieved from http://www.boston.com/news/globe/editorial_opinion/oped/articles/2005/07/17/personalized_medicine/

Collins, F., & Barker, A. (2007). Mapping the cancer genome. Pinpointing the genes involved in cancer will help chart a new course across the complex landscape of human malignancies. *Scientific American*, *296*(3), 50–57.

Collins, F. S. (2010). Has the revolution arrived? *Nature*, *464*(7289), 674–675.

Collins, F. S., Green, E. D., Guttmacher, A. E., & Guyer, M. S. (2003). A vision for the future of genomics research. *Nature*, *422*(6934), 835–847.

Collins, F. S., & McKusick, V. A. (2001). Implications of the Human Genome Project for medical science. *JAMA: The Journal of the American Journal Association*, *285*(5), 540–544.

Compton, C. (2007). Getting to personalized cancer medicine: Taking out the garbage. *Cancer*, *110*(8), 1641–1643.

Daly, A. K., & King, B. P. (2003). Pharmacogenetics of oral anticoagulants. *Pharmacogenetics*, *13*(5), 247–252.

Diekema, D. J., BootsMiller, B. J., Vaughn, T. E., Woolson, R. F., Yankey, J. W., Ernst, E. J., . . . Doebbeling, B. N. (2004). Antimicrobial resistance trends and outbreak frequency in United States hospitals. *Clinical Infectious Diseases*, *38*(1), 78–85.

Dietel, M., & Sers, C. (2006). Personalized medicine and development of targeted therapies: The upcoming challenge for diagnostic molecular pathology. A review. *Virchows Archiv: An International Journal of Pathology*, *448*(6), 744–755.

Doyle, J. M. (2006). What race and ethnicity measure in pharmacologic research. *Journal of Clinical Pharmacology*, *46*(4), 401–404.

Eckburg, P. B., & Relman, D. A. (2007). The role of microbes in Crohn's disease. *Clinical Infectious Diseases*, *44*(2), 256–262. http://dx.doi.org/10.1086/510385

Eckman, M. H., Rosand, J., Greenberg, S. M., & Gage, B. F. (2009). Cost-effectiveness of using pharmacogenetic information in warfarin dosing for patients with nonvalvular atrial fibrillation. *Annals of Internal Medicine*, *150*(2), 73–83. http://dx.doi.org/10.1059/0003-4819-150-2-200901200-00005

Edwards, A. O., Ritter, R., III, Abel, K. J., Manning, A., Panhuysen, C., & Farrer, L. A. (2005). Complement factor H polymorphism and age-related macular degeneration. *Science*, *308*(5720), 421–424. http://dx.doi.org/10.1126/science.1110189

ENCODE Project Consortium, Birney, E., Stamatoyannopoulos, J. A., Dutta, A., Guigó, R., Gingeras, T. R., . . . de Jong, P. J. (2007). Identification and analysis of functional elements in 1% of the human genome by the ENCODE pilot project. *Nature*, *447*(7146), 799–816.

ENCODE Project Consortium, Myers, R. M., Stamatoyannopoulos, J., Snyder, M., Dunham, I., Hardison, R. C., . . . Crawford, G. E. (2011). A user's guide to the encylopedia of DNA elements (ENCODE). *PLoS Biology*, *9*(4), 1–21. http://dx.doi.org/10.1371/journal.pbio.1001046

Evans, W. E., & McLeod, H. L. (2003). Pharmacogenomics–drug disposition, drug targets, and side effects. *The New England Journal of Medicine*, *348*(6), 538–549.

Evans, W. E., & Relling, M. V. (2004). Moving towards individualized medicine with pharmacogenomics. *Nature*, *429*(6990), 464–468.

FDA Press Conference on Warfarin Transcript. (2007). Transcript of FDA Press Conference on Warfarin. Retrieved from http://www.fda.gov/downloads/NewsEvents/Newsroom/MediaTranscripts/ucm123583.pdf

FDA Package Insert Label (2010). Coumadin. Retrieved from http://www.accessdata.fda.gov/drugsatfda_docs/label/2010/009218s108lbl.pdf

Fischer, O. M., Streit, S., Hart, S., & Ullrich, A. (2003). Beyond Herceptin and Gleevec. *Current Opinion in Chemical Biology*, *7*(4), 490–495.

Foster, M. W., & Sharp, R. R. (2002). Race, ethnicity, and genomics: Social classifications as proxies of biological heterogeneity. *Genome Research*, *12*(6), 844–850.

Fraser, C. M., & Rappuoli, R. (2005). Application of microbial genomic science to advanced therapeutics. *Annual Review of Medicine*, *56*, 459–474.

Frayling, T. M., Timpson, N. J., Weedon, M. N., Zeggini, E., Freathy, R. M., Lindgren, C. M., . . . McCarthy, M. I. (2007). A common variant in the FTO gene is associated with body mass index and predisposes to childhood and adult obesity. *Science*, *316*(5826), 889–894. http://dx.doi.org/10.1126/science.1141634

Gage, B. F., Eby, C., Milligan, P. E., Banet, G. A., Duncan, J. R., & McLeod, H. L. (2004). Use of pharmacogenetics and clinical factors to predict the maintenance dose of warfarin. *Thrombosis and Haemostasis*, *91*(1), 87–94.

Geisen, C., Watzka, M., Sittinger, K., Steffens, M., Daugela, L., Seifried, E., . . . Oldenburg, J. (2005). VKORC1 haplotypes and their impact on the inter-individual and inter-ethnical variability of oral anticoagulation. *Thrombosis and Haemostasis*, *94*(4), 773–779.

Ginsburg, G. S., Konstance, R. P., Allsbrook, J. S., & Schulman, K. A. (2005). Implications of pharmacogenomics for drug development and clinical practice. *Archives of Internal Medicine*, *165*(20), 2331–2336.

Goddard, K. A., Moore, C., Ottman, D., Szegda, K. L., Bradley, L., & Khoury, M. J. (2007). Awareness and use of direct-to-consumer nutrigenomic tests, United States, 2006. *Genetics In Medicine: Official Journal of the American College of Medical Genetics*, *9*(8), 510–517.

Goldstein, D. B., Tate, S. K., & Sisodiya, S. M. (2003). Pharmacogenetics goes genomic. *Nature Reviews Genetics*, *4*(12), 937–947.

Gollust, S. E., Wilfond, B. S., & Hull, S. C. (2003). Direct-to-consumer sales of genetic services on the Internet. *Genetics In Medicine: Official Journal of the American College of Medical Genetics*, *5*(4), 332–337.

Government Accountability Office. (2010). Direct to consumer genetic tests: Misleading test results are further complicated by deceptive marketing and other questionable practices. Retrieved from http://www.gao.gov/new.items/d10847t.pdf

Grice, E. A., Kong, H. H., Renaud, G., Young, A. C., Bouffard, G. G., Blakesley, R. W., . . . Serge, J. A. (2008). A diversity profile of the human skin microbiota. *Genome Research*, *18*(7), 1043–1050. http://dx.doi.org/10.1101/gr.075549.107

Gwinn, M., & Khoury, M. J. (2006). Genomics and public health in the United States: Signposts on the translation highway. *Community Genetics*, *9*(1), 21–26.

Haga, S. B., Thummel, K. E., & Burke, W. (2006). Adding pharmacogenetics information to drug labels: Lessons learned. *Pharmacogenetics and Genomics*, *16*(12), 847–854.

Haines, J. L., Hauser, M. A., Schmidt, S., Scott, W. K., Olson, L. M., Gallins, P., . . . Pericak-Vance, M. A. (2005). Complement factor H variant increases the risk of age-related macular degeneration. *Science*, *308*(5720), 419–421. http://dx.doi.org/10.1126/science.1110359

Harris, R. P., Helfand, M., Woolf, S. H., Lohr, K. N., Mulrow, C. D., Teutsch, S. M., Atkins, D. (2001). Current methods of the US Preventive Services Task Force: A review of the process. *American Journal of Preventive Medicine*, *20*(3 Suppl), 21–35.

He, Y. D. (2006). Genomic approach to biomarker identification and its recent applications. *Cancer Biomark*, 2(3–4), 103–133.

Helgason, A., Pálsson, S., Thorleifsson, G., Grant, S. F., Emilsson, V., Gunnarsdottir, S., . . . Stefánsson, K. (2007). Refining the impact of TCF7L2 gene variants on type 2 diabetes and adaptive evolution. *Nature Genetics*, *39*(2), 218–225. http://dx.doi.org/10.1038/ng1960

Higashi, M. K., Veenstra, D. L., Kondo, L. M., Wittkowsky, A. K., Srinouanprachanh, S. L., Farin, F. M., & Rettie, A. E. (2002). Association between CYP2C9 genetic variants and anticoagulation-related outcomes during warfarin therapy. *JAMA: The Journal of the American Medical Association*, *287*(13), 1690–1698.

Higgins, J. P., Little, J., Ioannidis, J. P., Bray, M. S., Manolio, T. A., Smeeth, L., . . . Khoury, M. J. (2007). Turning the pump handle: Evolving methods for integrating the evidence on gene-disease association. *American Journal of Epidemiology*, *166*(8), 863–866.

Hindorff, L. A., Sethupathy, P., Junkins, H. A., Ramos, E. M., Mehta, J. P., Collins, F. S., & Manolio, T. A. (2009). Potential etiologic and functional implications of genome-wide association loci for human diseases and traits. *Proceedings of the National Academy of Sciences of the United States of America*, *106*(23), 9362–9367. http://dx.doi.org/10.1073/pnas.0903103106

Hsiao, W. W., & Fraser-Liggett, C. M. (2009). Human Microbiome Project–Paving the way to a better understanding of ourselves and our microbes. *Drug Discovery Today*, *14*(7–8), 331–333.

Hudson, K., Javitt, G., Burke, W., & Byers, P. (2007). ASHG Statement* on direct-to-consumer genetic testing in the United States. *Obstetrics and Gynecology*, *110*(6), 1392–1395. http://dx.doi.org/10.1097/01.AOG.0000292086.98514.8b

Human Microbiome Project. (2007). Retrieved from https://commonfund.nih.gov/hmp /

Hunter, D. J., & Kraft, P. (2007). Drinking from the fire hose–Statistical issues in genome-wide association studies. *The New England Journal of Medicine*, *357*(5), 436–439. http://dx.doi.org/10.1056/NEJMp078120

Illumina, Inc. (2011). Personal Genome Sequencing Service. Retrieved from http://www.everygenome.com/

Institute of Medicine. (2001). *Crossing the quality chasm: A new health system for the 21st century*. Washington, DC: National Academy Press.

International HapMap Consortium. (2005). A haplotype map of the human genome. *Nature*, *437*(7063), 1299–1320.

International Human Genome Sequencing Consortium. (2001). Initial sequencing and analysis of the human genome. *Nature*, *409*(6822), 860–921.

International Human Genome Sequencing Consortium. (2004). Finishing the euchromatic sequence of the human genome. *Nature*, *431*(7011), 931–945. http://dx.doi.org/10.1038/nature03001

Janssens, A. C., Gwinn, M., Bradley, L. A., Oostra, B. A., van Duijn, C. M., & Khoury, M. J. (2008). A critical appraisal of the scientific basis of commercial genomic profiles used to assess health risks and personalize health interventions. *American Journal of Human Genetics*, *82*(3), 593–599. http://dx.doi.org/10.1016/j.ajhg.2007.12.020

Janssens, A. C., Ioannidis, J., van Duijn, C., Little, J., Khoury, M. J., & GRIPS Group. (2011). Strengthening the reporting of genetic risk prediction studies: The GRIPS statement. *Genome Medicine*, *3*(3), 16.

Janssens, A. C., Moonesinghe, R., Yang, Q., Steyerberg, E. W., van Duijn, C. M., & Khoury, M. J. (2007). The impact of genotype frequencies on the clinical validity of genomic profiling for predicting common chronic diseases. *Genetics In Medicine: Official Journal of the American College of Medical Genetics*, *9*(8), 528–535. http://dx.doi.org/10.1097/GIM.0b013e31812eece0

Jensen, T., Jacques, L., Ciccanti, M., Long, K., Eggleston, L., & Roche, J. (2009). *Decision Memorandum for Pharmacogenomic Testing to Predict Warfarin Responsiveness (ID# CAG-00400N)*. Retrieved from http://www.ncbi.nlm.nih.gov/sites/GeneTests/

Katsanis, S. H., Javitt, G., & Hudson, K. (2008). Public health. A case study of personalized medicine. *Science, 320*(5872), 53–54. http://dx.doi.org/10.1126/science.1156604

Khoury, M. J., Feero, W. G., Reyes, M., Citrin, T., Freedman, A., Leonard, D., . . . Terry, S. (2009). The genomic applications in practice and prevention network. *Genetics In Medicine: Official Journal of the American College of Medical Genetics, 11*(7), 488–494. http://dx.doi.org/10.1097/GIM.0b013e3181a551cc

Khoury, M. J., Gwinn, M., Burke, W., Bowen, S., & Zimmern, R. (2007). Will genomics widen or help heal the schism between medicine and public health? *American Journal of Preventive Medicine, 33*(4), 310–317. http://dx.doi.org/10.1016/j.amepre.2007.05.010

Khoury, M. J., Little, J., Gwinn, M., & Ioannidis, J. P. (2007). On the synthesis and interpretation of consistent but weak gene-disease associations in the era of genome-wide association studies. *International Journal of Epidemiology, 36*(2), 439–445.

Khoury, M. J., Yang, Q., Gwinn, M., Little, J., & Dana Flanders, W. (2004). An epidemiologic assessment of genomic profiling for measuring susceptibility to common diseases and targeting interventions. *Genetics In Medicine: Official Journal of the American College of Medical Genetics, 6*(1), 38–47.

Klein, M. L., Francis, P. J., Rosner, B., Reynolds, R., Hamon, S. C., Schultz, D. W., . . . Seddon, J. M. (2008). CFH and LOC387715/ARMS2 genotypes and treatment with antioxidants and zinc for age-related macular degeneration. *Ophthalmology, 115*(6), 1019–1025. http://dx.doi.org/10.1016/j.ophtha.2008.01.036

Klein, R., Peto, T., Bird, A., & Vannewkirk, M. R. (2004). The epidemiology of age-related macular degeneration. *American Journal of Ophthalmology, 137*(3), 486–495. http://dx.doi.org/10.1016/j.ajo.2003.11.069

Klein, R. J., Zeiss, C., Chew, E. Y., Tsai, J. Y., Sackler, R. S., Haynes, C., . . . Hoh, J. (2005). Complement factor H polymorphism in age-related macular degeneration. *Science, 308*(5720), 385–389. http://dx.doi.org/1109557 [pii] 10.1126/science.1109557

Kutz, G. (2006). *Nutrigenetic testing: Tests purchased from four web sites mislead consumers.* (GAO-06-977T). Washington, DC: United States Government Accountability Office. Retrieved from http://www.gao.gov/new.items/d06977t.pdf

Lam, J. S., Shvarts, O., Leppert, J. T., Figlin, R. A., & Belldegrun, A. S. (2005). Renal cell carcinoma 2005: New frontiers in staging, prognostication and targeted molecular therapy. *The Journal of Urology, 173*(6), 1853–1862. http://dx.doi.org/10.1097/01.ju.0000165693.68449.c3

Lander, E. S. (2011). Initial impact of the sequencing of the human genome. *Nature, 470*(7333), 187–197.

Lee, S. S. (2007). The ethical implications of stratifying by race in pharmacogenomics. *Clinical Pharmacology and Therapeutics, 81*(1), 122–125.

Leong, S. P. (2006). Paradigm shift of staging and treatment for early breast cancer in the sentinel lymph node era. *The Breast Journal, 12*(5 Suppl. 2), S128–S133. http://dx.doi.org/10.1111/j.1075-122X.2006.00326.x

Lin, B. K., Clyne, M., Walsh, M., Gomez, O., Yu, W., Gwinn, M., & Khoury, M. J. (2006). Tracking the epidemiology of human genes in the literature: The HuGE Published Literature database. *American Journal of Epidemiology, 164*(1), 1–4.

Little, J., Bradley, L., Bray, M. S., Clyne, M., Dorman, J., Ellsworth, D. L., & Weinberg, C. (2002). Reporting, appraising, and integrating data on genotype prevalence and gene-disease associations. *American Journal of Epidemiology, 156*(4), 300–310.

Loos, R. J., Lindgren, C. M., Li, S., Wheeler, E., Zhao, J. H., Prokopenko, I., & Mohlke, K. L. (2008). Common variants near MC4R are associated with fat mass, weight, and risk of obesity. *Nature Genetics, 40*(6), 768–775. http://dx.doi.org/10.1038/ng.140

Ludwig, J. A., & Weinstein, J. N. (2005). Biomarkers in cancer staging, prognosis, and treatment selection. *Nature Reviews. Cancer*, *5*(11), 845–856.

Mahajan, P., Meyer, K. S., Wall, G. C., & Price, H. J. (2011). Clinical applications of pharmacogenomics guided warfarin dosing. *International Journal of Clinical Pharmacy*, *33*(1), 10–19.

Manolio, T. A., Brooks, L. D., & Collins, F. S. (2008). A HapMap harvest of insights into the genetics of common disease. *Journal of Clinical Investigation*, *118*(5), 1590–1605. http://dx.doi.org/10.1172/JCI34772

McBride, C. M., & Brody, L. C. (2007). Point: Genetic risk feedback for common disease time to test the waters. *Cancer Epidemiology, Biomarkers & Prevention*, *16*(9), 1724–1726.

McGowan, J. E., Jr., & Tenover, F. C. (2004). Confronting bacterial resistance in health care settings: A crucial role for microbiologists. *Nature Reviews. Microbiology*, *2*(3), 251–258.

McGuire, A., Evans, B. J., Caulfield, T., & Burke, W. (2010). Science and regulation. Regulating direct-to-consumer personal genome testing. *Science*, *330*(6001), 181–182.

McGuire, A. L., & Majumder, M. A. (2009). Two cheers for GINA? *Genome Medicine*, *1*(1), 6. http://dx.doi.org/10.1186/gm6

McPherson, R., Pertsemlidis, A., Kavaslar, N., Stewart, A., Roberts, R., Cox, D. R., . . . Cohen, J. C. (2007). A common allele on chromosome 9 associated with coronary heart disease. *Science*, *316*(5830), 1488–1491.

Metzker, M. L. (2010). Sequencing technologies — The next generation. *Nature Reviews. Genetics*, *11*(1), 31–46.

Meyer, U. A. (2004). Pharmacogenetics — Five decades of therapeutic lessons from genetic diversity. *Nature Reviews. Genetics*, *5*(9), 669–676.

Millican, E. A., Lenzini, P. A., Milligan, P. E., Grosso, L., Eby, C., Deych, E., . . . Gage, B. F. (2007). Genetic-based dosing in orthopedic patients beginning warfarin therapy. *Blood*, *110*(5), 1511–1515. http://dx.doi.org/10.1182/blood-2007-01-069609

Moonesinghe, R., Khoury, M. J., & Janssens, A. C. (2007). Most published research findings are false-but a little replication goes a long way. *PLoS Medicine*, *4*(2), e28.

National Cancer Institute. (2011). *Dictionary of cancer terms*. Retrieved from http://www.cancer.gov/dictionary

National Center for Advancing Translational Sciences, Office of Rare Diseases Research. (2011). *Terms and definitions*. Retrieved from http://rarediseases.info.nih.gov/Glossary.aspx

National Center for Biotechnology Information, GeneReviews. (2011). *Illustrated glossary. Terms and definitions*. Retrieved from http://www.ncbi.nlm.nih.gov/books/NBK5191/

National Human Genome Research Institute. (2011a). *The ENCODE (ENCyclopedia Of Dna Elements)*. Retrieved from http://www.genome.gov/10005107

National Human Genome Research Institute. (2011b). *A catalog of published genome-wide association studies*. Retrieved from http://www.genome.gov/gwastudies/

National Human Genome Research Institute. (2011c). *Talking glossary of genetic terms*. Retrieved from http://www.genome.gov/glossary/index.cfm?

Nguyen, D. M., & Schrump, D. S. (2006). Lung cancer staging in the genomics era. *Thoracic Surgery Clinics*, *16*(4), 329–337.

NIH HMP Working Group, Peterson, J., Garges, S., Giovanni, M., McInnes, P., Wang, L., . . . Guyer, M. (2009). The NIH human microbiome project. *Genome Research*, *19*(12), 2317–2323.

Pepe, M. S., Etzioni, R., Feng, Z., Potter, J. D., Thompson, M. L., Thornquist, M., . . . Yasui, Y. (2001). Phases of biomarker development for early detection of cancer. *Journal of the National Cancer Institute*, *93*(14), 1054–1061.

Phillips, K. A., & Van Bebber, S. L. (2006). Regulatory perspectives on pharmacogenomics: A review of the literature on key issues faced by the United States Food and Drug Administration. *Medical Care Research and Review*, *63*(3), 301–326.

Phillips, K. A., Veenstra, D. L., Oren, E., Lee, J. K., & Sadee, W. (2001). Potential role of pharmacogenomics in reducing adverse drug reactions: A systematic review. *JAMA: The Journal of the American Medical Association*, *286*(18), 2270–2279.

Piccaluga, P. P., Califano, A., Klein, U., Agostinelli, C., Bellosillo, B., Gimeno, E., . . . Pileri, S. A. (2008). Gene expression analysis provides a potential rationale for revising the histological grading of follicular lymphomas. *Haematologica*, *93*(7), 1033–1038. http://dx.doi.org/10.3324/haematol.12754

Prokunina-Olsson, L., Welch, C., Hansson, O., Adhikari, N., Scott, L. J., Usher, N., . . . Hall, J. L. (2009). Tissue-specific alternative splicing of TCF7L2. *Human Molecular Genetics*, *18*(20), 3795–3804. http://dx.doi.org/10.1093/hmg/ddp321

Rai, A. J. (2007). Biomarkers in translational research: Focus on discovery, development, and translation of protein biomarkers to clinical immunoassays. *Expert Review of Molecular Diagnostics*, *7*(5), 545–553. http://dx.doi.org/10.1586/14737159.7.5.545

Ransohoff, D. F. (2004). Rules of evidence for cancer molecular-marker discovery and validation. *Nature Reviews. Cancer*, *4*(4), 309–314. http://dx.doi.org/10.1038/nrc1322

Raoult, D., Fournier, P. E., & Drancourt, M. (2004). What does the future hold for clinical microbiology? *Nature Reviews. Microbiology*, *2*(2), 151–159.

Relman, D. A. (2011). Microbial genomics and infectious diseases. *The New England Journal of Medicine*, *365*(4), 347–357.

Rieder, M. J., Reiner, A. P., Gage, B. F., Nickerson, D. A., Eby, C. S., McLeod, H. L., . . . Rettie, A. E. (2005). Effect of VKORC1 haplotypes on transcriptional regulation and warfarin dose. *The New England Journal of Medicine*, *352*(22), 2285–2293.

Roses, A. (2007). "Personalized medicine: Elusive dream or imminent reality?": A commentary. *Clinical Pharmacology and Therapeutics*, *81*(6), 801–805.

Roses, A. D. (2004). Pharmacogenetics and drug development: The path to safer and more effective drugs. *Nature Reviews. Genetics*, *5*(9), 645–656.

Schaumberg, D. A., Christen, W. G., Kozlowski, P., Miller, D. T., Ridker, P. M., & Zee, R. Y. (2006). A prospective assessment of the Y402H variant in complement factor H, genetic variants in C-reactive protein, and risk of age-related macular degeneration. *Investigative Ophthalmology and Visual Science*, *47*(6), 2336–2340. http://dx.doi.org/10.1167/iovs.05-1456

Schaumberg, D. A., Hankinson, S. E., Guo, Q., Rimm, E., & Hunter, D. J. (2007). A prospective study of 2 major age-related macular degeneration susceptibility alleles and interactions with modifiable risk factors. *Archives of Ophthalmology*, *125*(1), 55–62. http://dx.doi.org/10.1001/archopht.125.1.55

Scott, L. J., Mohlke, K. L., Bonnycastle, L. L., Willer, C. J., Li, Y., Duren, W. L., . . . Boehnke, M. (2007). A genome-wide association study of type 2 diabetes in Finns detects multiple susceptibility variants. *Science*, *316*(5829), 1341–1345.

Secretary's Advisory Committee on Genetics, Health, and Society. (2008). U.S. system of oversight of genetic testing: A response to the charge of the Secretary of Health and Human Services (pp. 1–276). Washington, DC: U.S. Department of Health & Human Services.

Seddon, J. M., Gensler, G., Milton, R. C., Klein, M. L., & Rifai, N. (2004). Association between C-reactive protein and age-related macular degeneration. *JAMA: The Journal of the American Medical Association*, *291*(6), 704–710. http://dx.doi.org/10.1001/jama.291.6.704

Seddon, J. M., George, S., Rosner, B., & Rifai, N. (2005). Progression of age-related macular degeneration: Prospective assessment of C-reactive protein, interleukin 6, and other cardiovascular biomarkers. *Archives of Ophthalmology*, *123*(6), 774–782. http://dx.doi.org/10.1001/archopht.123.6.774

Seminara, D., Khoury, M. J., O'Brien, T. R., Manolio, T., Gwinn, M. L., Little, J., . . . Ioannidis, J. P. (2007). The emergence of networks in human genome epidemiology: Challenges and opportunities. *Epidemiology*, *18*(1), 1–8.

Slaughter, L. M. (2008). The Genetic Information Nondiscrimination Act: Why your personal genetics are still vulnerable to discrimination. *The Surgical Clinics of North America*, *88*(4), 723–738, vi. http://dx.doi.org/10.1016/j.suc.2008.04.004

Smits, K. M., Schouten, J. S., Smits, L. J., Stelma, F. F., Nelemans, P., & Prins, M. H. (2005). A review on the design and reporting of studies on drug-gene interaction. *Journal of Clinical Epidemiology*, *58*(7), 651–654.

Snyderman, R., & Langheier, J. (2006). Prospective health care: The second transformation of medicine. *Genome Biology*, *7*(2), 104. http://dx.doi.org/10.1186/gb-2006-7-2-104

Srivastava, S. (2006). Molecular screening of cancer: The future is here. *Molecular Diagnosis & Therapy*, *10*(4), 221–230.

Sudmant, P. H., Kitzman, J. O., Antonacci, F., Alkan, C., Malig, M., Tsalenko, A., . . . Shendure, J. (2010). Diversity of human copy number variation and multicopy genes. *Science*, *330*(6004), 641–646.

Tenover, F. C. (2006). Mechanisms of antimicrobial resistance in bacteria. *American Journal of Infection Control*, *34*(5 Suppl. 1), S3–S10.

Teutsch, S. M., Bradley, L. A., Palomaki, G. E., Haddow, J. E., Piper, M., Calonge, N., . . . Berg, A. O. (2009). The Evaluation of Genomic Applications in Practice and Prevention (EGAPP) Initiative: Methods of the EGAPP Working Group. *Genetics In Medicine: Official Journal of the American College of Medical Genetics*, *11*(1), 3–14. http://dx.doi.org/10.1097/GIM.0b013e318184137c

The Cancer Genome Atlas. (2011). Retrieved from http://cancergenome.nih.gov/

Thompson, P. A. (2007). Counterpoint: Genetic risk feedback for common disease time to test the waters. *Cancer Epidemiology, Biomarkers & Prevention*, *16*(9), 1727–1729.

Ting, A. Y., Lee, T. K., & MacDonald, I. M. (2009). Genetics of age-related macular degeneration. *Current Opinion in Ophthalmology*, *20*(5), 369–376. http://dx.doi.org/10.1097/ICU.0b013e32832f8016

Tomlinson, I., Webb, E., Carvajal-Carmona, L., Broderick, P., Kemp, Z., Spain, S., . . . Houlston, R. (2007). A genome-wide association scan of tag SNPs identifies a susceptibility variant for colorectal cancer at 8q24.21. *Nature Genetics*, *39*(8), 984–988.

Turnbaugh, P. J., Ley, R. E., Hamady, M., Fraser-Liggett, C. M., Knight, R., & Gordon, J. I. (2007). The human microbiome project. *Nature*, *449*(7164), 804–810. http://dx.doi.org/10.1038/nature06244

Turnbaugh, P. J., Ley, R. E., Mahowald, M. A., Magrini, V., Mardis, E. R., & Gordon, J. I. (2006). An obesity-associated gut microbiome with increased capacity for energy harvest. *Nature*, *444*(7122), 1027–1031. http://dx.doi.org/10.1038/nature05414

U.S. Preventive Services Task Force. (2002a). Screening for colorectal cancer: Recommendation and rationale. *Annals of Internal Medicine*, *137*(2), 129–131.

U.S. Preventive Services Task Force. (2002b). Screening for depression: Recommendations and rationale. *Annals of Internal Medicine*, *136*(10), 760–764.

U.S. Preventive Services Task Force. (2003a). Screening for obesity in adults: Recommendations and rationale. *Annals of Internal Medicine*, *139*(11), 930–932.

U.S. Preventive Services Task Force. (2003b). Screening for type 2 diabetes mellitus in adults: Recommendations and rationale. *Annals of Internal Medicine*, *138*(3), 212–214.

U.S. Preventive Services Task Force. (2004). Screening for coronary heart disease: Recommendation statement. *Annals of Internal Medicine*, *140*(7), 569–572.

U.S. Preventive Services Task Force. (2005). Genetic risk assessment and BRCA mutation testing for breast and ovarian cancer susceptibility: Recommendation statement. *Annals of Internal Medicine*, *143*(5), 355–361.

Wadelius, M., Chen, L. Y., Eriksson, N., Bumpstead, S., Ghori, J., Wadelius, C., . . . Deloukas, P. (2007). Association of warfarin dose with genes involved in its action and metabolism. *Human Genetics*, *121*(1), 23–34.

Weinshilboum, R. (2003). Inheritance and drug response. *The New England Journal of Medicine, 348*(6), 529–537.

Weinshilboum, R. M., & Wang, L. (2006). Pharmacogenetics and pharmacogenomics: Development, science, and translation. *Annual Review of Genomics and Human Genetics*, 7, 223–245.

Wellcome Trust Case Control Consortium. (2007). Genome-wide association study of 14,000 cases of seven common diseases and 3,000 shared controls. *Nature*, 447(7145), 661–678.

Wen, L., Ley, R. E., Volchkov, P. Y., Stranges, P. B., Avanesyan, L., Stonebraker, A. C., . . . Chernovsky, A. V. (2008). Innate immunity and intestinal microbiota in the development of Type 1 diabetes. *Nature*, 455(7216), 1109–1113. http://dx.doi.org/10.1038/nature07336

Woodcock, J., Witter, J., & Dionne, R. A. (2007). Stimulating the development of mechanism-based, individualized pain therapies. *Nature Reviews. Drug Discovery*, 6(9), 703–710.

CHAPTER 3

Diseasome

An Approach to Understanding Gene–Disease Interactions

Kenneth Wysocki and Leslie Ritter

ABSTRACT

Using bioinformatics computational tools, network maps that integrate the complex interactions of genetics and diseases have been developed. The purpose of this review is to introduce the reader to new approaches in understanding disease–gene associations using network maps, with an emphasis on how the human disease network (HDN) map (or diseasome) was constructed. A search was conducted in PubMed using the years 1999–2011 and using key words diseasome, molecular interaction, interactome, protein–protein interaction, and gene. The information reviewed included journal reviews, open source and web-based databases, and open source computational tools.

A review of the literature revealed the complexity of molecular, genetic, and protein structures that contribute to cellular function and possible disease, and how network mapping can help the clinician and scientist gain a better understanding of this complexity. Using computational tools and databases of genetics, protein interactions, and diseases, scientists have developed a network map of human genes and human diseases referred to as a diseasome. The diseasome is composed of 22 disease classes represented in different colored circular

http://dx.doi.org/10.1891/0739-6686.29.55

nodes. Lines connecting nodes indicate shared genes among diseases. Thus, the diseasome map provides a colorfully visual display that helps the user conceptualize gene–disease relationships.

This review provides an overview of the use of network maps to understand the interrelationships of genomics and disease. One such map, the diseasome, could be used as a reference for biomedical researchers and multidiscipline health care providers, including nurse practitioners and genetic counselors, to enhance their conceptualization and understanding of the genetic origins of disease.

BACKGROUND

Currently it is estimated that there are more than 3,000 human genes associated with one or more of over 2,000 human disorders (Vidal, Cusick, & Barabási, 2011). Thus, there continues to be a need for innovative approaches to advance the understanding of the complexities of genetics and disease. Some of the questions commonly asked are what proteins are involved in a disease of interest, what possible genes might be involved in the production and regulation of these proteins, and how might genes be associated with a disease of interest. Over the past 11 years, scientists have combined new computational network-based tools with bioinformatics to create a novel approach to understand these complex relationships or "interrelated parts" between proteins, diseases, and genes. In general terms, a network is described as a system of interrelated parts. An emerging science, known as network science, simplifies complex systems by summarizing them into components (nodes) and describing the interactions (edges) between them. In the context of genetics and disease, network scientists have worked to map human molecular interactions, protein–protein interactions (PPI), and protein–disease relationships.

In this chapter, we present network mapping as a novel approach to understanding gene–disease interactions. The authors review approaches to understanding and categorizing molecular interactions (i.e., interactome), PPI, public databases and tools for functional genomics, and bioinformatics computational methods used to explore complex networks of human diseases and disease genes. The authors also present how the human disease network (HDN) map (or diseasome) was constructed and how network maps such as this can be used to enhance understanding of the complex links between genes and diseases.

The articles included in this review were retrieved from PubMed using the years 1999–2011 and key words diseasome, molecular interaction, interactome, protein–protein interaction, and gene. The literature review included more than 60 journal articles and review of web-based resources that included

25 frequently reported open source databases, and 15 open source prediction/computational tools.

INTERACTOME

As science evolves over time, our approach and interpretation of the complex patterns and processes within the human body also evolves. Similar to the intense focus on genome projects 20 years ago, in the past decade, there have been intense efforts toward mapping PPI (De Las Rivas & Fontanillo, 2010). In this section, fundamentals of the interactome and how it has been used as a template to explain PPI and, more recently, a way to express disease–gene interactions, such as the diseasome, will be discussed. In addition, an overview of the components and developmental approach of an interactome will be introduced in order to understand the scientific underpinnings of network maps. In subsequent sections, the molecular interactome as it relates to protein-to-protein interactions will be addressed to facilitate understanding of how the diseasome was developed.

Although the Human Genome Project was initiated more than 20 years ago to sequence the human genome, discover disease related genes, and develop new strategies for diagnosis, treatment, and prevention, our understanding of many molecular mechanisms remains incomplete. The term interactome was introduced in 1999 by a group of French scientists led by Bernard Jacq and defined as the complete set of molecular interactions in living organisms including carbohydrates, lipids, nucleic acids, and proteins and is typically displayed as a directed graph (Sanchez et al., 1999). The aim of using the term interactome is to describe the contents, structure, function, and behavior of these interactions in organisms and the human body. (Sanchez et al., 1999) Interactome networks have been described as global views of complex biological processes within an organism at the molecular level (Missiuro et al., 2009).

It is well understood that diseases are considered perturbed states of the molecular system involved in cell and organism operation (Kanehisa, Goto, Furumichi, Tanabe, & Hirakawa, 2010). Living organisms and associated phenotypes are composed of a complex web of interactions between tissue, cellular, genetic, and molecular networks. A complex phenotype, such as disease, has been described as genetic and environmental perturbations in the system (Sieberts & Schadt, 2007). Knowledge of these complex macromolecular interactions has fueled innovative approaches to understanding expression of disease phenotypes, progression of the disease process, and potential treatments. Although genomic research has driven the need for more computer space to store all this new information and need for computational power that can handle the analysis of this data, there is need for humans to be able to read, understand, and work with this information (Kanehisa, 2009). Resources have been developed that include (a) lists of molecules, genes, proteins, diseases, environmental factors, diagnostic markers, and therapeutic drug lists; (b) computational tools to process this information into associations and networks; and (c) network representations (i.e., network maps) in two dimensions or, in some cases, three dimensions. Studies of the interactome include networks of PPI and disease–gene interactions.

PROTEIN–PROTEIN INTERACTIONS

With respect to proteomics, the interactome is referred to as a protein–protein interaction (PPI) network. PPI is defined as the molecular docking between proteins occurring in a cell or in a living organism; the interface should be intentional and evolved for a specific purpose as compared to other generic functions like protein production and degradation (De Las Rivas & Fontanillo, 2010). In PPI network maps, nodes represent proteins and edges represent the interaction between two proteins. Other protein interactions with genetic material, with non-protein chemicals, or with binding properties to an ion or a molecule (such as protein-RNA, protein-DNA, protein-cofactor, or protein-ligand) may be important components of an interactome but are not to be confused with, or included in, a PPI data base (De Las Rivas & Fontanillo, 2010). Model organisms such as *Saccharomyces cerevisiae*, *Caenorhabditis elegans*, *Drosophila melanogaster*, and *Mus musculus* have been used in high throughput (HTP) experiments to determine PPI, complex interrelationships that serve as a network, and provide evidence of potential human PPIs. These experiments include large-scale yeast one-, two-, and three-hybrid screens, affinity chromatography, co-immunoprecipitation (Co-IP) coupled mass spectrometry, fluorescence resonance energy transfer (FRET), regular low-scale experiments, and next generation Stitch-seq that combines PCR stitching with sequencing (Souiai et al., 2011; Walhout, Boulton, & Vidal, 2000; Yang, Li, Wu, Kwoh, & Ng, 2011; Yu et al., 2011).

It is estimated that there are more than 100,000 protein interactions in the human body (Bonetta, 2010). Protein interactions are integral in signaling pathways that serve in the metabolic function of cells (Brown & Jurisica, 2005). There are different approaches to constructing a PPI network map. Methods that predict function and interaction between proteins (i.e., functional prediction methods) include Rosetta Stone method, neighbour genes method, and phylogenetic profiles method. The Rosetta Stone method is based on the concept that gene organization is conserved between organisms (Marcotte, Pellegrini,

Thompson, Yeates, & Eisenberg, 1999). Neighbour genes method incorporates the concept that gene order is conserved between organisms and encodes functionally related proteins (Dandekar, Snel, Huynen, & Bork, 1998). The phylogenetic profiles method is founded on the concept that, because gene content varies between organisms, coinheritance of proteins in different species may indicate their functional link (Brun et al., 2003; Pellegrini, Marcotte, Thompson, Eisenberg, & Yeates, 1999).

Based on the principal that two proteins are more likely to be functionally related if they share common molecular interactions or functional links, computational methods can be used in building a network map to illustrate these relationships (Brun et al., 2003). Protein interactions can be direct, such as a cytokine receptor interaction; or indirect, such as part of a complex cellular pathway. Using bioinformatics, mapping these PPI into networks can be used as a powerful way to compare, study, and predict protein function. High-quality PPI maps, such as molecular medicine GPS, have been used to develop molecular tools such as biomarker diagnostic tools and RNA interference (RNAi), controllers of gene activation, that help accelerate drug discovery for certain diseases (Coulombe, 2010).

Although computational methods are useful in analyzing PPIs, there are limitations when dealing with the large number of proteins in the human proteome (Brun et al., 2003). Proteins may share a functional contact with each other but not all proteins in close proximity interact. Because of the large volume of protein interactions within a cell or organism, it may be difficult to place all of the protein interactions in a two-dimensional network map; however, sophisticated computer programs allow these networks to be represented in a three dimensional space. Both a 2-D or 3-D perspective provides the scientist a unique approach to visualizing potential interactions and associations that may not have been previously considered.

Pathguide, http://www.pathguide.org/, is a web-based resource that provides a list of 325 biological pathways and molecular interactions including PPI, metabolic pathways, signaling pathways, pathway diagrams, transcription factors/gene regulatory networks, protein–compound interactions, genetic interaction networks, protein sequences, and other databases. There are more than 130 PPI resources and 69 resources related to genetic interactions and networks.

Minimum standards for molecular interactions have been set by the International Molecular Exchange Consortium (IMEx, http://www.imexconsortium.org/about-imex) and HUPO Proteomics Standards Initiative (PSI, http://www.psidev.info/). The purpose of these international standards is to improve data quality, collection, and exchange of molecular interaction data.

TOOLS TO EXAMINE FUNCTIONAL PROTEIN–PROTEIN INTERACTIONS

Public databases are a rich resource for the study of PPIs. Although several of these databases are described here, the savvy scientist may find favor with one database over another, depending on the focus of their unique research question. Large web-based PPI databases, such as Online Predicted Human Interaction Database (OPHID) and Interologous Interaction Database (I2D), are designed to integrate known, experimental, and predicted PPIs for model organisms and humans (Brown & Jurisica, 2005, 2007). User-friendly software helps the scientist to navigate through volumes of PPI data and build visual networks. Jurisica Lab developed the Network Analysis, Visualization, & Graphing TORonto (NAViGaTOR) software package to visualize and analyze OPHID and I2D databases. Import and export formats in this program support the Gene Ontology (GO), a standardized representation of gene and gene product attributes across species and databases initiated the National Human Genome Research Institute and Proteomics Standards Initiative (PSI) and a standardized representation of proteomics initiated by the Human Proteome Organization (Brown et al., 2009). Displays of NAViGaTOR networks can be viewed in either 2- or 3-dimension (http://ophid.utoronto.ca/navigator/) (Brown et al., 2009).

Using these tools, PPI network analysis reveals that human proteins have interacting partners and exhibit preferential evolutionary conservation leading to enrichment in protein complexes when transferred as a conservative protein interaction in species of similar evolutionary origin (Brown & Jurisica, 2007). Knowledge of these pathways can help with understanding normal cell function as well as abnormal cell function because of gene variation (i.e., cancer). Examples of open source and web-based databases related to in-silico protein functional relationships are presented in Table 3.1.

TABLE 3.1

Interactome and PPI Public Databases

	Database	Web Address
Alliance for Cellular Signaling (AfCS)	PPI and others	http://www.signaling-gateway.org (Gilman et al., 2002)
Biomolecular Object Network Databank (BOND) formerly known as Biomolecular Interaction Network Database (BIND)	PPI only	http://crystal.uvm.edu/index.php?id=128 (http://www.bind.ca)

TABLE 3.1
Interactome and PPI Public Databases (Continued)

	Database	Web Address
BioGRID	PPI and others	http://thebiogrid.org/
Database of Interacting Proteins (DIP)	PPI only	http://dip.doe-mbi.ucla.edu/dip/Main.cgi (Xenarios et al., 2000)
Disease Gene Networks (DisGeNET)	PPI and other	http://ibi.imim.es/DisGeNET/DisGeNETweb.html
HIV-1, Human Protein Interaction Database	PPI and HIV	http://www.ncbi.nlm.nih.gov/RefSeq/HIVInteractions/
Human Protein Reference Database (HPRD)	PPI only	http://www.hprd.org/ (Alfarano et al., 2005)
Information Hyperlinked over Proteins (iHOP)	PPI only	http://www.ihop-net.org/UniPub/iHOP/
IntAct molecular interaction database (IntAct)	PPI and others	http://www.ebi.ac.uk/intact/main.xhtml (Aranda et al., 2010)
Jena Protein–Protein Interaction	PPI only	http://ppi.fli-leibniz.de/
Kyoto Encyclopedia of Genes and Genomes (KEGG)	PPI and others	http://www.genome.jp/kegg/ (Kanehisa et al., 2010)
Mammalian Protein–Protein Interaction Database (MIPS)	PPI only	http://mips.helmholtz-muenchen.de/proj/ppi/ (Mewes et al., 2006)
MetaCore	PPI and others	http://www.genego.com/metacore.php
Molecular INTeraction database (MINT)	PPI only	http://mint.bio.uniroma2.it/mint/Welcome.do (Chatr-Aryamontri, Zanzoni, Ceol, & Cesareni, 2008)
Negatome	Database of unlikely protein–protein interaction	http://mips.helmholtz-muenchen.de/proj/ppi/negatome/ (Smialowski et al., 2010)

(Continued)

TABLE 3.1
Interactome and PPI Public Databases (Continued)

	Database	Web Address
Online Predicted Human Interaction Database (OPHID)	PPI only	http://ophid.utoronto.ca/ophidv2.201/ (Brown & Jurisica, 2005)
PDZBase	PPI only	http://icb.med.cornell.edu/services/pdz/start (Beuming, Skrabanek, Niv, Mukherjee, & Weinstein, 2005)
PROTein–protein INterface residues Data Base (ProtinDB)	PPI only	http://protindb.cs.iastate.edu/index.py
STRING	PPI only	http://string-db.org/ (Huang, Barker, Chen, & Wu, 2003)
Reactome	PPI and others	http://www.reactome.org (Vastrik et al., 2007)

Note. Although not exhaustive, this table exhibits common available online resources frequently reported in the literature over the past decade that provide human molecular, protein, and cellular data. For a comprehensive list see Pathguide (http://www.pathguide.org/).

BUILDING DISEASE–GENE NETWORKS FROM KNOWLEDGE OF PPI NETWORKS

In the past decade, PPI networks have served as a template to address the complex relationships between human disease and genes. Paired with PPI network computational tools, such as those presented in Table 3.2, disease and genomic data have served as a framework to identify gene–disease relationships. Computational tools analyze similar characteristics of disease genes and include data of literature descriptions, sequence features, functional annotations, physical interactions, and expression patterns. In the example of the Correlating protein Interaction network and PHEnotype network to pRedict disease genes (CIPHER) method, a model was constructed from a complete set of standardized phenotypes, a reliable complete set of physical disease–gene interactions, and known disease gene–phenotype associations (Wu, Jiang, Zhang, & Li, 2008). Through analysis of new network maps, researchers have demonstrated that genes associated with similar diseases exhibit a higher likelihood of physical interactions between their expressed proteins, supporting the existence of disease-specific functional systems or networks (Goh et al., 2007).

TABLE 3.2
Computational and Prediction Model Resources

	Resource	Web Address
Agile Protein Interaction Data Analyzer (APID)	Computational tools	http://bioinfow.dep.usal.es/apid/index.htm
BioNetBuilder Plugin	Computational tools	http://err.bio.nyu.edu/cytoscape/bionetbuilder/ (Avila-Campillo, Drew, Lin, Reiss, & Bonneau, 2007)
Cytoscape	Computational tools	http://www.cytoscape.org/
GeneMANIA	Prediction models	http://www.genemania.org/
MetaCore	Computational tools	http://www.genego.com/metacore.php
Michigan Molecular Interactions (MiMI)	Prediction models	http://mimi.ncibi.org/MimiWeb/main-page.jsp
Online Predicted Human Interaction Database (OPHID)	Prediction models	http://ophid.utoronto.ca/ophidv2.201/ (Brown & Jurisica, 2005)
Human Protein–Protein Interaction Prediction (PIPs)	Prediction models	http://www.compbio.dundee.ac.uk/www-pips/
Protein Interaction Network Analysis platform (PINA)	Computational tools	http://cbg.garvan.unsw.edu.au/pina/
Prodistin	Computational tools	http://crfb.univ-mrs.fr/webdistin/
ProLinks	Prediction models	http://prl.mbi.ucla.edu/prlbeta/prolinks.jsp (Bowers et al., 2004)
STRING	Prediction models	http://string-db.org/ (Huang et al., 2003)
Structural Prediction for pRotein fOlding UTility System (SPROUTS)	Prediction models	http://bioinformatics.eas.asu.edu/springs/Sprouts/projectsSprouts.html (Lonquety, Lacroix, Papandreou, & Chomilier, 2009)
VisANT, formerly known as Predictome	Computational tools and prediction models	http://visant.bu.edu (Hu et al., 2007; Mellor, Yanai, Clodfelter, Mintseris, & DeLisi, 2002)

Note. This table presents selected available online resources (active as of August 2011) for prediction models and molecular/protein computational tools frequently referred in the literature.

Computational tools are also used to combine information from these databases to create biological networks for organisms including model organisms and humans. Open-source client-server computational tools are available, such as BioNetBuilder that uses BIND, BioGrid, DIP, GO, HPRD, KEGG, Prolinks; and VisANT that uses BIND, Biogrid, HPRD, MIPS, and Predictome. Table 3.2 describes in more detail other molecular and cellular computational tools and prediction models.

Knowledge of PPIs provides a foundation in understanding the interactions of proteins as well as cellular pathways that may lead to disease. Developing a HDN provides a novel approach to understanding the origins and interactions of functional components of diseases (i.e., genes, proteins, and phenotypic expression). Approaching disease at the molecular level can be very complex for the clinician, however, computational tools can be used to allow the clinician and researcher to take a step back and visualize the global picture of associated genes and disease.

Over the past two decades, candidate gene association studies, genome-wide association studies (GWAS), positional cloning, epigenetic, and microarray gene expression studies have revealed volumes of data associating disease with genomic variation (Vidal et al., 2011). Efforts to map PPI and metabolism in humans have led to more detailed maps of disease genes (Bonetta, 2010). Research has revealed that few diseases are single gene disorders, but rather a sum of mutations in many genes (i.e., heterogeneity). Likewise, mutation in a particular gene may influence expression of many different diseases. Given this knowledge, approaching a disease/disorder phenotype as a silo is no longer in vogue, but rather, an approach that involves interrelationships of genes; and disease clusters may reveal important information that will help answer some of the important questions of these complex relationships, pathways, and expression, as well as provide novel approaches to hypothesis development, is favored.

DISEASOME

Equipped with the evolving knowledge of disease and the genome, Goh et al., set forth to develop a conceptual framework to link disease and gene relationships in a network map, built on the fundamental ideas of the interactome. Their network map was termed the “diseasome” or “diseasome map” (Goh et al., 2007). Key components in the development of the diseasome will be discussed first.

In developing the HDN, also known as diseasome map, databases were analyzed such as the Online Mendelian Inheritance in Man (OMIM; Goh et al., 2006b). McKusick suggested on average, a gene is related to 1.7 phenotypes (physical characteristics; McKusick, 2007). As discussed previously, developing a network map is a complex process and involves many components. From the morbid map (MM) of OMIM, 2,929 disease–gene association entries were chosen with at least one gene mutation known to be associated with a disease. These were further classified into 1,286 distinct diseases based on physiological system affected by the disease (Goh et al., 2006b).

In general, a network map typically includes a scatter of many nodes (i.e., focal hubs) exhibited as a circles or rectangles. In the HDN, each node is represented as a circle with the color corresponding to one of 22 disease classes in which the disease belongs (see Figure 3.1; Goh et al., 2007). The node size is proportional to the number of genes involved in the disease (Goh et al., 2006b). Links or connections between nodes are represented as lines and referred as edges. In the example of HDN (Figure 3.1), two circular disease nodes are linked if genes are implicated in both diseases, whereas the thickness of the link/edge is proportional to the number of genes implicated in both diseases. For example, in Figure 3.1, the obesity node is connected with the asthma node, illustrating the fact that there are shared gene(s) associated with these two distinct diseases. Diseases having multiple clinical features are assigned to the "multiple" class and those diseases with insufficient information

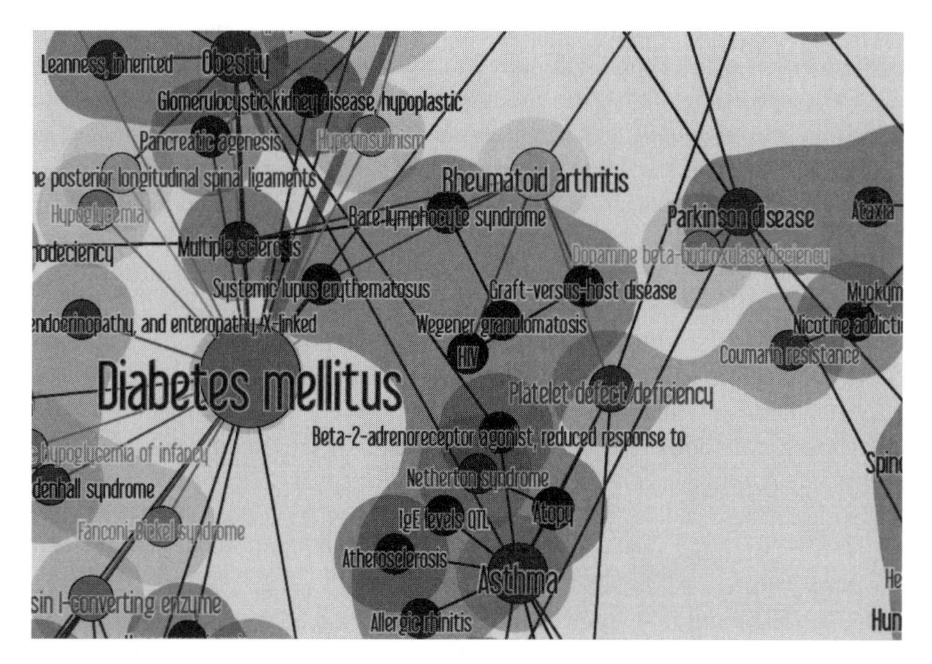

FIGURE 3.1 Illustration of the human diseasome network, also simply referred to as diseasome. Circles depicting diseases and width of lines correspond to number of shared genes associated with the connecting diseases. The illustration can be viewed online in further detail and disease specific node colors at http://diseasome.eu/. (Goh et al., 2006a). Reprinted with permission.

are categorized into the "unclassified" class (Goh et al., 2006b). Disease classes included in the HDN are cancer, endocrine, ear/nose/throat, ophthalmological, neurological, hematological, cardiovascular, muscular, immunological, dermatological, nutritional, connective tissue, renal, psychiatric, metabolic, bone, skeletal, developmental, gastrointestinal, respiratory, multiple, and unclassified. In the HDN, there are 516 nodes and 1,188 edges. Top diseases represented in the HDN include deafness, leukemia, colon cancer, retinitis pigmentosa, and diabetes mellitus exhibited by larger nodes. In the HDN, a single disease associated with a single gene appears disconnected or grouped in clusters (Goh et al., 2007).

To continue the example of Figure 3.1, the scientist may use this network to further explore other phenotypes not previously thought about that may have shared genes and consider the biomolecular pathways involved in these other disease/disorders as potential influences not previously considered. The biomolecular relationship between asthma and allergies or atopy may be intuitive to the clinician or genetic counselor, but a biomolecular relationship between asthma and obesity is typically not the first thought. Or the biomolecular relationship between diabetes and kidney disease may be intuitive; however, considering shared genes in diabetes and rheumatoid arthritis may not be as intuitive. After viewing a network map such as the diseasome, the scientist will easily confirm many of the connections; however, he or she may express, "I had not thought of this other disease process." Thus, the diseasome serves as the visual tool in framing the research, and the resources in Table 3.1 serve as the evidence that provides the parts in developing such a tool.

In a reverse approach to exhibiting disease gene association, the disease gene network (DGN), two gene nodes, exhibited as squares, are linked if they are associated with a disease; whereas the thickness of the link is proportional to the number of diseases in which the two genes are implicated (Goh et al., 2007). Top genes represented in the DGN include Tumor Protein p53 (TP53), Paired box 6 (PAX6), Fibroblast Growth Factor Receptor 2 (FGFR2), and MutS Homolog 2 (MSH2), a human mutator gene (Goh et al., 2006b).

Based on a simple force-directed algorithm, the HDN and DGN were laid out in a map format and rearranged for visual clarity (like a map of airport hubs and airlines flights across the United States; Goh et al., 2006a). The HDN and DGN maps were not as spread out as expected if randomly plotted because of functional clustering (Goh et al., 2006b).

The HDN and DGN projections depict known genetic diseases and known disease genes respectively. Through a graph-theoretical framework, a network of disease genes and known phenotype associations were illustrated as common genetic origins in many diseases (Goh et al., 2007). The map of the HDN reveals many interconnected nodes and links between disease classes suggesting shared genetic origins (http://diseasome.eu/) (Goh et al., 2006a).

USING INTERACTOME AND DISEASOME NETWORK MAPS

In disease–gene analysis, Goh et al. (2007) reported genes associated with common diseases are also associated with increased tendency for PPI, tend to be expressed together in specific tissues, have corresponding protein products participating in the same cellular pathway, function in synchronized expression as a group, and share common cellular and functional characteristics. Genes and proteins represented in network maps (e.g., diseasome or PPI respectively) and involved in regulatory and developmental functions are typically central; whereas, genes and proteins involved in tissue-specific interactions, such as organ function, are peripheral on a network map (Goh et al., 2007; Souiai et al., 2011). Nodes in the central portion of the diseasome and human interactome are known to be more likely involved in housekeeping and regulatory functions (Goh et al., 2007; Souiai et al., 2011).

In terms of the PPI, the central core was described as the largest common interactome network (LCIN) containing 4,200 interactions between 1,996 expressed proteins and devoted to housekeeping functions (Souiai et al., 2011). Souiai et al. report that 33% of LCIN interactions involve disease genes compared with 43.5% in the rest of the interactome supporting disease genes typically expressed as tissue specific or more physiological functions and thus found illustrated in the periphery of the interactome (Souiai et al., 2011). Essential and monogenic disease genes have been demonstrated to be located centrally and complex disease genes located on the periphery (Barrenas, Chavali, Holme, Mobini, & Benson, 2009). This is supported by research involving knockout of genes (e.g., in mouse models). Genes located in the central positions of organism network maps typically resulted in lethality compared to knockouts of genes typically located in the periphery (Missiuro et al., 2009). Using the example of an electrical circuit model, informational flow analysis has revealed that proteins involved in high information flow are located centrally within interactome networks, further supporting essential proteins involved in essential functioning are central in an organism interactome (Missiuro et al., 2009).

Kim, Korbel, and Gerstein (2007) reported that the interactome periphery was more likely to depict proteins with higher potential to evolve. This group also reported that the interactome appears to map to cellular organization with cell periphery (i.e., cell membrane or extracellular space) associated with the interactome periphery. The next step would be to evaluate genes in the periphery of a DGN and then study their potential to be associated with other diseases. This may help to demystify the pleiotropy of genes and the genetic heterogeneity of diseases.

Evidence of interconnected groups of proteins responsible for specific cellular functioning supports the network-based diseasome. Disease results when variations in genes and possible environmental exposure affect phenotypic expression of components of the functional system that in turn produce developmental or physiological abnormalities. In the HDN, the connection between diseases was supported by more than 500 human genetic disorders belonging to a single interconnected main component (Goh et al., 2007). Equipped with this knowledge, the scientist may want to consider using an interactome framework when researching human disease.

Because of the network of disease and genes, one or a low number of genetic variations can have wide spread repercussions. Here is where overlapping or scaffolding of different molecular maps through computational tools might help us with the discovery of new genes associated with disease and development of potential compounds that might interfere with a particular functional pathway leading to alleviation of disease and restoration of health.

In an effort to transcend the "knowledge pockets" covering only small pockets of all available knowledge, a European team designed the Disease Gene Network (DisGeNET) database to include the whole spectrum of human disease genes including mendelian, complex and environmental diseases, and human gene–disease associations (Bauer-Mehren et al., 2011). Results of this project confirmed that most human diseases are associated with more than one biological process (Bauer-Mehren et al., 2011). As scientific research moves forward to explore the multifaceted components of personalized medicine, the next logical step is to analyze this DGN information with data from the Pharmacogenomics Knowledge Base (PharmGKB), which focuses on genes involved in modulating drug response (Klein et al., 2001), and Comparative Toxicogenomics Database (CTD), focused on effects of environmental chemicals on human disease (Mattingly et al., 2006), in an effort to increase our molecular understanding of human disease and the influences of the environment factors, including drugs, on human health.

LIMITATIONS WITH INTERACTOME AND DISEASOME NETWORK MAPPING

Although network mapping provides a resource to illustrate relationships between biological components, limitations in this approach do exist and should be addressed when constructing an interactome. PPI can be influenced by protein binding strengths, regulatory ligands and microenvironmental factors of protein conformation, sensitivity to extraction and isolation in vitro, and intracellular/extracellular physiology. It has been argued that these factors have not been incorporated into explaining the interactome beyond the molecular level (Bossi & Lehner, 2009; Welch, 2009).

Networking mapping captures the fundamental nuances of interactome and provides a unique visual representation of the relationships between protein–protein or disease–gene; however, reporting bias in the literature may influence the development of a network map, such as the diseasome described in this chapter (Cai, Borenstein, & Petrov, 2010). Legrain, Wojcik, and Gauthier (2001) suggest the limitation in bioinformatics computational tools, such as those used in network mapping, involves the production of rich data but poor in hypothesis-driven scientific research. Thus, network maps may provide a display of complicated relationships but not necessarily test hypotheses. However, this approach to understanding associations between diseases and genes can serve as a mechanism of generating novel hypotheses and scientific discovery (Kell & Oliver, 2004). Benefits of carefully developed network maps to scientists, clinicians, counselors, and educators quickly outweigh the minor limitations.

SUMMARY

The enormity of data related to disease and genomics has served as a catalyst for scientists to take novel approaches to conceptualize associations and interactions between these components. This review covered the concept of interactome and how this evolved into the development of network maps of protein–protein and disease–gene interactions. It is suggested that network mapping of proteins, genes, and diseases may provide molecular explanations to diseases, reveal targets and cell regulators involved in disease cells and tissues, provide avenues for research follow-up, and be used for functional studies leading to biomarker and drug discovery.

Many web-based open-resource interaction databases and networks are available. In the example of a diseasome, the HDN can provide biomedical researchers, health care providers, and genetic counselors quick visual references and global perspectives of genetic links between diseases and disease genes. Identifying diseases of interest on a diseasome map can help the researcher, health care provider, and counselor to explore other interconnected diseases and consider these when searching for clues about underlying molecular influences as well as consider therapies that may have a larger impact on the health and well-being of the individual. Further research may render another generation of network maps that increase our understanding of environmental substance interaction and degradation as well as drug metabolism in the human body genome and metagenome.

REFERENCES

Alfarano, C., Andrade, C. E., Anthony, K., Bahroos, N., Bajec, M., Bantoft, K., . . . Hogue, C. W. (2005). The Biomolecular Interaction Network Database and related tools 2005 update. *Nucleic Acids Research*, *33*(Database issue), D418–D424.

Aranda, B., Achuthan, P., Alam-Faruque, Y., Armean, I., Bridge, A., Derow, C., . . . Hermjakob, H. (2010). The IntAct molecular interaction database in 2010. *Nucleic Acids Research*, *38*(Database issue), D525–D531.

Avila-Campillo, I., Drew, K., Lin, J., Reiss, D. J., & Bonneau, R. (2007). BioNetBuilder: Automatic integration of biological networks. *Bioinformatics*, *23*(3), 392–393.

Barrenas, F., Chavali, S., Holme, P., Mobini, R., & Benson, M. (2009). Network properties of complex human disease genes identified through genome-wide association studies. *Public Library of Science One*, *4*(11), e8090.

Bauer-Mehren, A., Bundschus, M., Rautschka, M., Mayer, M. A., Sanz, F., & Furlong, L. I. (2011). Gene-disease network analysis reveals functional modules in mendelian, complex and environmental diseases. *Public Library of Science One*, *6*(6), e20284.

Beuming, T., Skrabanek, L., Niv, M. Y., Mukherjee, P., & Weinstein, H. (2005). PDZBase: A protein-protein interaction database for PDZ-domains. *Bioinformatics*, *21*(6), 827–828.

Bonetta, L. (2010). Protein-protein interactions: Interactome under construction. *Nature*, *468*(7325), 851–854.

Bossi, A., & Lehner, B. (2009). Tissue specificity and the human protein interaction network. *Molecular Systems Biology*, *5*, 260.

Bowers, P. M., Pellegrini, M., Thompson, M. J., Fierro, J., Yeates, T. O., & Eisenberg, D. (2004). Prolinks: A database of protein functional linkages derived from coevolution. *Genome Biology*, *5*(5), R35.

Brown, K. R., & Jurisica, I. (2005). Online predicted human interaction database. *Bioinformatics*, *21*(9), 2076–2082.

Brown, K. R., & Jurisica, I. (2007). Unequal evolutionary conservation of human protein interactions in interologous networks. *Genome Biology*, *8*(5), R95.

Brown, K. R., Otasek, D., Ali, M., McGuffin, M. J., Xie, W., Devani, B., . . . Jurisica, I. (2009). NAViGaTOR: Network Analysis, Visualization and Graphing Toronto. *Bioinformatics*, *25*(24), 3327–3329.

Brun, C., Chevenet, F., Martin, D., Wojcik, J., Guénoche, A., & Jacq, B. (2003). Functional classification of proteins for the prediction of cellular function from a protein-protein interaction network. *Genome Biolology*, *5*(1), R6.

Cai, J. J., Borenstein, E., & Petrov, D. A. (2010). Distinct Properties of Human Disease Genes in Protein Interaction Networks. *Genome Biology and Evolution*.

Chatr-Aryamontri, A., Zanzoni, A., Ceol, A., & Cesareni, G. (2008). Searching the protein interaction space through the MINT database. *Methods in Molecular Biology*, *484*, 305–317.

Coulombe, B. (2010). Mapping the disease protein interactome: Toward a molecular medicine GPS to accelerate drug and biomarker discovery. *Journal of Proteome Research*, *10*(1), 120–125.

Dandekar, T., Snel, B., Huynen, M., & Bork, P. (1998). Conservation of gene order: A fingerprint of proteins that physically interact. *Trends in Biochemical Sciences*, *23*(9), 324–328.

De Las Rivas, J., & Fontanillo, C. (2010). Protein—protein interactions essentials: Key concepts to building and analyzing interactome networks. *Public Library of Science Computational Biology*, *6*(6), e1000807.

Gilman, A. G., Simon, M. I., Bourne, H. R., Harris, B. A., Long, R., Ross, E. M., . . . Sambrano, G. R. (2002). Overview of the Alliance for Cellular Signaling. *Nature*, *420*(6916), 703–706.

Goh, K. I., Cusick, M. E., Valle, D., Childs, B., Vidal, M., & Barabási, A. L. (2006a). *The Human Disease Network. Diseasome poster*. Retrieved from http://diseasome.eu/poster.html

Goh, K. I., Cusick, M. E., Valle, D., Childs, B., Vidal, M., & Barabási, A. L. (2006b). *Supporting information on the human disease network.* Retrieved from http://www.nd.edu/~alb/Publication06/145-HumanDisease_PNAS-14My07-Proc/Suppl/

Goh, K. I., Cusick, M. E., Valle, D., Childs, B., Vidal, M., & Barabási, A. L. (2007). The human disease network. *Proceedings of the National Academy of Sciences of the United States of America*, *104*(21), 8685–8690.

Hu, Z., Ng, D. M., Yamada, T., Chen, C., Kawashima, S., Mellor, J., . . . DeLisi, C. (2007). VisANT 3.0: New modules for pathway visualization, editing, prediction and construction. *Nucleic Acids Research*, *35*(Suppl 2), W625–W632.

Huang, H., Barker, W. C., Chen, Y., & Wu, C. H. (2003). iProClass: An integrated database of protein family, function and structure information. *Nucleic Acids Research*, *31*(1), 390–392.

Kanehisa, M. (2009). Representation and analysis of molecular networks involving diseases and drugs. *Genome Informatics*, *23*(1), 212–213.

Kanehisa, M., Goto, S., Furumichi, M., Tanabe, M., & Hirakawa, M. (2010). KEGG for representation and analysis of molecular networks involving diseases and drugs. *Nucleic Acids Research*, *38*(Database issue), D355–D360.

Kell, D. B., & Oliver, S. G. (2004). Here is the evidence, now what is the hypothesis? The complementary roles of inductive and hypothesis-driven science in the post-genomic era. *Bioessays*, *26*(1), 99–105.

Kim, P. M., Korbel, J. O., & Gerstein, M. B. (2007). Positive selection at the protein network periphery: Evaluation in terms of structural constraints and cellular context. *Proceedings of the National Academy of Sciences of the United States of America*, *104*(51), 20274–20279.

Klein, T. E., Chang, J. T., Cho, M. K., Easton, K. L., Fergerson, R., Hewett, M., . . . Altman, R. B. (2001). Integrating genotype and phenotype information: An overview of the PharmGKB project. Pharmacogenetics Research Network and Knowledge Base. *The Pharmacogenomics Journal*, *1*(3), 167–170.

Legrain, P., Wojcik, J., & Gauthier, J. M. (2001). Protein—protein interaction maps: A lead towards cellular functions. *Trends in Genetics*, *17*(6), 346–352.

Lonquety, M., Lacroix, Z., Papandreou, N., & Chomilier, J. (2009). SPROUTS: A database for the evaluation of protein stability upon point mutation. *Nucleic Acids Research*, *37*(Database issue), D374–D379.

Marcotte, E. M., Pellegrini, M., Thompson, M. J., Yeates, T. O., & Eisenberg, D. (1999). A combined algorithm for genome-wide prediction of protein function. *Nature*, *402*(6757), 83–86.

Mattingly, C. J., Rosenstein, M. C., Davis, A. P., Colby, G. T., Forrest, J. N., Jr., & Boyer, J. L. (2006). The comparative toxicogenomics database: A cross-species resource for building chemical-gene interaction networks. *Toxicological Sciences*, *92*(2), 587–595.

McKusick, V. A. (2007). Mendelian Inheritance in Man and its online version, OMIM. *American Journal of Human Genetics*, *80*(4), 588–604.

Mellor, J. C., Yanai, I., Clodfelter, K. H., Mintseris, J., & DeLisi, C. (2002). Predictome: A database of putative functional links between proteins. *Nucleic Acids Research*, *30*(1), 306–309.

Mewes, H. W., Frishman, D., Mayer, K. F., Münsterkötter, M., Noubibou, O., Pagel, P., . . . Stümpflen, V. (2006). MIPS: Analysis and annotation of proteins from whole genomes in 2005. *Nucleic Acids Research*, *34*(Database issue), D169–D172.

Missiuro, P. V., Liu, K., Zou, L., Ross, B. C., Zhao, G., Liu, J. S., & Ge, A. H. (2009). Information flow analysis of interactome networks. *Public Library of Science Computational Biology*, *5*(4), e1000350.

Pellegrini, M., Marcotte, E. M., Thompson, M. J., Eisenberg, D., & Yeates, T. O. (1999). Assigning protein functions by comparative genome analysis: Protein phylogenetic profiles. *Proceedings of the National Academy of Sciences of the United States of America*, *96*(8), 4285–4288.

Sanchez, C., Lachaize, C., Janody, F., Bellon, B., Röder, L., Euzenat, J., . . . Jacq, B. (1999). Grasping at molecular interactions and genetic networks in Drosophila melanogaster using FlyNets, an Internet database. *Nucleic Acids Research*, *27*(1), 89–94.

Sieberts, S. K., & Schadt, E. E. (2007). Moving toward a system genetics view of disease. *Mammalian Genome*, *18*(6–7), 389–401.

Smialowski, P., Pagel, P., Wong, P., Brauner, B., Dunger, I., Fobo, G., . . . Ruepp, A. (2009). The Negatome database: A reference set of non-interacting protein pairs. *Nucleic Acids Research*, *38*(Suppl 1), D540–D544.

Souiai, O., Becker, E., Prieto, C., Benkahla, A., De las Rivas, J., & Brun, C. (2011). Functional integrative levels in the human interactome recapitulate organ organization. *Public Library of Science One*, *6*(7), e22051.

Vastrik, I., D'Eustachio, P., Schmidt, E., Gopinath, G., Croft, D., de Bono, B., . . . Stein, L. (2007). Reactome: A knowledge base of biologic pathways and processes. *Genome Biology*, *8*(3), R39.

Vidal, M., Cusick, M. E., & Barabási, A. L. (2011). Interactome networks and human disease. *Cell*, *144*(6), 986–998.

Walhout, A. J., Boulton, S. J., & Vidal, M. (2000). Yeast two-hybrid systems and protein interaction mapping projects for yeast and worm. *Yeast*, *17*(2), 88–94.

Welch, G. R. (2009). The 'fuzzy' interactome. *Trends in Biochemical Sciences*, *34*(1), 1–2; author reply 3.

Wu, X., Jiang, R., Zhang, M. Q., & Li, S. (2008). Network-based global inference of human disease genes. *Molecular Systems Biology*, *4*, 189.

Xenarios, I., Rice, D. W., Salwinski, L., Baron, M. K., Marcotte, E. M., & Eisenberg, D. (2000). DIP: The database of interacting proteins. *Nucleic Acids Research*, *28*(1), 289–291.

Yang, P., Li, X., Wu, M., Kwoh, C. K., & Ng, S. K. (2011). Inferring gene-phenotype associations via global protein complex network propagation. *Public Library of Science One*, *6*(7), e21502.

Yu, H., Tardivo, L., Tam, S., Weiner, E., Gebreab, F., Fan, C., . . . Vidal, M. (2011). Next-generation sequencing to generate interactome datasets. *Nature Methods*, *8*(6), 478–480.

CHAPTER 4

The State of Genomic Health Care and Cancer

Are We Going Two Steps Forward and One Step Backward?

Karen E. Greco and Suzanne M. Mahon

ABSTRACT

As the application of genomic information and technology crosses the horizon of health care into our everyday lives, expanding genomic knowledge continues to affect how health care services are defined and delivered. Genomic discoveries have led to enhanced clinical capabilities to predict susceptibility to common diseases and conditions such as cancer, diabetes, cardiovascular disease, and Alzheimer's disease. Hundreds of genetic tests are now available that can identify individuals who carry one or more gene mutations that increase their risk of developing cancer or other common diseases. Increased availability and direct-to-consumer marketing of genetic testing is moving genetic testing away from trained genetics health professionals and into the hands of primary care providers and consumers. Genetic tests available on the Internet are being directly marketed to individuals, who can order these tests and receive a report of their risk for numerous health conditions and diseases. Health care providers are expected to interpret these test results, evaluate their accuracy, address the psychosocial consequences of those distressed by receiving their results, and translate genomic information into effective care. However, as we move two steps forward,

http://dx.doi.org/10.1891/0739-6686.29.73

we are also moving one step backward because many health care providers are unprepared for this genomic revolution. A number of international education, practice, and policy efforts are underway to address the challenges health care providers face in providing competent genomic health care in the context of unprecedented access to information, technology, and global communication. Efforts to integrate standard of care guidelines into electronic medical records increases health care providers' access to information for individuals at risk for or diagnosed with a genomic condition. Development of genomic competencies for health care providers has led to increased genomic content in academic programs. These and other efforts will keep the state of genomic health care stepping forward as we face the challenges of health care in the genomic era.

INTRODUCTION

Scientific advances in genetics and genomics are redefining our understanding of health and illness worldwide as the application of genomic information and technology crosses the horizon of health care into our everyday lives. The paradigm has shifted from genetics, the study of single genes and their effects on relatively rare single gene disorders, to the broader focus of genomics, the study of all the genes in the human genome together, including their interactions with each other and the environment (Guttmacher, Collins, & Drazen, 2004). Genomic discoveries have led to enhanced clinical capabilities to predict susceptibility to common diseases and conditions caused by both hereditary and nonhereditary factors such as cancer, diabetes, cardiovascular disease, Alzheimer's disease, asthma, glaucoma, and obesity. Hundreds of genetic tests are now commercially available that can identify individuals who carry one or more gene mutations that increase their risk of developing cancer or other common diseases, giving health care providers the ability to provide highly personalized health care tailored to an individual's genomic make up.

Expanding genomic knowledge will continue to affect how health care services are defined and delivered. The public increasingly expects health care providers, including nurses, to translate genomic science into effective care, including personalized health care that integrates genomic information related to hereditary disease susceptibility. However, as we move two steps forward, we are also moving one step backward because many health care providers are unprepared for this genomic revolution.

Increased commercial availability of genetic testing, including marketing to non-genetics professionals (especially primary care providers) and direct-to-consumer marketing (DTC) of genetic testing, is rapidly moving genetic testing away from trained genetics health professionals and into the hands of health care providers and consumers, resulting in a potentially higher risk of negative outcomes for patients and families. Genetic testing companies market genetic tests directly to primary care providers, suggesting that the tests are simple to order. The reality is that the tests are easily ordered and simple to submit (usually a blood or buccal sample); the challenge is in completing an accurate assessment to order the best test, comprehensive and accurate interpretation of the results, and assuring coordinated follow-up for other at-risk family members. Similarly, genetic tests are available on the Internet, directly marketed to individuals who can order these tests from the privacy of their own home and receive a report of their risk for numerous health conditions and diseases. Health care providers are expected to interpret these test results, evaluate their accuracy, and address the psychosocial consequences of those distressed by receiving their results. Unfortunately, many health care providers might not have the academic preparation to provide these interventions.

THE CURRENT STATE OF CANCER GENETIC TESTING IN CLINICAL PRACTICE

Over a thousand genetic tests are currently available from testing laboratories. Unfortunately there are no regulations in the United States to evaluate the accuracy and reliability of genetic testing (U.S. Department of Energy, 2011). Some of these genetic tests provide information about whether or not an individual has a relatively rare genetic disease or condition. Much more common are the increasing numbers of genetic susceptibility tests now available that provide information about the future probability that an individual will be diagnosed the with a common adult-onset disease, such as cancer or Alzheimer's disease, that is caused by a combination of hereditary and nonhereditary risk factors (U.S. Department of Energy, 2011). Because of the expanding availability of genetic testing, primary care providers are increasingly being expected to provide this testing, interpret the findings, and tailor patient care based on genetic test results. However, these susceptibility tests are much more difficult to interpret and the clinical implications for patient care are not always clear.

As this movement toward mainstream genetic testing in primary care is most evident with genetic testing for cancer, the focus of examples in this chapter will follow suit. Increased availability of cancer risk assessment tools and the direct marketing of cancer genetic tests to primary care providers rather than genetics professionals are major contributors to the movement of cancer genetic testing into primary care. In a study of primary care physicians, 83% reported they routinely assessed hereditary cancer risk; however, only 33% reported including a three generation pedigree and only 14% used a cancer risk assessment tool such as the Gail Model as part of their assessment. About half reported referring to a genetics specialist (Vig et al., 2009).

The clinical implications of genetic testing potentially can be significant. About 5%–10% of cancers are thought to be caused by hereditary factors resulting from inherited gene mutations (Lindor, McMaster, Lindor, & Greene, 2008; Sweet, Bradley, & Westman, 2002). Current cancer screening and risk reduction guidelines are often based on an individual's level of cancer risk, which includes whether or not they carry a cancer susceptibility gene. For example, more than half of hereditary breast cancer is thought to be related to mutations in two genes, *BRCA1* and *BRCA2*. Annual MRI for breast cancer screening is now recommended for women who carry a *BRCA* mutation, have a first-degree blood relative with a *BRCA* mutation, or have a strong family history of breast or ovarian cancer with an approximately 20%–25% or greater lifetime risk of breast cancer (Saslow et al, 2007). This is an example of how known or suspected hereditary risk can dramatically alter screening recommendations and risk reduction guidelines.

Systematic collection of family history information is necessary for identification of individuals with hereditary cancer risk (Qureshi et al., 2009). Obtaining a family history using a systematic approach enables primary care providers to more accurately identify an individual's level of cancer risk, determine whether cancer genetic testing is indicated, and facilitates decision-making regarding appropriate cancer screening and risk reduction options.

Despite the importance of family history information, a study found more than half of the individuals at increased risk for breast or colorectal cancer based on their family history did not have documentation of this risk within their medical record, and the age of relatives at diagnosis was frequently missing (that is often a critical indicator in identifying suspected hereditary risk; Murff, Greevy, & Syngal, 2007). Another study by Sweet et al. (2002), comparing participant completed online cancer risk assessment surveys to medical records, found 15% of participants had no cancer family history information documented in their medical record. Family history information was most often completed on new patients and not routinely updated; changes in family history are important to document because as changes occur hereditary risk may be identified. Twenty eight percent of the study participants were assigned to a high cancer risk category because of a personal or family history of cancer, however, only 14% had documentation in their medical record of being at increased cancer risk and only 7% had documentation of receiving a referral for genetics consultation. Another study using unannounced standardized patients found physicians collected sufficient family history for breast cancer risk only if there was a strong maternal breast cancer family history. Only 48% collected sufficient family history in both a 36-year-old female with a strong paternal family history of breast and ovarian cancer and in a 33-year-old female presenting with a breast lump with a mother and paternal great aunt with breast cancer (Burke et al., 2009). Another limitation with family history assessment is that the reported family history may be inaccurate or incomplete, especially for second degree relatives (Mai et al., 2011). Primary care providers may not have the time or resources to obtain pathology reports and other medical records to confirm diagnoses, which is an important component of pedigree assessment.

Although individuals should receive genetic counseling by a genetics professional prior to genetic testing, in clinical practice, an increasing number of individuals are not receiving genetic counseling prior to testing and therefore may not fully understand the implications of their genetic test results. Many professional organizations advocate for genetic counseling by a trained credentialed genetics professional prior to and after testing (International Society of Nurses in Genetics, 2009; Oncology Nursing Society, 2009a, 2009b; Trepanier et al., 2004). Health care providers often lack the educational preparation to provide adequate counseling and may not fully understand the complexity of genetic testing and interpretation of results. They may also have limited experience providing care to individuals with a gene mutation that predisposes them to a disease and may not be familiar with screening, risk reduction guidelines, implications for other family members, and the psychosocial needs of these individuals.

DIRECT-TO-CONSUMER MARKETING MOVEMENT

Many genetic testing laboratories have begun to market their tests directly to consumers. This is similar to the marketing that has been done by pharmaceutical companies since the 1980s. Direct-to-consumer testing (DTC) is accomplished largely through print, television advertisements, and the Internet. Traditionally, genetic tests have been available only through health care providers such as physicians, nurse practitioners, and genetic counselors; DTC genetic testing allows a consumer to completely circumvent health care providers when accessing genetic information (Weitzel, Blazer, MacDonald, Culver, & Offit, 2011). Typically a health care provider orders the appropriate test from a laboratory, collects and sends the sample, and when available, interprets the test results.

Currently, there are at least 35 DTC genetic testing companies doing business through the Internet (Marietta & McGuire, 2009). The types of information tested for can be grouped into four general categories: (a) susceptibility to disease, (b) nutritional and metabolic assessments, (c) individual traits or characteristics such as athletic capability or earwax type, and (d) ancestry information (Caulfield, Ries, Ray, Shuman, & Wilson, 2010). The cost can run from several hundred dollars to more than $2,000 depending on the panels selected and it is not covered by insurance (Direct-to-consumer genetic testing kits, 2010). DTC genetic testing raises numerous questions involving privacy and confidentiality, the nature of what constitutes a medical test, who should regulate access to genomic information, and what is the best way to impart complex technical information to different individuals so they understand and can act appropriately on such information (Evans & Green, 2009).

DTC is viewed by some as an enabling tool for exercising one's autonomous quest for personal health information (McCabe & McCabe, 2004; Tamir, 2010). Proponents of DTC advocate that DTC could provide a mechanism to address the limited availability of genetic service providers across the United States (Hock et al., 2011). Most credentialed genetics professionals are concentrated in larger cities; in some rural areas, patients must travel more than a hundred miles to access a credentialed provider. Proponents also emphasize that DTC genetic testing may be less costly to consumers or insurers, especially because a clinic visit is not necessary (Caulfield et al., 2010). DTC genetic tests may also be an attractive option to consumers who are concerned about genetic discrimination and confidentiality. Because DTC occurs outside of a health care provider's office, genetic test results are not automatically part of medical record, however, concerns have been raised about how DTC companies will keep records confidential (Lynch, Parrott, Hopkin, & Myers, 2011). Finally, the advertisement of DTC tests for genetic conditions could raise public awareness and knowledge about genetic tests and empower consumers to be better advocates for their health care (Bloss et al., 2010; Farkas & Holland, 2009). In some cases the tests have alerted patients and doctors to certain conditions (Hauskeller, 2011).

Some DTC companies use privacy as a marketing tool, emphasizing the benefits of obtaining genetic testing outside the health care system that avoids the risks of having genetic information contained in a medical record (Hudson, Javitt, Burke, Byers, & American Society of Human Genetics Social Issues, 2007). Unfortunately, often these companies do not necessarily disclose their privacy policies or explain that a patient's subsequent disclosure of the test results to a health care may lead to the information becoming part of his or her medical record. Some companies say where data is kept and comment on the security of the website, but most companies are not clear about whether data will be kept anonymous. Further, some websites are not clear that the data may be used for future research. Genome-wide testing of hundreds of thousands of individuals for thousands or even millions of single-nucleotide polymorphisms (SNPs) generates a lot of data that can provide data for numerous genomic studies; it is not clear if consumers understand that their samples in some cases may be used for additional research (Howard, Knoppers, & Borry, 2010).

Most DTC testing occurs in the context of a for-profit company. It is a business venture. Proponents of DTC emphasize the possibility of analyzing a large amount of genomic data encourages individuals to be aware of their risks for developing diseases and to adopt preventative measures. Opponents of DTC testing emphasize that these tests offer little value as the scientific validity may be questionable and the clinical utility limited. Additional concerns include that in the absence of a trained professional, many individuals may not be able to correctly interpret the meaning of the results (Parthasarathy, 2010). Further, supporters of DTC testing argue that direct access to genetic testing reduces health care costs by eliminating the need for consultation with trained professionals (McCabe & McCabe, 2004). However, this argument does not consider potential costs resulting from the uninformed and inappropriate use of genetic tests and from the potentially inappropriate responses and health behaviors resulting from test results. For example, if a test result states a woman is at increased risk for breast cancer, and high penetrance genes such as *BRCA1/2* have not been completely sequenced, the risk perception may be inaccurate. Consequently, there may be unnecessary anxiety or inappropriate assessment of risk and the woman may engage in costly screening procedures that are not needed.

The impact of the advertising message on a DTC genetic test website should not be underestimated. A recent study that included 284 women with a personal or family history of breast/ovarian cancer were randomly assigned to view a simulated DTC commercial website that was the control condition or the same simulated website that included information on the potential risks of obtaining genetic testing online (Gray et al., 2009). Participants completed an online survey. Women exposed to risk information had lower intentions to get *BRCA* testing than women in the control group. The researchers concluded that exposing women to information on the potential risks of online *BRCA* testing altered their intentions, beliefs, and preferences for *BRCA* testing.

There are an estimated 480 different tests offered in the combined testing options of current DTC genetic testing companies (Vashlishan Murray, Carson, Morris, & Beckwith, 2010). These tests are not limited to diagnostic testing for genetic syndromes or predisposition to illness; there are tests for the creative, musical, linguistic, and shyness, as well as for intelligence, athletic aptitude, and difficult behavior. One of the most obvious problems is the packaging together of tests for seemingly trivial traits, such as variations for hair type along with potentially life-changing tests for predicting risk for serious, life-threatening illnesses (Udesky, 2010). Some DTC companies offer testing for highly penetrant diseases, weak associations that are well-supported, as well as highly questionable associations. A recent study of 29 health-related DTC companies found that only 11 of their websites provide any scientific evidence to support the markers being tested and, of those, only 6 reference the scientific literature (Lachance, Erby,

Ford, Allen, & Kaphingst, 2010). One company offers 166 tests in one of its testing packages where approximately 60% of the tests (99 tests) are categorized as "preliminary research" because the genetic association data have not yet been replicated (Vashlishan Murray et al., 2010). The scientific evidence provided for many of the genetic associations in this category raises the question of whether these tests should even be included in a genetic-testing package (Imai, Kricka, & Fortina, 2011).

A report of a year-long investigation of DTC genetic testing firms by the U.S. General Accounting Office (GAO) stated that test results were misleading and of limited or no practical use. Because of this report, the U.S. Food and Drug Administration is currently investigating and tightening the regulations that apply to DTC genetic tests (Udesky, 2010).

There are several approaches to DTC genetic testing. Most analyze SNPs but the decision of which SNPs to offer varies from company to company. For example, 23andMe adds a SNP to its testing platform as soon as researchers have established its association with a particular trait or health condition; for this reasons, there are a large number of SNPs on its testing menu. Navigenics only adds a SNP once it has been approved by its own advisory boards that look at and consider research studies and other information prior to adding the SNP (Parthasarathy, 2010). These varying approaches influence the price of the testing package and the clinical utility. Although a genetics professional might recognize what it means, the public might not; for example, SNPs in the *FGFR2* gene that have been associated with a 1.2-fold increased risk of developing breast cancer (Offit, 2008). This small increased risk is similar to the risk of developing breast cancer in women whose first pregnancy occurs after age 35 years. In either case, the risk is very small and there is no clear course of action to reduce the risk. The clinical utility of knowing one has this particular SNP is very small.

Marketing of DTC genetic testing appeals to human emotion. Websites sites offering DTC often appear authoritative and scientific. Most encourage consumers to discuss findings with their personal health care provider and have the consumer sign informed consent forms that appear to add credibility. However, communication in DTC genetic testing often occurs using web-based methods instead of personal interaction with credentialed genetics professional; often the primary care provider is unaware that such testing has been completed (Speicher, Geigl, & Tomlinson, 2010). Of concern is what occurs when someone receives a result that suggests increased risk for a significant health problem such as colorectal cancer. In an undercover investigation, an investigator received a report suggesting increased risk of colorectal cancer. The investigator contacted the company for additional information and clarification and the response from the "expert" was that the investigator should be familiar with the signs and symptoms of colorectal cancer such as blood in the stool, but there was little else that could be done because of the silent nature of colorectal cancer. Knowledgeable genetics professionals would have instructed a person at risk for colorectal cancer about possible screening modalities such as colonoscopy and recommended an appropriate interval for screening based on risk (Udesky, 2010).

In general, once test results are available, DTC testing companies send the consumer an email with a link to open the results. Some companies provide a scientific or research report that accompanies the test results, whereas others compare the consumer's risk for a particular condition to that of the general public. In a study of 29 DTC genetic testing websites, researchers noted that DTC websites tend to have a very high reading level, employ relatively few simple sentences, and lack common language or clear explanations for technical terms (Lachance et al., 2010). The availability of a credentialed genetics professional to help interpret the results is quite variable. Some sites offer social networking so persons with similar results can learn more and provide peer support (Parthasarathy, 2010). Interpretation of DTC genetic test results can be confusing. Although consumers expect to receive a definitive meaningful answer from a medical test, in the case of genetic testing, the results suggest probability of developing a disease and it is often difficult to understand results in the absence of a trained professional (McCabe & McCabe, 2004).

Until governmental regulators and professional bodies sort out the appropriate use and oversight of DTC genetic tests, health care providers will continue to encounter and counsel patients who present with the results from DTC genetic testing and need to be prepared with appropriate responses (Caulfield et al., 2010; Kuehn, 2008). For example, if a patient expresses concern that their test results indicate they are at elevated risk of developing cancer, an appropriate response might be, "I'm glad you are thinking about risk for cancer. Here is what we know about preventing and detecting cancer in its earliest stages."

A study of 3,640 individuals who had freely chosen to use DTC testing approximately half expressed concerns about the process and the meaning of the experience (Bloss et al., 2010). The researchers concluded that if the clinical validity and utility of these tests are demonstrated, tailored genetic education and counseling services may be of benefit. Current approaches to education may be inadequate.

Confidentiality of information is another concern with DTC testing. Surreptitious DNA testing is probably occurring with DTC genetic testing (Udesky, 2010). In this scenario, a person's DNA is submitted for analysis without their knowledge. The extent to which this occurs is not known. It is unlikely to occur when a person must submit a blood or buccal swab. Some companies are willing and have the capability to analyze DNA left on discarded items such as chewing gum, cigarette butts, or strands of hair. In this case, surreptitious testing is possible and some consumers may consider it fun to submit a sample as a birthday present or joke, for example. The potential for problems occurs if the person learns something about their genetic make-up that they did not want to know.

Another related concern is that some companies also store DNA so that customers can use their genetic information again at a later stage or that parents could test minors and later withhold information (Borry, Howard, Sénécal, & Avard, 2010). Often there is little or no discussion of the privacy and consent issues in relation to the storage and analysis of samples (Patch, Sequeiros, & Cornel, 2009). This raises the issue of how samples belonging to minors are handled, given that there are concerns expressed relating to the appropriate consent for genetic tests for adult onset conditions in asymptomatic minors where there is no clinical benefit at present.

There is a need for scientists and knowledgeable health care professionals, at the individual level, to publicly speak out against misrepresentations of the science behind the tests (Vashlishan Murray et al., 2010). Until governmental regulators and professional bodies sort out the appropriate use and oversight of DTC genetic tests, health care providers will continue to encounter and counsel patients who present with the results from DTC genetic testing and need to be prepared with appropriate responses (Kuehn, 2008).

CASE STUDY EXAMPLES OF WHAT CAN GO WRONG WITH GENETIC TESTING

Case: Issues With Coordination of Care for the Entire Family

A 25-year-old female presented with metastatic breast cancer. A genetics professional was consulted and the pedigree drawn. There was no significant history of breast or ovarian cancer with exception of one aunt diagnosed age 62 on the paternal side. Genetic testing was completed and she was found to have a *BRCA1* mutation. Her father was tested and found to be negative. Next, her mother was tested and found to have the mutation. Her mother was estranged from her siblings but instructed to try to contact her siblings about their potential risk and the availability of testing. As the proband's mother reestablished contact, she learned her sister had breast cancer approximately 7 years earlier, genetic testing had been ordered by her oncologist and the same mutation had been identified in her. Her sister had not informed any of the siblings; had the proband's mother known of her sister's mutation, testing could have been arranged and perhaps a different outcome for her daughter who eventually died from her metastatic disease. Additionally, the mother learned she was an obligate carrier (which was confirmed by genetic testing) and was confronted with many difficult decisions in an already stressful point of her life. This case illustrates the importance of coordinated care when dealing with genetics issues. A genetics professional will typically coordinate all care and make every effort to assist the proband with providing information to at-risk family members and provide anticipatory guidance on the implications of testing especially in the case of potential obligate carriers. Although this oncologist correctly tested the proband's aunt, no testing was offered to any of the other aunts or uncles of the proband; this often occurs when testing occurs in primary care or other subspecialties. The primary care or subspecialist cares for the individual; the genetics professional cares for the entire family.

Case: Impact of Incomplete Testing

A 32-year-old self-referred for genetic counseling to better understand the implications of her testing. She had breast cancer at age 30 and the surgical oncologist had ordered *BRCA* testing to determine if the patient should consider prophylactic mastectomies. Because the result was negative, she opted for lumpectomy and radiation therapy after chemotherapy. The woman was concerned about her risks for second cancers, for her children, and siblings.

At the appointment, a pedigree was constructed and her testing results reviewed. The testing had not included testing for the large rearrangement panels. The large deletions, duplications, or large genomic rearrangements in the *BRCA* genes were described in 1997 (Rodríguez, Torres, Borràs, Salvat, & Gumà, 2010). The precise clinical implications of these mutations are not completely known but it has been reported that they may account for a low percentage of all *BRCA1* and *BRCA2* mutations. The frequency of large rearrangements of these genes varies among populations but is probably 1%–2%. Although these rearrangements are not common, they do occur. The genetics professional decided to order the additional testing because of the proband's young age at diagnosis and significant family history of breast cancer and wrote the necessary letters of medical necessity for coverage. The results showed a mutation in *BRCA1*.

This woman now faced several difficult tasks. First, her risk of a second malignancy in the breast and/or ovaries was now substantially increased. Initially she had done testing to make a decision about mastectomy. She thought she had made a reasonable decision and was confronting with having a bilateral mastectomy after radiation, which is more complicated. She opted for the bilateral mastectomy and reconstruction and is undergoing regular ovarian cancer monitoring until she decides to have an oophorectomy. Had the correct test been ordered from the beginning, the radiation therapy could have been avoided. This woman also needed to recontact her relatives informing them they were potentially at increased risk for developing cancer, and thus additional testing was coordinated by the genetics professional.

Case: Management of a Noninformative Negative Result

A 49-year-old woman with a diagnosis of breast cancer 10 years earlier moved to a new city and located a new oncologist. The oncologist wanted her to be seen by the genetics professional for assessment and she told him it was not necessary as she had already tested negative with her previous provider and did not have hereditary risk. The oncologist wanted the history reviewed and she saw the genetics professional. A complete pedigree was completed for all first, second, and third degree relatives. Significant additional findings in the proband included a head circumference of 58 cm, hyperthyroidism, and a uterine fibroid necessitating a hysterectomy at age 42. There were other cases of uterine cancer, hyperplastic polyps, and hyperthyroidism in the family. The proband was subsequently found to have a *PTEN* mutation. Although it was appropriate to consider *BRCA1/2* mutations in this woman, because of her young age of onset (less than age 40), it was not the only diagnosis to consider. When there is not a known mutation in the family, a negative result is often noninformative and does not eliminate the possibility of hereditary predisposition as this patient incorrectly assumed. A health care provider without training in genetics may not consider all of the other possibilities in the differential diagnosis.

Case: Issues in the Case of a Variant of Unknown Significance

A 58-year-old woman presented with a diagnosis of colorectal cancer at age 45 and bladder cancer at age 54. There was an extensive history of other family members with cancers associated with Lynch syndrome. Testing was completed and the results showed two variants of unknown significance. The genetics professional provided interpretation of the results, recommendations, and offered to help the proband with follow-up. A pedigree was submitted to the testing company and 26 relatives were identified for testing for the variant. The genetics professional worked with the patient to try to contact as many relatives from the very large extended family and arrange for testing for the variant. Many hours of professional time were invested in this process, but almost half of the identified relatives were contacted. Eventually, one of the variants was reclassified as deleterious; the genetics professional provided follow-up education and made recommendations based on the new information. Had this family not been managed by a genetics professional, the necessary follow-up and testing to get the variant reclassified resulting in more meaningful information might not have been completed.

Case: Impact of Direct-to-Consumer Testing

A 33-year-old Ashkenazi Jewish (AJ) woman presented to a cancer genetics professional on the suggestion of a coworker with concern about genetic testing results she had obtained from a DTC genetic test. The proband and her husband had decided to do the testing as many of their close friends had done it and found it entertaining. She came with a result showing she had one of the three common *BRCA* mutations in the AJ population (185delAG *BRCA1*, exon. 2).

The proband and her husband were completely unprepared for the results and wanted an interpretation. The results suggested she was at increased risk for breast and gynecologic cancer but this couple had no idea of the magnitude of the risk. The genetics professional began by constructing a family tree and identified that there was possible paternal transmission. The proband had two sisters and one brother who were also potentially at risk. The family received information on the implications of what a positive mutation means and the three common mutations were reordered and confirmed in a certified lab.

Once the mutation was confirmed, the proband and her husband required extensive counseling and information about how to manage the risk. She needed support on making a decision about breast cancer prevention and the importance of ovarian prophylaxis in the next few years. This required multiple visits and phone calls. This was followed by arranging follow-up for her siblings; one sister also tested positive.

Although the DTC genetic testing company offered access to a genetic counselor for an additional fee, the proband did not really understand the importance of genetic counseling. In retrospect, she stated she would not have done the DTC testing had she known of the possible results and discussed it with her primary care provider. This woman was fortunate in that she ultimately received appropriate care and prophylaxis as well as counseling, but when she received the results she had no idea of the significance of the results. There is a very real risk that others will misinterpret the results from DTC testing or not be emotionally or intellectually prepared for the results.

STEPPING FORWARD: MOVING TOWARDS COMPETENT GENOMIC HEALTH CARE

A number of practice, education, and policy efforts are underway to address the challenges nurses and other health care providers face in providing competent genomic health care in the context of unprecedented access to information, technology, and global communication (see Table 4.1). Efforts to integrate standard of care guidelines into electronic medical records increases health care providers' access information to help care for individuals at risk for or diagnosed with a

TABLE 4.1
Genomics Resources for Health Care Providers

Resource	Services/Tools Available
National Coalition of Health Professionals in Genetics, http://nchpeg.org/	• Family History Tool: Information on family history collection and a family history collection form. • Genetic Red Flags: Six genetic red flags that indicate there might be increased genetic risk in an individual or family. • GeneFacts provides decision support for non-geneticist clinicians at the point-of-care by providing concise, accurate, fact sheets on genetic conditions.
National Institutes of Health: National Human Genome Research Institute, http://www.genome.gov	• Health section provides resources for patients and the public and for health professionals. • Education section provides genetics/genomics educational materials. • Some resources are available in Spanish.
National Genetics Education and Development Centre, http://www.geneticseducation.nhs.uk/	• Genetics education resources for learning genetics, teaching genetics, and genetics in practice
National Genetics Education and Development Centre: Telling Stories, http://www.tellingstories.nhs.uk/index.asp	• Telling stories about an individual with a genetic condition is used as the framework to teach real life genetics to patients, family members, and professionals.
World Health Organization Human Genetics Programme, http://www.who.int/genomics/en/	• Online resources and educational tools for health professionals. • Information on the genetics of common diseases. • Ask the Expert is a resource that allows an individual to query a group of health professionals in genetics and related disciplines about genetics and related issues.

genomic condition. Development of essential genomic nursing competencies has resulted in increased genomic content in academic programs. Policy efforts include development of regulations for direct-to-consumer testing. These and other efforts will keep the state of genomic health care stepping forward as we face the challenges of health care in the genomic era.

GENOMIC COMPETENCIES FOR HEALTH PROFESSIONALS

The development of genomic competencies by various health professional organizations and the integration of these competencies into academic programs is an essential step toward assuring health care providers have the knowledge and skills needed to provide competent genomic health care in the clinical setting. A key effort in the United States is the National Coalition for Health Professional Education in Genetics (NCHPEG), *Core Competencies in Genetics Essential for all Health-Care Professionals*, originally published in 2001, which were revised in 2007. This document is widely endorsed and has influenced the development of other genetic/genomic competencies by specific health professional groups (NCHPEG, 2001, 2007). NCHPEG is a coalition of organizations established in 1996 by the American Medical Association, the American Nurses Association, and the National Human Genome Research Institute. As an interdisciplinary group of leaders worldwide from more than 50 diverse health professional organizations, consumer and volunteer groups, government agencies, private industry, managed care organizations, and genetics professional societies, NCHPEG is committed to promoting health professional education and access to information about advances in human genetics.

Genomic competencies for all public health professionals in the United States, including nurses, were published by the Centers for Disease Control and Prevention (CDC), also in 2001 (CDC, 2001). In 1996, the CDC began a long-range program to develop the capacity to integrate genomic and genetic discoveries into public health to ensure the appropriate use of genomic information to promote health and prevent disease. Working groups, including 50 state and local public health, academic and private sector participants, were convened to create competencies for the entire public health workforce, including all public health professionals, clinicians, and health educators. These genomic competencies were later endorsed in an Institute of Medicine public health report (Institute of Medicine, 2005); however, they have not been as influential as the NCHPEG competencies.

In the United Kingdom, workforce competencies have been developed for genetics in clinical practice for non-genetics health care staff, including nurses, with the goal of ensuring that patients have access to genetic advances now and in the future. These competencies are presented in an easy to navigate framework, which includes the performance criteria, and knowledge and understanding needed for a provider to achieve each competency (National Genetics Education and Development Centre, 2007).

GENETICS AND GENOMICS COMPETENCIES IN NURSING

Significant efforts are being made to facilitate the integration of genetics and genomics into nursing education. A recent study surveyed 10 countries: Brazil, Israel, Italy, Japan, Netherlands, Oman, Pakistan, South Africa, United Kingdom, and the United States regarding the current state of the integration of genetics/genomics into nursing education and practice and the incorporation into nursing regulation in each country (Kirk, Calzone, Arimori, & Tonkin, 2011). Establishment of the genetics nurse specialist role and the status of genetics services within health care were factors related to level of incorporation of genetics/genomics competencies into nursing regulatory standards. Other critical factors include the formation of the specialist professional community as a recognized body and recognition and support from policymakers, the public, and other professionals.

The *Essentials of Genetic and Genomic Nursing: Competencies, Curricula Guidelines, and Outcome Indicators, 2nd Edition* (American Nurses Association Consensus Panel on Genetic/Genomic Nursing Competencies, 2009) define the minimum genetic and genomic competencies for all nurses in the United States. These widely accepted competencies, first published in 2006, have been endorsed by about 50 different nursing organizations and have significantly affected recent efforts to increase genomics in undergraduate nursing programs. This document includes strategies for incorporating these competencies into nursing curriculum. In addition, to assist faculty in teaching the essential genetic and genomic nursing competencies, the Genetics/Genomics Competency Center for Education was developed to provide nursing educators with curriculum materials and resources to teach genomics content (G2C2, 2011). *Essential Genetic and Genomic Competencies For Nurses With Graduate Degrees* have just been published (Greco, Tinley, & Seibert, 2012). Discussion of these new competencies and their development can be found in Chapter Nine of this issue.

Fit for Practice in the Genetics Era: A Competence Based Education Framework for Nurses, Midwives and Health Visitors was published in 2003, which identified seven key genetics competency standard statements for nurses in the United Kingdom (Kirk, McDonald, Anstey, & Longley, 2003). This document has become the foundation for the integration of genetics and genomics into nursing education in the United Kingdom. Another key effort is the National Health Service (NHS) Genetics Education and Development Centre that provides a rich array of genetics and genomics resources targeted to all levels of health professionals.

Competencies of genetic nursing practice in Japan for both the basic and advanced levels were first published in 2004 (Arimori et al., 2007). Efforts are underway in Japan to integrate these competencies into academic nursing curricula.

REGULATION OF DIRECT-TO-CONSUMER GENETIC TESTING

Internationally, there are only a few examples of governmental regulatory agencies specifically designed to deal with DTC genetic tests. The Advisory Committee on Genetic Testing (ACGT) in the United Kingdom was established in 1996 for the purpose of considering public health and consumer protection issues around genetic testing including DTC testing with the establishment of the Code to regulate policies (Hogarth, Javitt, & Melzer, 2008). The ACGT was disbanded in 1999, and its responsibilities were assumed by the Human Genetics Commission (HGC), the U.K. government's strategic advisory body on developments in human genetics. However, after enforcing the code once in the case of the nutrigenetics company, the HGC concluded its position as a regulator was incompatible with its primary mission to offer independent strategic policy guidance to the government; thus, the code is no longer enforced. In the absence of a system designed specifically to regulate DTC testing, regulation of these tests falls under the existing regulatory mechanisms that cover clinical laboratories, medical devices, and fair trade and advertising practices similar to what occurs in the United States. Regulations vary by country (Borry et al., 2012). In France, Germany, Portugal and Switzerland laws specify that genetic tests can only be ordered by a physician following sufficient counseling and patient consent. In the Netherlands, some DTC genetic tests may not be allowed if the test is deemed scientifically unsound or might not be beneficial. Both Belgium and the UK allow the provision of DTC genetic tests.

Regulation of genetic tests is complicated. In the United States, genetic tests must go through a premarket review by the FDA. At present, manufacturers and private laboratories have been able to avoid the routine FDA review process for diagnostic tests and compliance with applicable federal regulations (Farkas & Holland, 2009; Gniady, 2008). They do this by manufacturing and using their own reagents in-house and selling these proprietary testing services. These tests are typically regulated by the Clinical Laboratory Improvement Amendments (CLIA), which requires that the test have analytic validity. This means that each test identifies the polymorphism, mutation, or variant that it states it will. Laboratory tests also ideally should have clinical validity, which assures that the test provides useful clinical information that can guide care or screening recommendations. CLIA authorizes regulation of laboratories that conduct genetic testing, but does not regulate the individuals who order the tests or who can receive test results. CLIA regulations do not differentiate between facilities performing DTC genetic testing and facilities performing provider ordered testing (Clinical Laboratory Improvement Amendments, 2011). All facilities that meet the definition of "laboratory" under CLIA must obtain CLIA certification prior to conducting patient testing. This includes provider ordered and DTC genetic testing. CLIA certification must be maintained and the CLIA laboratory procedures must be followed throughout all phases of testing. In contrast to the United States, Canada, and Australia, regulation is much stronger and genetic tests have been placed into a higher risk category requiring greater oversight, most tests within Europe are classified as low risk, meaning that claims are not reviewed before tests are marketed (Patch et al., 2009).

In the United States, the Federal Trade Commission (FTC) Act monitors for unlawful, unfair, or deceptive acts or practices in or affecting commerce. The statute specifically prohibits the dissemination of false advertising or misleading advertising to induce or encourage the purchase of drugs, devices, food, or cosmetics and includes misrepresentations as well as omissions of information. Technically, the FTC has the legal authority to bring enforcement action to DTC testing companies that make claims of clinical validity without adequate scientific evidence, but the FTC has not pursued such enforcement to date (Hogarth et al., 2008).

Many professional organizations are genuinely concerned about the impact of DTC and have issued guidelines and recommendations regarding use of DTC genetic testing (see Table 4.2). Issues addressed in these statements include the appropriateness of DTC testing, strengths and limitations of testing, informed consent, involvement of a genetics professional, and usefulness of test results.

WORLD HEALTH ORGANIZATION GENOMIC EFFORTS

The World Health Organization (WHO) has a Human Genetics Programme that is actively involved in several efforts aimed at facilitating the provision of genomics services in developing countries. An Internet-based Genomic Resource Centre is dedicated to providing reliable information on genomics and health to developing countries on topics such as genetics and common diseases, craniofacial anomalies, genetic research, and ethical legal and social implications of genetics. WHO efforts also include promoting equitable access of affordable

TABLE 4.2
Position Statements on Direct-to-Consumer Genetic Testing

Professional Organization	Summary of Position of DTC Genetic Testing
American Congress of Obstetricians and Gynecologists (ACOG): Direct-to-Consumer Marketing of Genetic Testing (2008) [Committee Opinion No. 409] http://www.acog.org/~/media/Committee%20Opinions/Committee%20on%20Genetics/co409.pdf?dmc=1&ts=20120329T1002401749	• Discourages direct-to-consumer testing
American College of Medicine Genetics (ACMG) Statement on Direct-to-Consumer Genetic Testing (2004) [policy statement] http://www.acmg.net/StaticContent/StaticPages/Direct_Consumer.pdf	• Advises consumers to involve a genetics expert in the process of genetic testing
American Society of Human Genetics (ASHG) Statement on Direct-to-Consumer Genetic Testing in the United States (2007) http://ashg.org/pdf/dtc_statement.pdf	Recommendations: • Companies must provide all relevant information about offered tests in a readily accessible and understandable manner. • To ensure analytic and clinical validity as well as truthful claims of DTC genetic tests offered, relevant agencies of the federal government should take appropriate and targeted regulatory action.
The American College of Clinical Pharmacology (ACCP) Direct-to Consumer/Patient Advertising of Genetic Testing: A Position Statement of the American College of Clinical Pharmacology (Ameer & Krivoy, 2009) http://www.accp1.org/pdf/DirectToConsumerPatient.pdf	Recommends those seeking genetic testing to seek advice from a genetics professional, recognize the strengths and limitations of genetic testing, and realize that in most cases, the companies marketing genetic tests to the consumer do not provide interpretation of the results.

(Continued)

TABLE 4.2

Position Statements on Direct-to-Consumer Genetic Testing (continued)

Professional Organization	Summary of Position of DTC Genetic Testing
International Society of Nurses in Genetics (ISONG) Direct-to-Consumer Marketing of Genetic Tests (2009) http://www.isong.org/ISONG_PS_direct_consumer_marketing_genetic_tests.php	DTC genetic testing be undertaken once the consumer has considered the following: • mechanisms to ensure confidentiality • test purpose • how results will be used • clinical value as well as strengths and limitations of the test • concerns posed in testing minors • ability of test results to provide scientifically based information • reputation of the company offering testing • accuracy and adequacy of results interpretation for individual and family, • fate of the genetic material after the test is complete
The National Society of Genetic Counselors (NSGC) Direct-to-Consumer Testing (2007) [Position Statement] http://www.nsgc.org/Advocacy/PositionStatements/tabid/107/Default.aspx#DTC	Supports the rights of consumers to access high-quality genetic services and strongly encourages the involvement of appropriately trained clinical genetics professionals in the genetic testing process
The European Society of Human Genetics (ESHG). Statement of the ESHG on Direct-to-Consumer GeneticTesting for Health-Related Purposes (2010) https://www.eshg.org/fileadmin/www.eshg.org/documents/PPPC/2010-ejhg2010129a.pdf	• DTC genetic testing and the advertisement of genetic tests of unproven benefit or without adequate independent genetic counseling are in opposition to the professional standards the ESHG sustains. • DTC testing might negatively influence the perception of genetic testing and the tests' usefulness for health care.
The Oncology Nursing Society (ONS) Direct-to-Consumer Marketing of Genetic and Genomic Tests (2010) http://www.ons.org/Publications/Positions/DTCMarketing	• Recommends that patients have pretest and posttest counseling, informed consent, and post disclosure follow-up.

genetic tests, screening, diagnostics and other technology, and integrating genetic approaches into primary health care in addition to other key areas (World Health Organization [WHO], 2011).

WHO has also published several reports addressing a broad range of topics related to genomic health. WHO's report of Genomics and World Health (WHO, 2002) discussed the potential of genomic health care in developing countries, and made recommendations concerning how to facilitate developing health care systems in these countries so they have the opportunity and resources to benefit from genomic health care.

A 2007 WHO report addressed the ethical, legal, and social implications of pharmacogenomics in developing countries. Conclusions included that education programs for health professionals and the public needed to be developed and implemented addressing the benefits and risks of genetics, genomics, and pharmacogenomics to facilitate informed decision making and appropriate use of genomic medications and treatments (WHO, 2007).

In 2010, WHO published a report on community genetics services in low- and middle-income services focusing on reducing the prevalence and health impact of congenital disorders and genetic diseases. This report concluded that genetics services were inadequate, most likely because of issues such as lack of resources, low genetics literacy, misconceptions that control of genetic disorders is too expensive, fear of stigmatization, and an insufficient number of trained health professionals. Report recommendation examples include training health professionals in basic genetics concepts and their application to community genetics services; promoting the use of family history to identify genetic risks; public education to avoid alcohol, tobacco, and other potential teratogens during pregnancy; and newborn screening for congenital conditions in which early intervention is effective. An action plan is underway to address these issues (WHO, 2010).

THE STATE OF GENOMIC HEALTH CARE MOVING FORWARD

Increased availability of genomic information and technology will continue to drive health care toward highly personalized care with the ability to predict disease risk and tailor health care interventions based on an individual's genomic make-up. As genomic advances continue to traverse the horizon of health care into our everyday lives, consumers are faced with an abundance of their genomic information and the challenge of finding the gold nuggets that impact health care decisions and discarding the fool's gold. We live in an era of unprecedented technology and instant global communication. As the ability to sequence the human genome for $1,000 becomes a reality in the near future, it will be less expensive than the analysis of one or two genes. This raises the question of who is responsible for interpreting variants of uncertain significance or other test results not related to the reason the test was ordered (Pyeritz, 2011). Nurses must be prepared to be leaders in addressing these and other challenges in the clinical application of genetic science to the care of the individuals, families, and communities we serve.

REFERENCES

Ameer, B., & Krivoy, N. (2009). Direct-to-consumer/patient advertising of genetic testing: A position statement of the American College of Clinical Pharmacology. *Journal of Clinical Pharmacology*, *49*(8), 886–888.

American College of Medicine Genetics Board of Directors. (2004). ACMG statement on direct-to-consumer genetic testing. *Genetics in Medicine*, *6*(1), 60.

American Nurses Association Consensus Panel on Genetic/Genomic Nursing Competencies. (2009). *Essentials of genetic and genomic nursing: Competencies, curricula guidelines, and outcome indicators* (2nd ed.). Silver Spring, MD: American Nurses Association.

Arimori, N., Nakagomi, S., Mizoguchi, M., Morita, M., Ando, H., Mori, A., . . . Holzemer, W. L. (2007). Competencies of genetic nursing practise in Japan: A comparison between basic and advanced levels. *Japan Journal of Nursing Science*, *4*(1), 45–55.

Bloss, C. S., Ornowski, L., Silver, E., Cargill, M., Vanier, V., Schork, N. J., & Topol, E. J. (2010). Consumer perceptions of direct-to-consumer personalized genomic risk assessments. *Genetics in Medicine*, *12*(9), 556–566.

Borry, P., Howard, H. C., Sénécal, K., & Avard, D. (2010). Health-related direct-to-consumer genetic testing: A review of companies' policies with regard to genetic testing in minors. *Familial Cancer*, *9*(1), 51–59.

Borry, P., van Hellemondt, R. E., Sprumont, D., Jales, C. F., Rial-Sebbag, E., Spranger, T. M., . . . Howard, H. (2012). Legislation on direct-to-consumer genetic testing in seven European countries. *European Journal of Human Genetics*.

Burke, W., Culver, J., Pinsky, L., Hall, S., Reynolds, S. E., Yasui, Y., & Press, N. (2009). Genetic assessment of breast cancer risk in primary care practice. *American Journal of Medical Genetics. Part A*, *149A*(3), 349–356.

Caulfield, T., Ries, N. M., Ray, P. N., Shuman, C., & Wilson, B. (2010). Direct-to-consumer genetic testing: Good, bad or benign? *Clinical Genetics*, *77*(2), 101–105.

Centers for Disease Control and Prevention. (2001). *Genomic competencies for all public health professionals*. Retrieved from http://www.cdc.gov/genomics/translation/competencies/index.htm

Clinical Laboratory Improvement Amendments. (2011). Direct access testing (DAT) and the Clinical Laboratory Improvement Amendments (CLIA) regulations. Retrieved from https://www.cms.gov/CLIA/Downloads/directaccesstesting.pdf

Committee on Genetics, American College of Obstetricians and Gynecologists, & Committee on Ethics, American College of Obstetricians and Gynecologists. (2008). ACOG Committee Opinion No. 409: Direct-to-consumer marketing of genetic testing. *Obstetrics & Gynecology*, *111*(6), 1493–1494.

Core Competency Working Group of the National Coalition for Health Professional Education in Genetics. (2001). Recommendations of core competencies in genetics essential for all health professionals. *Genetics in Medicine*, *3*(2), 155–159.

Direct-to-consumer genetic testing kits. You send in a sample and get your results online. But is it worth the price? (2010). *Harvard Women's Health Watch*, *18*(1), 1–3.

Direct-to-consumer marketing of genetic and genomic tests. (2010). Retrieved from http://www.ons.org/Publications/Positions/DTCMarketing

European Society of Human Genetics. (2010). Statement of the ESHG on direct-to-consumer genetic testing for health-related purposes. *European Journal of Human Genetics*, *18*(12), 1271–1273.

Evans, J. P., & Green, R. C. (2009). Direct to consumer genetic testing: Avoiding a culture war. *Genetics in Medicine*, *11*(8), 568–569.

Farkas, D. H., & Holland, C. A. (2009). Direct-to-consumer genetic testing: Two sides of the coin. *The Journal of Molecular Diagnostics*, *11*(4), 263–265.

Gniady, J. A. (2008). Regulating direct-to-consumer genetic testing: Protecting the consumer without quashing a medical revolution. *Fordham Law Review*, *76*(5), 2429–2475.

Gray, S. W., O'Grady, C., Karp, L., Smith, D., Schwartz, J. S., Hornik, R. C., & Armstrong, K. (2009). Risk information exposure and direct-to-consumer genetic testing for BRCA mutations among women with a personal or family history of breast or ovarian cancer. *Cancer Epidemiology, Biomarkers & Prevention*, *18*(4), 1303–1311.

Greco, K. E., Tinley, S., & Seibert, D. (2012). *Essential genetic and genomic competencies for nurses with graduate degrees*. Silver Spring, MD: American Nurses Association and International Society of Nurses in Genetics. Retrieved from http://www.nursingworld.org/MainMenuCategories/EthicsStandards/Genetics-1/Essential-Genetic-and-Genomic-Competencies-for-Nurses-With-Graduate-Degrees.pdf

Guttmacher, A. E., Collins, F. S., & Drazen, J. M. (2004). Genomic medicine: A primer. In A. E. Guttmacher, F. S. Collins, & J. M. Drazen (Eds.), *Genomic medicine: Articles from the New England Journal of Medicine* (pp. 3–13). Baltimore, MD: John Hopkins University Press.

G2C2: Genetics/Genomics Competency Center for Education. (2011). Retrieved from http://www.g-2-c-2.org/

Hauskeller, C. (2011). Direct to consumer genetic testing. *British Medical Journal*, *342*, d2317. http://dx.doi.org/10.1136/bmj.d2317

Hock, K. T., Christensen, K. D., Yashar, B. M., Roberts, J. S., Gollust, S. E., & Uhlmann, W. R. (2011). Direct-to-consumer genetic testing: An assessment of genetic counselors' knowledge and beliefs. *Genetics in Medicine*, *13*(4), 325–332. http://dx.doi.org/10.1097/GIM.0b013e3182011636

Hogarth, S., Javitt, G., & Melzer, D. (2008). The current landscape for direct-to-consumer genetic testing: Legal, ethical, and policy issues. *Annual Review of Genomics & Human Genetics*, *9*, 161–182.

Howard, H. C., Knoppers, B. M., & Borry, P. (2010). Blurring lines. The research activities of direct-to-consumer genetic testing companies raise questions about consumers as research subjects. *European Molecular Biology Organization Reports*, *11*(8), 579–582.

Hudson, K., Javitt, G., Burke, W., & Byers, P.; American Society of Human Genetics Social Issues Committee. (2007). ASHG Statement* on direct-to-consumer genetic testing in the United States. *Obstetrics and Gynecology*, *110*(6), 1392–1395.

Hudson, K. Javitt, G., Burke, W., & Byers, P.; American Society of Human Genetics Social Issues Committee. (2007). ASHG Statement* on direct-to-consumer genetic testing in the United States. *The American Journal of Human Genetics*, *81*(3), 635–637. Retrieved from http://ashg.org/pdf/dtc_statement.pdf

Imai, K., Kricka, L. J., & Fortina, P. (2011). Concordance study of 3 direct-to-consumer genetic-testing services. *Clinical Chemistry*, *57*(3), 518–521.

Institute of Medicine. (2005). *Implications of genomics for public health: Workshop summary*. Washington, DC: The National Academies Press.

International Society of Nurses in Genetics. (2009). *Position statements: Direct-to-consumer marketing of genetic tests*. Retrieved from http://www.isong.org/ISONG_PS_direct_consumer_marketing_genetic_tests.php

Kirk, M., Calzone, K., Arimori, N., & Tonkin, E. (2011). Genetics-genomics competencies and nursing regulation. *Journal of Nursing Scholarship, 43*(2), 107–116.

Kirk, M., McDonald, K., Anstey, S., & Longley, M. (2003). *Fit for practice in the genetics era: A competence based education framework for nurses, midwives and health visitors*. University of Glamornan, Wales: National Health Services Genetics Team.

Kuehn, B. M. (2008). Risks and benefits of direct-to-consumer genetic testing remain unclear. *JAMA: The Journal of the American Medical Association*, *300*(13), 1503–1505.

Lachance, C. R., Erby, L. A., Ford, B. M., Allen, V. C., Jr., & Kaphingst, K. A. (2010). Informational content, literacy demands, and usability of websites offering health-related genetic tests directly to consumers. *Genetics in Medicine*, *12*(5), 304–312.

Lindor, N. M., McMaster, M. L., Lindor, C. J., & Greene, M. H. (2008). Concise handbook of family cancer susceptibility syndromes—Second edition. *Journal of the National Cancer Institute. Monographs*, *38*, 1–93.

Lynch, J., Parrott, A., Hopkin, R. J., & Myers, M. (2011). Media coverage of direct-to-consumer genetic testing. *Journal of Genetic Counseling*, *20*(5), 486–494.

Mai, P. L., Garceau, A. O., Graubard, B. I., Dunn, M., McNeel, T. S., Gonsalves, L., . . . Wideroff, L. (2011). Confirmation of family cancer history reported in a population-based survey. *Journal of the National Cancer Institute*, *103*(10), 788–797.

Marietta, C., & McGuire, A. L. (2009). Currents in contemporary ethics. Direct-to-consumer genetic testing: Is it the practice of medicine? *The Journal of Law, Medicine & Ethics*, *37*(2), 369–374.

McCabe, L. L., & McCabe, E. R. (2004). Direct-to-consumer genetic testing: Access and marketing. *Genetics in Medicine*, *6*(1), 58–59.

Murff, H. J., Greevy, R. A., & Syngal, S. (2007). The comprehensiveness of family cancer history assessments in primary care. *Community Genetics*, *10*(3), 174–180.

National Coalition for Health Professional Education in Genetics. (2001). Recommendations of core competencies in genetics essential for all health professionals. *Genetics in Medicine*, 3(2), 155–159.

National Coalition for Health Professional Education in Genetics. (2007). *Core competencies in genetics essential for all health-care professionals*. Retrieved from http://www.nchpeg.org/index.php?option=com_content&view=article&id=237&Itemid=84

National Genetics Education and Development Centre. (2007). *Enhancing patient care by integrating genetics in clinical practice: UK workforce competences for genetics in clinical practice for non-genetics health care staff*. Retrieved from http://www.geneticseducation.nhs.uk/media/16686/Competence_Framework.pdf

National Society of Genetic Counselors. (2007). Direct to consumer genetic testing. Retrieved from http://www.nsgc.org/Advocacy/PositionStatements/tabid/107/Default.aspx#DTC

Offit, K. (2008). Genomic profiles for disease risk: Predictive or premature. *JAMA: The Journal of the American Medical Association*, *299*(11), 1353–1355.

Oncology Nursing Society. (2009a). Cancer predisposition genetic testing and risk assessment counseling. Retrieved from http://www.ons.org/Publications/Positions/Predisposition

Oncology Nursing Society. (2009b). Role of the oncology nurse in cancer genetic counseling. Retrieved from http://www.ons.org/Publications/Positions/GeneticCounseling

Parthasarathy, S. (2010). Assessing the social impact of direct-to-consumer genetic testing: Understanding sociotechnical architectures. *Genetics in Medicine*, *12*(9), 544–547.

Patch, C., Sequeiros, J., & Cornel, M. C. (2009). Genetic horoscopes: Is it all in the genes? Points for regulatory control of direct-to-consumer genetic testing. *European Journal of Human Genetics, 17*(7), 857–859.

Pyeritz, R. E. (2011). The coming explosion in genetic testing—Is there a duty to recontact? *The New England Journal of Medicine, 365*(15), 1367–1369.

Qureshi, N., Carroll, J. C., Wilson, B., Santaguida, P., Allanson, J., Brouwers, M., & Raina, P. (2009). The current state of cancer family history collection tools in primary care: A systematic review. *Genetics in Medicine, 11*(7), 495–506.

Rodríguez, M., Torres, A., Borràs, J., Salvat, M., & Gumà, J. (2010). Large genomic rearrangements in mutation-negative BRCA families: A population-based study. *Clinical Genetics, 78*(4), 405–407.

Saslow, D., Boetes, C., Burke, W., Harms, S., Leach, M. O., Lehman, C. D., . . . Russel, C. A.; American Cancer Society Breast Cancer Advisory Group. (2007). American Cancer Society guidelines for breast screening with MRI as an adjunct to mammography. *CA: A Cancer Journal for Clinicians*, 57(2), 75–89.

Speicher, M. R., Geigl, J. B., & Tomlinson, I. P. (2010). Effect of genome-wide association studies, direct-to-consumer genetic testing, and high-speed sequencing technologies on predictive genetic counselling for cancer risk. *The Lancet Oncology, 11*(9), 890–898.

Sweet, K. M., Bradley, T. L., & Westman, J. A. (2002). Identification and referral of families at high risk for cancer susceptibility. *Journal of Clinical Oncology, 20*(2), 528–537.

Tamir, S. (2010). Direct -to-Consumer genetic testing: Ethical-legal perspectives and practical considerations. *Medical Law Review, 18*(2), 213–238.

Trepanier, A., Ahrens, M., McKinnon, W., Peters, J., Stopfer, J., Grumet, S. C., . . . Vockley, C. W.; National Society of Genetic Counselors. (2004). Genetic cancer risk assessment and counseling: Recommendations of the National Society of Genetic Counselors. *Journal of Genetic Counseling, 13*(2), 83–114.

Udesky, L. (2010). The ethics of direct-to-consumer genetic testing. *Lancet, 376*(9750), 1377–1378.

U.S. Department of Energy. (2011). Human Genome Project Information: Gene testing. Retrieved from http://www.ornl.gov/sci/techresources/Human_Genome/medicine/genetest.shtml

Vashlishan Murray, A. B., Carson, M. J., Morris, C. A., & Beckwith, J. (2010). Illusions of scientific legitimacy: Misrepresented science in the direct-to-consumer genetic-testing marketplace. *Trends in Genetics, 26*(11), 459–461.

Vig, H. S., Armstrong, J., Egleston, B. L., Mazar, C., Toscano, M., Bradbury, A. R., . . . Meropol, N. J. (2009). Cancer genetic risk assessment and referral patterns in primary care. *Genetic Testing and Molecular Biomarkers, 13*(6), 735–741.

Weitzel, J. N., Blazer, K. R., Macdonald, D. J., Culver, J. O., & Offit, K. (2011). Genetics, genomics, and cancer risk assessment: State of the art and future directions in the era of personalized medicine. *CA: A Cancer Journal for Clinicians, 61*(5), 327–359.

World Health Organization. (2002). *Genomics and world health.* Retrieved from http://whqlibdoc.who.int/hq/2002/a74580.pdf

World Health Organization. (2007). *The ethical, legal and social implications of pharmacogenomics in developing countries.* Retrieved from http://whqlibdoc.who.int/publications/2007/9789241595469_eng.pdf

World Health Organization. (2010). *Community genetics services: Report of a WHO consultation on community genetics in low-and middle-income countries.* Retrieved from http://whqlibdoc.who.int/publications/2011/9789241501149_eng.pdf

World Health Organization. (2011). *Human genetics programme.* Retrieved from http://www.who.int/genomics/en/

CHAPTER 5

From Mouse to Man

The Efficacy of Animal Models of Human Disease in Genetic and Genomic Research

Cynthia L. Renn and Susan G. Dorsey

ABSTRACT

Animal models are a critical component of biomedical and biobehavioral research and have contributed to the exponential expansion of our understanding of human disease. Now, as we move onward into the era of genetics and genomics research, the importance of animal models to the research process will become even more acute as we explore the significance of genetic differences that are found in the presence and absence of disease. The decision to use an animal model is not one that can be taken lightly; but, rather, requires careful thought and consideration. In this review, we will address (a) why we should consider using animal models, (b) several caveats that are associated with using animals for research, and (c) some of the common genetic tools that are used in animal research.

INTRODUCTION

In recent decades, our understanding of the physiological mechanisms underlying the development and progression of many diseases has increased significantly, leading to the development of many new pharmacological therapies and therapeutic strategies. This increase in our knowledge base has been aided greatly by the use of animal models of human disease, which will continue to be vital tools

http://dx.doi.org/10.1891/0739-6686.29.99

in our new era of research into the genetics and genomics of disease. This chapter will discuss the use of animal models in genetics and genomics research, with examples coming from the pain literature.

WHY USE AN ANIMAL MODEL

Ideally, all research into human disease would be done using human subjects. However, there are several mitigating factors that make this goal unrealistic. First, human subject research can be challenging from a practical standpoint (Mogil, 2009), largely because of the genetic heterogeneity of humans, requiring the use of tens, hundreds, or thousands of subjects to achieve significance in the research findings. For example, in pain research, one of the major challenges in experimental design is the extensive variability in pain sensitivity and analgesic response between individuals (Lacroix-Fralish & Mogil, 2009; Levine, Gordon, Smith, & Fields, 1981; Walker, Sheather-Reid, Carmody, Vial, & Day, 1997; Wolff, Kantor, Jarvik, & Laska, 1966). Using standard pain assays, inter-individual variability in pain threshold, pain tolerance, and pain scale ratings has been well-documented in the literature (Kim et al., 2004; Lacroix-Fralish & Mogil, 2009; Nielsen, Price, Vassend, Stubhaug, & Harris, 2005; Nielsen et al., 2008). Animal models, on the other hand, allow the researcher to minimize variability between subjects by using a standardized genetic background and living environment (Mogil, Davis, & Derbyshire, 2010). Therefore, the difference between subjects is because of the experimental condition and significance can be achieved with a small sample size. For example, in an experiment to test the behavioral response to a thermal stimulus of mice with an inflamed hind paw (Figure 5.1), we used 8-week-old adult male C57BL/6J mice that were housed in pairs on a 12-hour light–dark cycle with food and water available ad libitum. The plantar surface of the left hind paw was injected subcutaneously with the inflammatory agent complete Freund's adjuvant (CFA) in the experimental group ($n = 6$) or with saline in the vehicle control group ($n = 6$). A heat stimulus was applied to the plantar surface of the injected hind paw and the time latency for the mouse to withdraw his paw from the heat was recorded (Hargreaves, Dubner, Brown, Flores, & Joris, 1988). The mice that received the CFA injection had significantly shorter withdrawal latencies than the saline-injected mice by 3 hours after injection, which persisted for 4 weeks and gradually resolved. By using all male mice, we controlled for the possibility of gender differences and the effects of hormonal fluctuations associated with the 4-day estrous cycle of female mice. The C57BL/6J is an inbred strain, which produces genetically identical mice and minimizes the effects of genetic variability. Lastly, the mice were all housed in identical conditions to minimize the behavioral variability that might

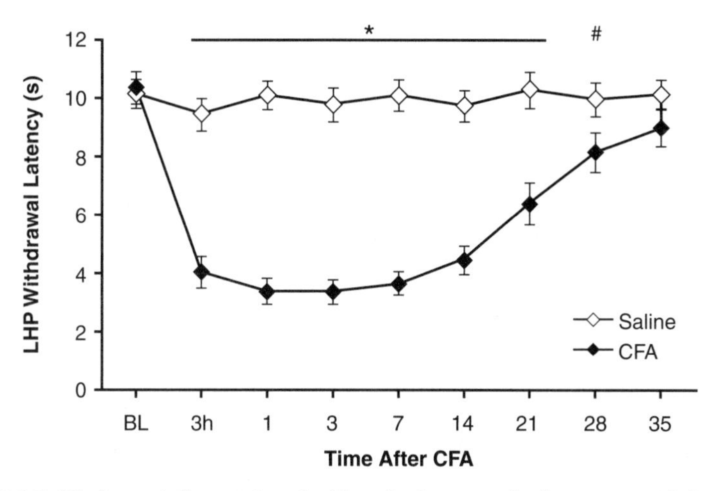

FIGURE 5.1 Hind paw inflammation significantly decreases the latency to withdraw from a heat stimulus. CFA-treated mice ($n = 6$) withdraw the inflamed hind paw from a heat stimulus significantly faster than the non-inflamed, saline-treated control mice ($n = 6$). Mean ± SEM; * = $p < .001$; # = $p < .05$.

occur because of differences in the living environment. All of the mice in both groups experienced a needle stick and injection of the same fluid volume in the left hind paw. Thus, we can conclude that the behavioral difference between the groups is because of the presence of inflammation and not likely because of other mitigating factors.

Research on human subjects is also inherently subjective (Mogil, 2009), with the potential for threats to internal validity being introduced, such as the Hawthorne effect, selection, and diffusion (Slack & Draugalis, 2001). With animal models, there is minimal or no subjectivity involved and many, though not all, threats to validity are minimized. Several threats to validity, such as history, maturation, instrumentation, and mortality are potential confounds for both human and animal research but, with careful experimental design and execution, these can be minimized. However, when animal models are used, the potential for other threats is absent. For example, there is negligible potential for the Hawthorne effect to be a factor because it is unlikely that animals will alter their behavior because they know they are being studied in an experiment. Animals may alter their behavior for other reasons, such as stress; however, with a good experimental design and the proper control groups, all of the animals will be exposed to the same intervening factor or condition, allowing it to be a covariate for all groups.

As we continue to work toward a better understanding of the physiological processes that underlie the development and persistence of human diseases, much of this research involves studying whole organs and organ tissue or causing damage to organs and nerves, as is the case in pain research (Bennett & Xie, 1988; Honore & Mantyh, 2000; Kim & Chung, 1992; Ren, Hylden, Williams, Ruda, & Dubner, 1992; Seltzer, Dubner, & Shir, 1990; Wesselmann, Czakanski, Affaitati, & Giamberardino, 1998). Thus, conducting these types of studies in human subjects is ethically self-limiting (Mogil, 2009) in healthy volunteers. Although there are situations where tissue can be collected from human subjects who are undergoing surgery that involves a diseased organ or tissue, the spectrum of experimental assays that this tissue can be used for is limited and often there is not a healthy control tissue specimen for comparison. Many of these ethical concerns can be obviated by using animal models, which have been shown to provide an excellent depiction of the anatomy, physiology, and neurochemistry involved in pain processing (Mogil et al., 2010). Animal models also allow the researcher to conduct electrophysiological recordings of neurons in and harvest protein and mRNA from key pain-related structures throughout the peripheral and central nervous systems (Table 5.1; Cross, 1994; Willis & Westlund, 1997), which is not possible in human subjects.

The physiology of human disease is complex and can be difficult to study. Recently, a movement has developed to require that research using animal models be replaced with ethically conducted studies on human patients and healthy volunteers whenever feasible (Langley et al., 2008). However, as discussed previously, this is often not possible, especially in studies of the cellular mechanisms that are involved in a disease process. Animal models in various species (Table 5.2) are still widely used and will continue to play a vital role in the discovery and understanding of the physiological mechanisms that underlie human diseases (Holden, 2011; Langley et al., 2008).

CAVEATS ASSOCIATED WITH ANIMAL RESEARCH

There are many benefits to using animal models to study human diseases and society has granted the research community the privilege of using animals for this purpose. However, this privilege comes with the expectation that the animals will not be used for trivial purposes and that significant knowledge will be gained from the research, which will improve both human and animal well-being (McCarthy, 1999; Perry, 2007). The decision to use animal models cannot be taken lightly, instead requiring careful thought and consideration, and the trust granted by society mandates the humane care and responsible use of all research animals (Institute for Laboratory Animal Research, 2011). Therefore, the researcher must accept the responsibility and every effort must be made to ensure the humane treatment of all research animals in all aspects of the research

TABLE 5.1
List of Nervous System Tissue that can be Manipulated, Removed, and Examined in an Animal Model but not in Human Research

Region	Structure	Sub-Structures
Peripheral nervous system	Peripheral sensory nerves	Sciatic nerve Tibial nerve Peroneal nerve Dorsal root ganglia
Central nervous system	Spinal cord	Spinal dorsal horn Superficial (I & II) and deep laminae (IV & V) of the dorsal horn Ascending pain transmission tracts (spinothalamic, spinomesencephalic, spinoreticular) Descending pain modulatory tracts
	Brainstem	Medulla (nucleus raphe magnus, gigantocellularis, lateral reticular nucleus) Pons (parabrachial nucleus, locus coeruleus) Midbrain (periaqueductal grey, superior colliculus, red nucleus, nucleus cuneiformis)
	Brain	Thalamus (midline nuclei, centrolateral nuclei, vental posterolateral nuclei) Hypothalamus Anterior cingulate gyrus Insular cortex Primary sensory cortex (S I) Secondary sensory cortex (S II)

TABLE 5.2
List of the Most Common Animal Species That Are Used for Research Purposes

Animal Species Commonly Used in Research	
Mouse	Rabbit
Rat	Cat
Gerbil	Dog
Hamster	Pig
Guinea pig	Non-human primates

venture, including a humane end point for the animals (Institute for Laboratory Animal Research [ILAR], 2011). A lot of factors are involved when planning a research study. However, it is not reasonable to consider using an animal model simply as a means of avoiding the regulations associated with human research. As with human research, animal researchers must comply with a strict set of regulations and guidelines set forth by the Federal Government (Office of Laboratory Animal Welfare National Institutes of Health, 2002; United States Department of Agriculture, 2010) and the research institution's Institutional Animal Care and Use Committee (IACUC; ILAR, 2011). All studies involving the use of animals must have a written protocol that has been reviewed and approved by the IACUC (Holden, 2011). The researcher is then legally obligated to adhere to the approved protocol, thus ensuring the humane care and use of all animals.

As the researcher designs the experiments and writes the protocol for a study that will use an animal model, one key component in the process is following the principle of the "*Three R's*," which stands for "*replacement, refinement, and reduction*" (Russell & Burch, 1959). Replacement means employing methods or systems (e.g., computer modeling or cell culture with immortalized cell lines) that negate the need for using animals to answer the research question. Another option is to use "relative replacement" by replacing vertebrate with invertebrate animals that are lower on the phylogenetic ladder (Institute for Laboratory Animal Research, 2011). From an ethical standpoint, the *replacement principle* is especially important because it advocates obtaining scientific findings without the use of animals (Olsson, Robinson, & Sandoe, 2011). The decision about what experimental system to use is based on the research question and the best experimental assays to answer that question. For example, to test the interaction between two specific proteins, the best method might be to take a very reductionist approach and study the protein–protein interaction in cultured cells using biochemical, molecular biological, and immunocytochemical assays. However, if the goal is to test the effect of agonizing or antagonizing a protein on the function of a physiological system or behavioral response, then cultured cells will not work and an animal model must be used. The same design principles apply when deciding between species on the phylogenetic ladder. There can be widely divergent physiological mechanisms that underlie the function of organ systems between species. Further, the spectrum of available behavioral assays varies greatly between species, such as drosophila, worms, fish, and mammals.

The *refinement principle* refers to taking measures to optimize husbandry and experimental practices, which minimizes stress and improves the overall well-being of the animals (Institute for Laboratory Animal Research, 2011; Olsson et al., 2011). It is the responsibility of the researcher to ensure that all staff members are carefully trained in the proper care and handling of experimental animals. Meticulously adhering to this principle will reduce extraneous variability between animals and sound scientific results can be obtained with fewer animals being used. For example, in an experiment (similar to the one described in Figure 5.1) that involved transporting mice between research buildings just prior to behavioral testing (Figure 5.2A; $n = 6$ per group; mean ± SE), the standard error of the data was quite large. When that experiment was repeated, doubling the N (Figure 5.2B; $n = 12$ per group; mean ± SE), the standard error was reduced but still remained fairly large. In a subsequent experiment that was identical to the first two, except the mice did not undergo transport or excessive handling immediately before behavioral testing (Figure 5.2C; $n = 6$ per group; mean ± SE), the standard error was markedly decreased. Of note, in this example, the treatment effect was quite robust and the difference between the groups was large enough that the excessive variability would not preclude achieving significance in the results. However, in many experiments, the treatment effect is more modest, resulting in a smaller difference between groups, and significance could be masked because of variability within the experimental groups. The example given here illustrates the importance of refining the experimental procedure (minimizing stress-inducing activity before testing) and the effect it has on data variability. But, this is just one piece of the refinement process. Every aspect of the care and handling of the animals (Table 5.3) must be tightly controlled to ensure that all of the experimental animals are treated identically and the only difference between the animals is because of the experimental condition.

The reduction principle has close ties with refinement. Reduction refers to taking measures that allow the researcher to use the fewest number of animals in a study that will yield maximal information from the experimental results (Institute for Laboratory Animal Research, 2011; Olsson et al., 2011). This principle relies on a careful analysis of the literature to avoid unnecessary repetition of experiments, a well-planned experimental design with all of the proper control conditions and an a priori power analysis to guide sample size decisions. Further, it is critical that the data are analyzed using the most appropriate statistical methods to ensure that the conclusions drawn from the study are valid (Institute for Laboratory Animal Research, 2011).

One last question that must be asked prior to conducting a study is, "Does the chosen animal model actually '*model*' the human condition?" The answer to this question is important when interpreting the experimental results. Animal models have been vital tools for mechanistic studies to understand the pathophysiology of diseases. However, what is learned from animal studies does not always translate to new treatment modalities that are efficacious, and may in fact be dangerous, when applied to humans (Kola & Landis, 2004). For example, in pain research, problems with using animal models can be found in the literature (Mogil, 2009)

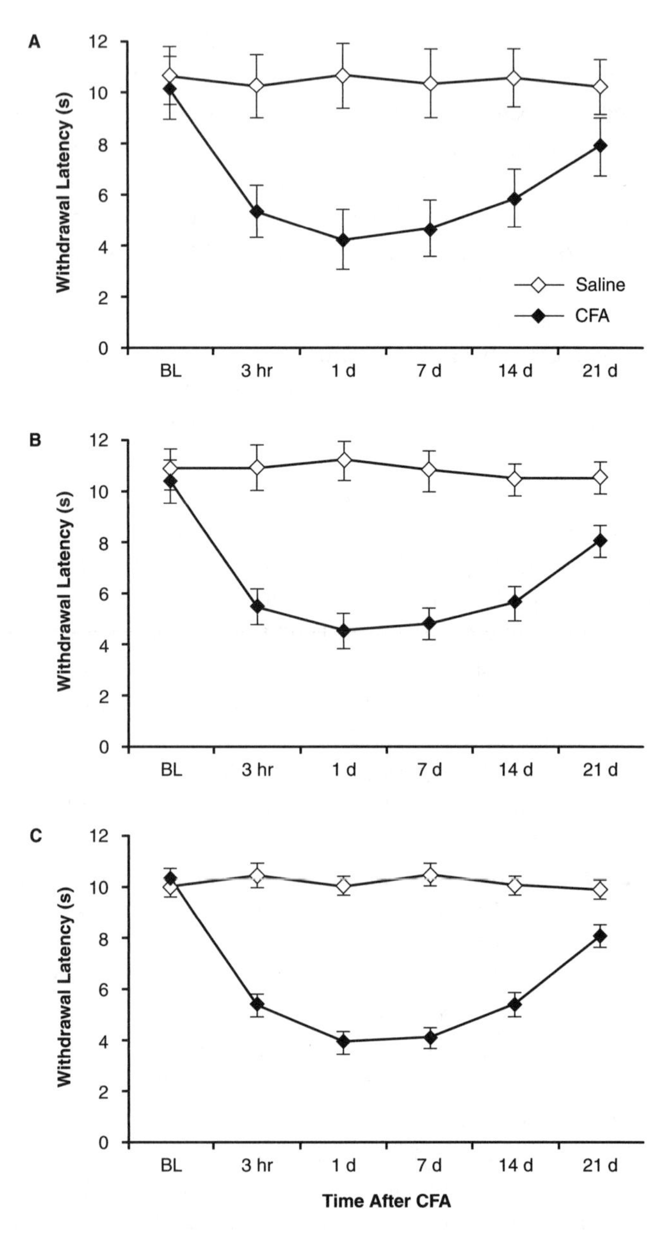

FIGURE 5.2 Stress-inducing handling of mice increases the variability of the data. A. Mice that underwent excessive handling during transport ($n = 6$ per group) exhibited large variability in the behavioral data. B. Increasing the number of mice ($n = 12$ per group) decreases the variability. C. Mice ($n = 6$ per group) that are transported to the testing facility and allowed to acclimate with minimal stress-inducing handling before the behavioral assay is conducted exhibit the least amount of variability in the data. The data shown are the mean ± SEM.

TABLE 5.3
List of Husbandry and Care Practices That Can Affect the Variability Between Experimental Animals

Variables That Can Affect Animal Data
Cage size
Number of animals per cage
Type of food
Source of water
Enrichment of the cage environment
Cage location in the housing room
Cage temperature and humidity
Ambient light
Ambient noise
Variation in the light: dark cycle
Handling by staff
Time of day
Proximity of females to males
Estrous cycle of females
Smell of non cage-mate animals
Hearing stress vocalizations from others

such as a lack of face validity (Rice et al., 2008), where animal models that are considered to be the standard in the field for testing analgesics often don't replicate the real clinical situation (International Association for the Study of Pain, 2010). Some steps that can be taken to improve the validity of animal models include refining the protocols for developing models and increasing the complexity of a study's outcome measures (Festing, 2003; Mogil, 2009; Quessy, 2009, 2010).

With all of the associated caveats, the careful use of animal models will continue to be the foundation of research into the physiological processes that underlie human disease. The use of these models will allow researchers to further elucidate the mechanistic processes of disease development, thus leading to the development of new and better therapeutic modalities.

GENETIC TOOLS TO STUDY DISEASES

The variability of a particular trait in humans has been well documented and attributed to the genetic diversity in the human genome between subjects (Storey et al., 2007). Comparing differences between inbred mouse strains is one frequently used method to study the role of genetic diversity in the variability of a selected trait (Lariviere et al., 2002; Mogil et al., 1999a, 1999b; Smith, Crager, &

Mogil, 2004). Inbred strains are virtually genetically identical (isogenic) because of sibling mating for at least 20 generations (Lyon & Searle, 1989). The use of this methodology showed that a genetic correlation existed between allodynic states after several different types of injury, suggesting that the degree of allodynia that develops after an injury is dependent on genetic factors rather than the type of injury (Lariviere et al., 2002; Mogil et al., 1999a, 1999b; Smith et al., 2004). Similar types of studies have also been done in sets of twins with varying results, though there is evidence of some degree of heritability of pain traits (Arguelles et al., 2006; Fejer, Hartvigsen, & Kyvik, 2006; Nielsen et al., 2008; Norbury, MacGregor, Urwin, Spector, & McMahon, 2007). Another method of studying the heritability of traits is through specific breeding strategies that are designed to produce high- or low-responding mice for a particular pain trait (Belknap, Haltli, Goebel, & Lamé, 1983; Devor & Raber, 1990; Panocka, Marek, & Sadowski, 1986). With the advent of new technologies that allow us to engineer mutations in mice, selective breeding studies have become less favorable (Lacroix-Fralish & Mogil, 2009).

In recent decades, the ability to genetically alter mice has increased and the use of mutant mice has greatly expanded, across various fields of study, to clarify the physiological function of a broad range of proteins and model human diseases (Picciotto & Wickman, 1998). There are several different forms of mutants that can be created: transgenics, inducible transgenics, knockouts, and inducible knockouts. Transgenic mice are created by inserting a fragment of DNA encoding a particular gene of interest into the mouse DNA. This causes the mouse to either express a gene that normally isn't found in the mouse or over-express higher quantities of a gene that is normally present (Picciotto & Wickman, 1998). By taking advantage of tissue-specific promoters, the transgenic mouse can be engineered to express the gene of interest in specific tissues or at specific developmental time points rather than ubiquitously (Brenner, Kisseberth, Su, Besnard, & Messing, 1994; Forss-Petter et al., 1990; Goujet-Zalc et al., 1993; Mayford, Wang, Kandel, & O'Dell, 1995). As the use of transgenic technology increased, it became apparent that many genes of interest play temporally critical roles during development. Researchers found that mice chronically expressing a transgene often exhibited embryonic or early postnatal lethality and developmental abnormalities that might distort the role of the gene/protein of interest (Picciotto & Wickman, 1998). This led to the development of various systems that allowed the transgene to be temporally inducible, such that the gene expression could be turned on at a postnatal time point. The generation of inducible transgenic mice is more complex, involving the crossing of two separate lines of transgenic mice, and results in a system where the expression of the transgene is controlled by exposing the mouse to an activating drug such as tetracycline (Saez, No, West, & Evans, 1997). Thus, the function of the gene of interest can be explored specifically in juvenile, young adult, middle aged, or older adult mice, depending on the disease process that is being studied.

The ability to express genes and study the functions of their protein products in vivo is critical to increasing our understanding of the pathophysiology of disease. Equally important is the generation of knockout mice, in which a gene is deleted from the mouse DNA to examine the consequences of losing the protein product of the gene of interest. This can be accomplished by either random or targeted mutagenesis. Random mutagenesis occurs when the organism is exposed to an agent that randomly attacks its genome, such as ultraviolet radiation, viruses, or mutagenic chemicals and causes random genomic lesions and inactivation of genes. This must be followed by in vivo assays that select for the expected function of the protein product of the gene of interest, which can be time consuming and difficult if no overt phenotype is present in the knockout (Picciotto & Wickman, 1998). On the contrary, a specific genomic lesion to remove a particular gene can be achieved through targeted mutagenesis, which takes advantage of homologous recombination to remove a section of DNA that encodes the gene of interest (Capecchi, 1989a, 1989b; Waldman, 1992). The process of targeted mutagenesis is complex in the early stages; however, once the knockout mouse has been generated, breeding strategies can be used to establish a knockout colony. Also, as was found with transgenic mice, knockout mice can have developmental abnormalities that alter the expected phenotype. Therefore, it was necessary to develop the capability of generating tissue-specific and inducible knockouts where the gene is removed from only the target tissue or at a temporally appropriate time (Picciotto & Wickman, 1998).

CONCLUSION

Animal models have been critical to gaining an understanding of anatomy, physiology, and the pathophysiology of human diseases, and they will continue to play a prominent role as we move forward with a greater exploration of the genetic and genomic underpinnings of diseases. As more genes are identified as playing a role in the diseases processes, we will continue to create genetic mutants to clarify the roles that genes and their protein products play. These transgenic and knockout mutants can be subjected to a broad array of experimental assays and paradigms that are not possible with human subjects, thus allowing the careful dissection of the cellular processes that are impacted by the addition or deletion of a specific protein. The overarching goal of all of this research is to discover new and better therapeutic modalities and pharmaceuticals for the management of human diseases.

REFERENCES

Arguelles, L. M., Afari, N., Buchwald, D. S., Clauw, D. J., Furner, S., & Goldberg, J. (2006). A twin study of posttraumatic stress disorder symptoms and chronic widespread pain. *Pain, 124*(1–2), 150–157.

Belknap, J. K., Haltli, N. R., Goebel, D. M., & Lamé, M. (1983). Selective breeding for high and low levels of opiate-induced analgesia in mice. *Behavioral Genetics, 13*(4), 383–396.

Bennett, G. J., & Xie, Y. K. (1988). A peripheral mononeuropathy in rat that produces disorders of pain sensation like those seen in man. *Pain, 33*(1), 87–107.

Brenner, M., Kisseberth, W. C., Su, Y., Besnard, F., & Messing, A. (1994). GFAP promoter directs astrocyte-specific expression in transgenic mice. *Journal of Neuroscience, 14*(3, Pt. 1), 1030–1037.

Capecchi, M. R. (1989a). Altering the genome by homologous recombination. *Science, 244*(4910), 1288–1292.

Capecchi, M. R. (1989b). The new mouse genetics: Altering the genome by gene targeting. *Trends in Genetics, 5*(3), 70–76.

Cross, S. A. (1994). Pathophysiology of pain. *Mayo Clinic Proceedings, 69*(4), 375–383.

Devor, M., & Raber, P. (1990). Heritability of symptoms in an experimental model of neuropathic pain. *Pain, 42*(1), 51–67.

Fejer, R., Hartvigsen, J., & Kyvik, K. O. (2006). Heritability of neck pain: A population-based study of 33,794 Danish twins. *Rheumatology (Oxford), 45*(5), 589–594.

Festing, M. F. (2003). Principles: The need for better experimental design. *Trends in Pharmacological Sciences, 24*(7), 341–345.

Forss-Petter, S., Danielson, P. E., Catsicas, S., Battenberg, E., Price, J., Nerenberg, M., & Sutcliffe, J. G. (1990). Transgenic mice expressing beta-galactosidase in mature neurons under neuron-specific enolase promoter control. *Neuron, 5*(2), 187–197.

Goujet-Zalc, C., Babinet, C., Monge, M., Timsit, S., Cabon, F., Gansmüller, A., . . . Mikoshiba, K. (1993). The proximal region of the MBP gene promoter is sufficient to induce oligodendroglial-specific expression in transgenic mice. *European Journal of Neuroscience, 5*(6), 624–632.

Hargreaves, K., Dubner, R., Brown, F., Flores, C., & Joris, J. (1988). A new and sensitive method for measuring thermal nociception in cutaneous hyperalgesia. *Pain, 32*(1), 77–88.

Holden, J. E. (2011). Putting the bio in biobehavioral: Animal models. *Western Journal of Nurings Research, 33*(8), 1017–1029.

Honore, P., & Mantyh, P. W. (2000). Bone cancer pain: From mechanism to model to therapy. *Pain Medicine, 1*(4), 303–309.

Institute for Laboratory Animal Research. (2011). *Guide for the care and use of laboratory animals* (8th ed.). Washington, DC: National Academy Press.

International Association for the Study of Pain. (2010). Do animal models tell us about human pain? *Pain Clinical Updates, 18*, 1–6.

Kim, H., Neubert, J. K., San Miguel, A., Xu, K., Krishnaraju, R. K., Iadarola, M. J., . . . Dionne, R. A. (2004). Genetic influence on variability in human acute experimental pain sensitivity associated with gender, ethnicity and psychological temperament. *Pain, 109*(3), 488–496.

Kim, S. H., & Chung, J. M. (1992). An experimental model for peripheral neuropathy produced by segmental spinal nerve ligation in the rat. *Pain, 50*(3), 355–363.

Kola, I., & Landis, J. (2004). Can the pharmaceutical industry reduce attrition rates? *Nature Reviews. Drug Discovovery, 3*(8), 711–715.

Lacroix-Fralish, M. L., & Mogil, J. S. (2009). Progress in genetic studies of pain and analgesia. *Annual Review of Pharmacology and Toxicology, 49*, 97–121.

Langley, C. K., Aziz, Q., Bountra, C., Gordon, N., Hawkins, P., Jones, A., . . . Tracey, I. (2008). Volunteer studies in pain research—opportunities and challenges to replace animal experiments: The report and recommendations of a Focus on Alternatives workshop. *NeuroImage, 42*(2), 467–473.

Lariviere, W. R., Wilson, S. G., Laughlin, T. M., Kokayeff, A., West, E. E., Adhikari, S. M., . . . Mogil, J. S. (2002). Heritability of nociception. III. Genetic relationships among commonly used assays of nociception and hypersensitivity. *Pain, 97*(1–2), 75–86.

Levine, J. D., Gordon, N. C., Smith, R., & Fields, H. L. (1981). Analgesic responses to morphine and placebo in individuals with postoperative pain. *Pain, 10*(3), 379–389.

Lyon, M., & Searle, A. (1989). *Genetic variants and strains of the laboratory mouse*. Oxford, United Kingdom: Oxford University Press.

Mayford, M., Wang, J., Kandel, E. R., & O'Dell, T. J. (1995). CaMKII regulates the frequency-response function of hippocampal synapses for the production of both LTD and LTP. *Cell, 81*(6), 891–904.

McCarthy, C. R. (1999). Bioethics of laboratory animal research. *ILAR Journal, 40*(1), 3–11.

Mogil, J. S. (2009). Animal models of pain: Progress and challenges. *Nature Reviews. Neuroscience, 10*(4), 283–294.

Mogil, J. S., Davis, K. D., & Derbyshire, S. W. (2010). The necessity of animal models in pain research. *Pain, 151*(1), 12–17.

Mogil, J. S., Wilson, S. G., Bon, K., Lee, S. E., Chung, K., Raber, P., . . . Devor, M. (1999a). Heritability of nociception I: Responses of 11 inbred mouse strains on 12 measures of nociception. *Pain, 80*(1–2), 67–82.

Mogil, J. S., Wilson, S. G., Bon, K., Lee, S. E., Chung, K., Raber, P., . . . Devor, M. (1999b). Heritability of nociception II. 'Types' of nociception revealed by genetic correlation analysis. *Pain, 80*(1–2), 83–93.

Nielsen, C. S., Price, D. D., Vassend, O., Stubhaug, A., & Harris, J. R. (2005). Characterizing individual differences in heat-pain sensitivity. *Pain, 119*(1–3), 65–74.

Nielsen, C. S., Stubhaug, A., Price, D. D., Vassend, O., Czajkowski, N., & Harris, J. R. (2008). Individual differences in pain sensitivity: Genetic and environmental contributions. *Pain, 136*(1–2), 21–29.

Norbury, T. A., MacGregor, A. J., Urwin, J., Spector, T. D., & McMahon, S. B. (2007). Heritability of responses to painful stimuli in women: A classical twin study. *Brain, 130*(Pt. 11), 3041–3049.

Office of Laboratory Animal Welfare National Institutes of Health. (2002). *Public health service policy on humane care and use of laboratory animals*. Retrieved from http://grants1.nih.gov/grants/olaw/references/PHSPolicyLabAnimals.pdf

Olsson, A. S., Robinson, P., & Sandoe, P. (2011). Ethics of animal research. In J. Hau & S. J. Shapiro (Eds.), *Handbook of laboratory animal science* (3rd ed., pp. 21–37). New York, NY: CRC Press.

Panocka, I., Marek, P., & Sadowski, B. (1986). Inheritance of stress-induced analgesia in mice. Selective breeding study. *Brain Research, 397*(1), 152–155.

Perry, P. (2007). The ethics of animal research: A UK perspective. *ILAR Journal, 48*(1), 42–46.

Picciotto, M. R., & Wickman, K. (1998). Using knockout and transgenic mice to study neurophysiology and behavior. *Physiological Reviews, 78*(4), 1131–1163.

Quessy, S. N. (2009). Comment on: Animal models and the prediction of efficacy in clinical trials of analgesic drugs: A critical appraisal and a call for uniform reporting standards. *Pain, 142*(3), 284–285.

Quessy, S. N. (2010). Two-stage enriched enrolment pain trials: A brief review of designs and opportunities for broader application. *Pain, 148*(1), 8–13.

Ren, K., Hylden, J. L., Williams, G. M., Ruda, M. A., & Dubner, R. (1992). The effects of a non-competitive NMDA receptor antagonist, MK-801, on behavioral hyperalgesia and dorsal horn neuronal activity in rats with unilateral inflammation. *Pain, 50*(3), 331–344.

Rice, A. S., Cimino-Brown, D., Eisenach, J. C., Kontinen, V. K., Lacroix-Fralish, M. L., Machin, I., . . . Stöhr, T. (2008). Animal models and the prediction of efficacy in clinical trials of analgesic drugs: A critical appraisal and call for uniform reporting standards. *Pain, 139*(2), 243–247.

Russell, W. M. S., & Burch, R. L. (1959). *The principles of humane experimental technique*. London, United Kingdom: Methuen & Co.

Saez, E., No, D., West, A., & Evans, R. M. (1997). Inducible gene expression in mammalian cells and transgenic mice. *Current Opinion in Biotechnology*, *8*(5), 608–616.

Seltzer, Z., Dubner, R., & Shir, Y. (1990). A novel behavioral model of neuropathic pain disorders produced in rats by partial sciatic nerve injury. *Pain*, *43*(2), 205–218.

Slack, M. K., & Draugalis, J. R. (2001). Establishing the internal and external validity of experimental studies. *American Journal of Health-System Pharmacy*, *58*(22), 2173–2181.

Smith, S. B., Crager, S. E., & Mogil, J. S. (2004). Paclitaxel-induced neuropathic hypersensitivity in mice: Responses in 10 inbred mouse strains. *Life Sciences*, *74*(21), 2593–2604.

Storey, J. D., Madeoy, J., Strout, J. L., Wurfel, M., Ronald, J., & Akey, J. M. (2007). Gene-expression variation within and among human populations. *American Journal of Human Genetics*, *80*(3), 502–509.

United States Department of Agriculture. (2010). *Animal Welfare Act*. Retrieved from http://www.aphis.usda.gov/animal_welfare/publications_and_reports.shtml

Waldman, A. S. (1992). Targeted homologous recombination in mammalian cells. *Critical Reviews in Oncology/Hematology*, *12*(1), 49–64.

Walker, J. S., Sheather-Reid, R. B., Carmody, J. J., Vial, J. H., & Day, R. O. (1997). Nonsteroidal antiinflammatory drugs in rheumatoid arthritis and osteoarthritis: Support for the concept of "responders" and "nonresponders." *Arthritis and Rheumatism*, *40*(11), 1944–1954.

Wesselmann, U., Czakanski, P. P., Affaitati, G., & Giamberardino, M. A. (1998). Uterine inflammation as a noxious visceral stimulus: Behavioral characterization in the rat. *Neuroscience Letters*, *246*(2), 73–76.

Willis, W. D., & Westlund, K. N. (1997). Neuroanatomy of the pain system and of the pathways that modulate pain. *Journal of Clinical Neurophysiology*, *14*(1), 2–31.

Wolff, B. B., Kantor, T. G., Jarvik, M. E., & Laska, E. (1966). Response of experimental pain to analgesic drugs. 1. Morphine, aspirin, and placebo. *Clinical Pharmacology and Therapeutics*, *7*(2), 224–238.

CHAPTER 6

Newborn Screening

Ethical, Legal, and Social Implications

Rebecca Anderson, Erin Rothwell, and Jeffrey R. Botkin

ABSTRACT

Newborn Dried Blood Spot Screening (NBS) is a core public health service and is the largest application of genetic testing in the United States. NBS is conducted by state public health departments to identify infants with certain genetic, metabolic, and endocrine disorders. Screening is performed in the first few days of life through blood testing. Several drops of blood are taken from the baby's heel and placed on a filter paper card. The dried blood, on the filter cards, is sent from the newborn nursery to the state health department laboratory, or a commercial partner, where the blood is analyzed. Scientific and technological advances have lead to a significant expansion in the number of tests—from an average of 6 to more than 50—and there is a national trend to further expand the NBS program. This rapid expansion has created significant ethical, legal, and social challenges for the health care system and opportunity for scholarly inquiry to address these issues. The purpose of this chapter is to provide an overview of the NBS programs and to provide an in-depth examination of two significant concerns raised from expanded newborn screening, specifically false-positives and lack of information for parents. Implications for nursing research in managing these ethical dilemmas are discussed.

http://dx.doi.org/10.1891/0739-6686.29.113

BACKGROUND

Newborn Dried Blood Spot Screening (NBS) is conducted primarily by state public health departments for the early identification of infants with certain genetic, metabolic, and endocrine disorders (March of Dimes, n.d.). Without early identification and prompt treatment, these disorders can result in significant morbidity and mortality (Centers for Disease Control and Prevention, 2011). In all states and most developed countries, a newborn's blood is obtained from a heel-prick within the first few days after birth to screen for numerous disorders. The blood is placed on filter paper as part of the postpartum care in the birthing facility. In some states, a second screen is conducted during one of the infant's first well-child visits with their primary care provider. The filter paper is then mailed to the states' department of health laboratory for analysis and the results are then sent to the infant's primary care provider. Most results are normal and received within 2 weeks. If the result is abnormal, indicating that the infant may have a disorder, the health care provider is contacted by the health department and specific action steps are recommended to the health care provider to complete additional testing. The additional testing is to confirm the infant as affected or unaffected with a disorder targeted by the NBS program.

In recent years, NBS programs have undergone significant changes, in part because of the counsel of advocacy groups and recommendations of national organizations (American Academy of Pediatrics; Botkin et al., 2006; Eunice Kennedy Shriver National Institute of Child Health and Human Development, 2009; March of Dimes, n.d.; Save Babies Through Screening Foundation, Inc., n.d.). Technological advances that allow the screening of several disorders in one test (tandem mass spectrometry), and research evidence of benefit for affected individuals when disorders are identified during the newborn period, have also driven these changes (Botkin, Anderson, Staes, & Longo, 2009; Botkin, 2005; Botkin et al., 2006; New York State Task Force on Life and the Law, 2000; Wilcken, Wiley, Hammond, & Carpenter, 2003). However, the significant expansion of disorders screened for challenges the assumptions that supported the original mandate for screening of all infants under state legislation. This chapter provides a brief overview of the history of NBS and the recent expansion and assumptions governing this important public health program. Ethical dilemmas raised from screening and the growing role nursing will play within NBS will be discussed.

HISTORY OF NEWBORN SCREENING

Newborn screening (NBS) began in the early 1960s with one disorder, phenylketonuria (PKU), that has an incidence of approximately 1 in 20,000–25,000 infants (Brosco, 2011; Guthrie & Susi, 1963; Therrell & Adams, 2007). PKU is a serious metabolic disease caused by a genetic condition that results in the absence of an enzyme that breaks down an amino acid found in food proteins, specifically phenylalanine to tyrosine. After birth when the phenylalanine builds up within the body, the high levels are toxic to the central nervous system, resulting in profound intellectual disability. If the disease is diagnosed in the newborn period, the infant can be placed on a low phenylalanine diet that can prevent significant morbidity.

PKU is the archetypical condition on which current newborn screening programs are built. The primary assumption of the early programs was that identification of the disorder in the newborn period would enable early treatment that would reduce significant morbidity in virtual all affected infants. The reduction in morbidity provides clear benefits for the affected infants, families, and society. At the time NBS was first instituted, a large proportion of individuals with severe intellectual impairments were provided care in state-run facilities; therefore, reduction in morbidity from PKU represented a direct reduction in state expenses and responsibilities (Grosse, 2005; Lord et al., 1999).

In the 1950s when PKU was identified as a cause of preventable intellectual disability, parents and disability organizations advocated for the screening of all infants for this condition (Guthrie, 1996; Paul, 1998). In 1963, Robert Guthrie developed a technique for testing of dried blood on filter paper for PKU, and this discovery enabled population screening with a method that facilitated ease of collection and mailing of the samples to off-site laboratories that had not been possible (Guthrie & Susi, 1963). The ability to screen all infants born in a state at a central laboratory guided state public health departments' development of the infrastructure for the screening. Because of in large part due to the work of parent advocacy groups and organizations such as the National Association for Retarded Citizens (Berry & Wright, 1967), state legislatures began mandating PKU screening for all infants born in their state. Massachusetts was the first state to establish a state health department newborn screening program in 1962 (Guthrie, 1996). By 1975, 43 other states had legislated mandatory newborn screening of all infants for PKU (President's Council on Bioethics, 2008).

EXPANSION

Since the initial implementation for screening of PKU, there recently has been an exponential expansion of the number of disorders for which states mandate newborns to be screened. In the 1990s, most U.S. states were screening newborns for less than 10 conditions and by 2005, many states were screening for at least 20 disorders but with significant variability of the type of tests (American Academy of Pediatrics, 2000; President's Council on Bioethics, 2008;

Therrell, Johnson, & Williams, 2006). In 2005, the U.S. Health Resources and Services Administration (HRSA), part of U.S. Health and Human Services, recognized disparities among the states screening programs and commissioned the American College of Medical Genetics (ACMG) to develop recommendations for a "uniform" screening panel to provide guidance and consistency across all state newborn screening programs (American College of Medical Genetics, 2005). Twenty-nine conditions were recommended in the ACMG report to HRSA as a "core panel" (American College of Medical Genetics, 2005). However, the methods used to develop and the level of evidence on which the recommendations are based have been criticized (Botkin et al., 2006; Moyer, Calonge, Teutsch, & Botkin, 2008). The ACMG process lacked an analytical framework and was based on limited empirical evidence (Botkin et al., 2006). Despite the criticisms, this report was broadly accepted in the professional newborn screening community. The Department of Health and Human Services, Secretary's Advisory Committee on Heritable Disorders and Genetic Diseases in Newborns and Children, the American Academy of Pediatrics, and the March of Dimes endorsed the report and supported the recommendation that all states screen for the "core panel" (American Academy of Pediatrics, 2006; March of Dimes, n.d.; Secretary's Advisory Commitee on Heritable Disorders in Newborns and Children, 2010). Currently, every state in the United States has expanded their newborn screening programs to screen for more than 30 conditions (National Newborn Screening & Genetics Resource Center, 2011). Newborn screening programs diagnose an estimated 4,000 infants each year (National Newborn Screening & Genetics Resource Center, 2011). The expansion of this state mandated program makes newborn screening the largest application of genetic testing in health care.

However, there are questions if all the disorders on the screening panel are justified for states to mandate under the original assumptions that established newborn screening programs (Botkin et al., 2006; Natowicz, 2005). Several of the conditions are extremely rare, poorly understood, and without treatment that effectively prevents morbidity and mortality. The governing principle for any population-screening program is to effectively prevent morbidity or mortality from the targeted disorders through earlier treatment and with limited harm to unaffected infants (Wilson & Jungner, 1968). This raises concerns whether mandated population screening by states is justified when the affected infant receives limited or uncertain benefit, the cost-effectiveness (the acceptability of additional cost to achieve the desired goal) is unknown, and risks associated with false-positives screening results are increased. With inadequate evidence available, it is challenging to determine the balance of the benefits and risks for several specific conditions for the population.

PARENS PATRIAE

An important aspect of these state-based newborn screening programs from their inception is that mandatory screening is conducted on virtually every infant born in the United States under the power of "*parens patriae*" (Holtzman, 1997; Pollitt, 2004; Tarini, 2007). The translation of "parens patriae" is parent of the country. *Parens patriae* is a legal doctrine that gives states the right to assume certain responsibility of parents if it benefits the child and society (Olson & Berger, 2010). It gives states the power to defend the interests of those that are unable to protect their own interest, such as newborns (Gostin, 2008).

All states and territories of the United States, and most developed countries, have state legislation requiring that all infants born in their state undergo newborn screening under the power of parens patriae. Currently, only Wyoming and District of Columbia have laws requiring signed parental permission for the conduct of NBS (Lewis, Goldenberg, Anderson, Rothwell, & Botkin, 2011). In most states, parents are allowed to refuse NBS for religious or philosophical reasons, but parents may not be aware that screening is conducted and, furthermore, are often not adequately informed of their right to refuse (Rothwell, Anderson, & Botkin, 2010; United States General Accounting Office, 2003). The justification for mandatory NBS is that the benefits to children and society are substantial and the state should use its power to protect the interests of children. In an extreme example, state officials in Nebraska removed an infant from his parents to conduct NBS after the parents refused screening. The parents claimed NBS violated their free exercise of religion. The state justified its action in that they were protecting the infant's health ("Douglas City versus Anaya," 2005).

As NBS evolves and expands, the justification for mandated screening under the power of parens patriae needs to be re-examined (Pollitt, 2004; President's Council on Bioethics, 2008; Tarini, 2007). The NBS paradigm has shifted from screening for PKU to the identification of numerous inherited disorders, many of which are poorly understood and lack treatments that prevent significant morbidity. In some cases, early detection may delay, but not prevent, significant disability (Powell et al., 2010). The discussion about benefits has evolved from prevention of morbidity for the affected infant and cost savings for public health to secondary benefits to parents and others (President's Council on Bioethics, 2008). Some of the secondary benefits from NBS include information to the parents for reproductive planning, prevention of a "diagnostic odyssey" (the relatively long period and multiple tests one usually has to undergo to be diagnosed with a rare genetic condition), identification of maternal disease, and benefits to research from the ability to conduct studies with the children identified with rare disorders (Harrell, 2009; President's Council on Bioethics, 2008). In this genomic era, where we can screen for a wide range of disorders that may or may not benefit from early detection and treatment, NBS raises ethical questions if the expanded screening should be justified under a parens patrie (Pollitt, 2004).

ETHICAL CONSIDERATIONS WITH THE EXPANSION OF NEWBORN SCREENING

Newborn screening programs continue to expand, most recently with a national recommendation of screening for critical cyanotic congenital heart disease (Secretary's Advisory Commitee on Heritable Disorders in Newborns and Children, 2011). More disorders will be possible when DNA-based testing, such as sequencing of the individual genome, platforms become feasible. As NBS programs expand their scope, especially when there is limited evidence of direct benefit for affected newborns, there is a need for parents to be adequately informed and provided choices about medical interventions that go beyond the justification of parens patriae. Inadequate provision of information to parents about NBS may also increase the adverse consequences, mainly from the impact of false positive results. These two issues, false-positives and inadequate education, are discussed later.

FALSE-POSITIVE RESULTS

For screening programs in general (such as for breast cancer, prostate cancer, vision screening, and NBS) to be effective, the sensitivity of the tests used to screen for the disorders must approach 100%. This ensures that almost all affected individuals are identified and will potentially receive the benefits from the early detection. However, there is often a trade-off of high sensitivity for low specificity. *Specificity* is the ability of a test to differentiate those that have the disease and those that do not. When the specificity is low, many healthy individuals will initially have abnormal results that with further testing are found to not have the suspected disordor (Tarini, Christakis, & Welch, 2006). As with most population-based screening programs, the newborn screening test is extremely sensitive, allowing virtually all affected infants to be detected, but it is not necessarily very specific. The ratio of true positive to all positive screens is the positive predictive value. For NBS, most of those with an initial positive screening test are found to be healthy after additional testing, a false-positive screen. The positive predictive values for many newborn screening tests is generally between 1% and 10%, so there may be 10–100 unaffected infants with false-positive results for every affected infant identified (Kwon & Farrell, 2000; National Newborn Screening & Genetics Resource Center, 2011). Those unaffected will include

(1) genetic carriers, (2) premature and sick infants with transient abnormalities, (3) infants with benign variants of a condition, (4) healthy infants without a known explanation for the out-of-range result, and (5) those with results from a laboratory error. This results in tens of thousands of false-positive NBS test every year in the United States (Tarini et al., 2006).

There has been a long-standing concern about the psychological effects of false-positive NBS screenings on parents (Holtzman, 2003). An abnormal NBS results can be a time of crisis for parents with a newborn, particularly when they may not have been aware that the screening was conducted or what it entails. For many parents, a false-positive result causes distress and worry that may not be resolved even after further testing confirms the infant is not affected with a targeted disorder (Gurian, Kinnamon, Henry, & Waisbren, 2006; Tluczek et al., 1991; Waisbren, 2006; Waisbren et al., 2003). However, confirmatory testing takes time. Resolution of the crisis situation induced by a positive screen with a normal follow-up test depends on whether parents are convinced that the child is not ill or does not have the suspected disease. The newborns with false-positive results and their families suffer this disadvantage of the program, resulting in a risk for psychosocial harm with little compensatory benefit from the NBS system.

Mothers who received false-positive results reported higher levels of stress (Gurian et al., 2006; Waisbren et al., 2003), depression (Tluczek, Koscik, Farrell, & Rock, 2005; Tluczek, Mischler, & Bowers, 1991), anxiety (Bodegård, Fyrö, & Larsson, 1983; Clemens, Davis, & Bailey, 2000; Fyrö & Bodegård, 1987; Sorenson, Levy, Mangione, & Sepe, 1984), disturbances of the parent–child relationship (Gurian, Kinnamon, Henry, & Waisbren, 2006; Waisbren et al., 2003), altered perceptions and worry about the child's health (Beucher et al., 2010; Moran, Quirk, Duff, & Brownlee, 2007; Tluczek et al., 2010), persistent worry about the child's future (Tluczek et al., 2010), and increased use of health care services by parents for the child (Waisbren et al., 2003). Furthermore, one study found about one-half ($N = 12$) of families following a false-positive NBS result exhibited an impaired parent–child relationship (Fyrö, 1988; Fyrö & Bodegård, 1987, 1988). Additionally, false-positive results and repeat testing have far-reaching effects for the family unit (Sobel & Cowan, 2000). One example is on the parents' reproductive choices. Parents who received a false-positive result were negative influenced on future reproductive decisions (Mischler et al., 1998; Tluczek et al., 1992). However, it has been found by some that a false-positive NBS does not increase use of medical services (Prosser, Kong, Rusinak, & Waisbren, 2010; Tarini et al., 2011).

The management of expanding NBS programs presents numerous challenges because of the need to respond quickly to positive results while reducing the burdens of false-positive results. Those directly involved must respond quickly and efficiently on behalf of affected infants including clinicians, parents, and public health officials. Although concerns about the psychosocial distress some parents experience following a false-positives screening results have been raised since the 1960s with the description of the PKU anxiety syndrome (Rothenberg & Sills, 1968), what methods of NBS education might reduce the negative impacts of false-positives have not been empirically tested.

INFORMATION FOR PARENTS

NBS is unique from other health care services because it is a state-mandated public health program. There is no requirement to communicate with parents about the screening test, and/or to ask for permission to perform the heel-prick prior to taking the blood sample. Although most states provide educational materials, a significant percentage of parents are not aware the screening is done (Bonhomme, 2009). When a positive result requires further testing, this may be the first point in time that parents are aware that the screening test was performed by the state. Following confirmatory testing, affected children may be referred to subspecialty clinics, depending on the condition, for initial education, further testing, counseling, management plans, and subsequent monitoring. Communication of genetic information, especially screening results, is complex and many health care providers have difficulty fully understanding and effectively communicating the results (Kemper, Uren, Moseley, & Clark, 2006).

The structure and timing of NBS in the postpartum period is a significant contributor to why many parents are unaware that NBS was conducted or what it entails (Davis et al., 2006). Furthermore, recent findings suggest that primary care providers would prefer for someone else to inform families of positive NBS, and a significant portion of providers do not feel competent to discuss the disorders (Kemper et al., 2006). The information for and the active engagement of parents in newborn screening programs remain a major challenge and national priority. Parents want information on this topic, but research indicates that the current educational approaches are largely ineffective (Arnold et al., 2006; Davis et al., 2006). Despite the significant expansion in scope and complexity of these important public health programs, there have been no innovative, broad-scale, or evidence-based approaches to improve the education of new parents about newborn screening.

The lack of adequate education for parents about NBS is a problem for several reasons. First, parents are key participants and consumers in state newborn screening programs, prompt and efficient collaboration between professionals and parents is imperative to providing initial screening, confirmatory testing and evaluation, and follow-up services for affected children. Infants with disorders can be harmed if parents do not act as directed in a quick and appropriate way to the initial screening results when notified. Second, as recognized by the American Academy of Pediatrics (2000), parents have a right to basic information about medical interventions conducted on their children, regardless of the mandatory nature of newborn screening. Third, surveys and focus groups document parents have a strong preference for better education about NBS and for this information to be delivered prenatally (Davis et al., 2006; Rothwell et al., 2010). Fourth, 21 states require parental education through their newborn screening legislation and all states prepare educational materials for parents (Lewis et al., 2011). Because of the fact that most states do not require informed permission for the conduct of NBS, the legislative requirements can be met without actively engaging the parent or assuring they understand the information or even read the brochures. Finally, key professional organizations, federal agencies, and lay advocacy groups have specifically called for enhanced parental education regarding newborn screening with consistent suggestions that this intervention should be provided prenatally (American Academy of Pediatrics, 2000; International Society of Nurses in Genetics, 2011).

Recognizing this problem, the American Academy of Pediatrics Task Force on Newborn Screening outlined a national agenda for strengthening state newborn screening systems and specifically called for the development and assessment of new educational tools for parents and professionals (American Academy of Pediatrics, 2000). The task force's recommendations regarding newborn screening education were that public health agencies (federal and state), in partnership with health professionals, should address the following strategies (American Academy of Pediatrics, 2000):

- Design and evaluate tools and strategies to inform families and the general public more effectively.
- Public screening programs should not be implemented until they have first demonstrated their value in well-conducted pilot studies.
- Prospective parents should receive information about newborn screening during the prenatal period.
- Studies should be done to broaden understanding of the ways in which communication can be performed more effectively for the benefit of consumers.
- Develop and provide family educational materials about newborn screening in each state or region, with input from families who have children with special needs and/or parent information centers.

There is an emerging consensus that education of parents regarding newborn screening should occur as a part of prenatal care, rather than in the postpartum period alone (Campbell & Ross, 2004; Kim, Lloyd-Puryear, & Tonniges, 2003; Larson, 2002). Childbirth is a busy, confusing, and exhausting time that does not lend itself to education about topics that are not immediately relevant to the care of the baby. Davis et al. (2006) conducted 22 focus groups of English and Spanish speaking parents of infants less than 1 year of age. They found that parents were rarely given information about newborn screening and that education provided during childbirth was ineffective because of the demands of a newborn (Davis et al., 2006). Diem (2004) emphasized the potential effectiveness of education in the prenatal period when many parents are eager to learn about anything related to the health of their future child.

In 2009, the Consumer Task Force on Newborn Screening of the Genetic Alliance, a consumer advocacy organization, conducted a survey of 2,266 women age 18–45 regarding their attitudes on various issues relevant to newborn screening (Bonhomme, 2009). Two of the conclusions of the study were that the public needs and wants more information about NBS, and second, prospective and recent mothers want information about NBS before their baby is born. Results showed that 86% of recent mothers wanted information before or during pregnancy and 90% of prospective mothers wanted it during their pregnancy (Bonhomme, 2009). Prenatal education about NBS is further supported by the American College of Obstetrics and Gynecology (ACOG). In a 2011 opinion, the ACOG Committee on Genetics recommended, that prenatal care providers make resources about NBS available to patients during pregnancy ("Committee Opinion No. 481: Newborn screening," 2011). The support of ACOG is of critical importance in fostering this intervention by obstetricians.

The lack of adequate information for parents about NBS has become an issue of social justice that is being tested in state and federal courts (Fan, Chen, Lai, Chen, & Chen, 2010; Lilley et al., 2010; Lim et al., 2011). There is a growing dialogue on the importance of transparency that may alter the assumptions under which NBS programs have operated (Botkin, Anderson, & Rothwell, 2011). Furthermore, parents are demanding to know more about NBS and expect to be informed of the testing. If NBS program activities go beyond the states' legitimate *parens patriae* authority, parents may be motivated to protect their children from perceived risks or wrongs. But without accurate information, parents' often lack the tools to make informed decisions, potentially creating health risks for their children.

RESEARCH SUPPORT

There are two primary sources of support for scholarly nursing research in the domain of NBS, the National Institutes of Health (NIH, n.d.) and the Health

Resource Service Administration (HRSA, n.d.), both are part of the United States Health and Human Services. The NIH has two institutes with specific interests in NBS: National Human Genome Research Institute (NHGRI, n.d.a) and the Eunice Kennedy Shriver National Institute of Child Health & Human Development (NICHD, n.d.). The NHGRI's, Ethical Legal and Social Research Program is an important component of the institute that has funded NBS research (NHGRI, n.d.b). HRSA's Genetic Diseases in Newborns and Children Branch supports that the Secretary's Advisory Committee on Heritable Disorders and the Regional Genetics and Newborn Screening Collaborative, the National Newborn Screening, and Genetic Resource Center all provide funding for clinical research and education programs (Health Resource Service Administration, n.d.).

The National Human Genome Research Institute is one of the leaders in the science and support of ethical, legal, and social research at the National Institutes of Health. The Ethical Legal and Social Implications (ELSI) program was established as part of the Human Genome project in 1990 to study the implications of genetic and genomic research (NHGRI, n.d.b). From its inception the ELSI program has a fixed percent of the center's extramural funding budget for extramural research. Originally 3% and in FY 1991, it was increased to 5% where it is today (NHGRI, n.d.c). These dedicated resources offer stable funding for NBS research. The ELSI program recently posted program announcements for regular research applications (R01), exploratory/developmental grants (R21), and small research grants (R03; NHGRI, n.d.b). Furthermore, NHGRI seeks to address ELSI issues through public and community engagement through the support of six Centers of Excellence in ELSI research (CEERs) at universities across the nation (National Human Genome Research Institute, Center for Excellence in ELSI Research). All these funding mechanisms offer excellent opportunities for nursing research to anticipate, analyze, and address the ethical, legal, and social implication of the discovery of new genetic technologies and the use of genetic information.

In 2008, the Newborn Screening Saves Lives Act sponsored by Sen. Christopher Dodd was passed into law. The act authorized the U.S. Health and Human Services to award grants to (1) provide services to newborns and children having or at risk for heritable disorders, (2) provide screening laboratory personnel, (3) develop educational programs, and (4) to assess and coordinate long-term follow-up ("Newborn Screening Save Lives Act," 2008). This act supports the NIH's National Institute of Child Health and Develop (NICHD) to develop methods for adding conditions to the recommend panel, develop, and evaluate interventions to improve outcomes, develop systems to communicate with and educate health care providers, and support research

(National Institutes of Health, 2011). One of the main research interests supported by NICHD, through ACMG's National Coordinating Center (NCC) is the Newborn Screening Translational Research Network (NBSTRN). The NBSTRN is to develop infrastructure in which long-term follow-up and outcome data can be collected for use by investigators (Newborn Screening Translational Network). This dedicated funding has moved the system of newborn screening forward, improving coordination, and providing empirical evidence for the development of NBS interventions and policies. The Newborn Screening Saves Lives Act further charged the Secretary's Advisory Committee on Heritable Disorders and Genetic Diseases in Newborns and Children to recommend disorders to be screened for and to ensure that states had the capacity to screen for the recommended disorders (Eunice Kennedy Shriver National Institute of Child Health and Human Development, 2009). It required the establishment of a clearinghouse of information, which is being developed by the Genetic Alliance called "Baby's First Tests" (Genetic Alliance, n.d.). The Centers for Disease Control was to provide laboratory quality assurance/control and establish a national plan in the event of public health emergency (Department of Health and Human Services, 2008). Finally, the act established Hunter Kelly Newborn Screening Research Program within NICHD to identify new screening technologies and research management strategies for conditions the technologies can detect (Eunice Kennedy Shriver National Institute of Child Health and Human Development, 2009).

The third federal component in the support of NBS systems, projects, and research is the Health Resource Service Administrations (HRSA) Genetics Branch. HRSA aims to support the infrastructure for state NBS and genetic programs and integrate the programs with other community services such as the Medical Home ("Health Resource Service Administration," n.d.). Activities supported by HRSA have primarily included state genetic needs assessment and state genetic plans, evaluation of new technologies, and development of uniform guidelines for state genetic and NBS programs ("Health Resource Service Administration," n.d.). HRSA has accomplished this in large part through the development and support of Regional Genetics and Newborn Screening Services Collaboratives ("National Coordinating Center for the Regional Genetic and Newborn Screening Service Collaboratives," n.d.). The aim of the Collaboratives is to improve the ability and capacity of states to provide and expand genetic and NBS services ("National Coordinating Center for the Regional Genetic and Newborn Screening Service Collaboratives," n.d.). Finally, HRSA has contracted with the American College of Medical Genetics to operate the National Coordinating Center (NCC) for the regional

Collaboratives ("National Coordinating Center for the Regional Genetic and Newborn Screening Service Collaboratives," n.d.).

Furthermore, there is support for NBS research from other organizations such as the Hastings Center, March of Dimes, and the Greenwall Foundation. Laboratorians, physicians, bioethicist, psychologist, and genetic counselors drive much of the research; only a small portion is guided by nursing research. Nursing researchers have much to contribute to this domain of genetic research, especially bringing together transdisciplinary teams to address pressing ethical, legal, and social issues of this rapidly evolving domain of NBS that can serve as an exemplar for other domains of genetics in health care.

IMPLICATIONS FOR NURSING PRACTICE

Nursing scholarship will play a critical role in addressing current and future challenges facing NBS and may provide a mechanism for markedly changing how it is conducted. The most obvious and immediate impact nursing will have on maintaining and improving this important public health program is through empirically based policy development and recommendations for educational interventions. The lack of a permission requirement, in part, has resulted in poor communication about NBS with new parents and limited awareness about it by the general public. In addition, health care providers, who obtain the specimens, are given little or no information on how to communicate when, what, and why this medical procedure is conducted on infants. This responsibility often falls on nurses who oversee, manage, and staff NBS programs.

Many states have been reticent to provide more information about NBS to parents, in part, because of fear of parental decisions to decline screening that may result in missing a child with a disorder (Rothwell et al., 2011). Lack of adequate understanding of these programs and their substantial value by the general public has left health departments vulnerable to criticism and lawsuits, particularly over the management of samples after testing is completed (Maschke, 2009). As a result, some parents are now being encouraged to opt out of newborn screening with misinformation, promoting distrust, and misunderstanding about the programs (Brase, 2009).

As newborn screening continues to evolve, it is important for nurses to be leaders in research that will inform policies and educational interventions (Kemper, Fant, & Clark, 2005). Previous research found that more than half of the nurses involved with the birth of a child talked to parents about NBS, whereas physicians rarely or never discussed NBS with parents (Hayeems et al., 2009). This may be in part because nurses are the ones who interact most with the parents before, during, and after the birth of their child and during NBS specimen collection. Nurses have more opportunities to interact and engage with parents during this important time about the medical procedures conducted on their infants. This experience can provide critical insight into the policies and practices that will be the most efficacious. For example, nurses already provided leadership in several aspects associated with NBS ("International Society of Nurses in Genetics," n.d.). Nurses were responsible for national policy changes for NBS in the newborn intensive care units to ensure all infants received NBS and to improve accuracy of testing (Southeast Regional NBS & Genetics Collaborative). In addition, nurses have provided key insight into the clinical applications of screening and provided input on national guidelines for newborn hearing screening (Balk, 2007; Brennan, 2004).

It will be important to educate parents effectively about the process of NBS and the risk of not participating. Much of this education is likely to occur prenatally and at the hospital level and by nurses who have multiple interactions with patients (Hayeems et al., 2009; Kemper et al., 2005). Education of parents is likely to also improve follow-up with children with initial positive screens and minimize the harm of false-positives (Waisbren et al., 2003). Furthermore, patient education is a vital component to the nurse–patient relationship and nurses provide most of all education in the clinical setting including the conduct of research (Bastable, 2006). Thus, nursing will be critical to leading efforts to develop valid empirically supported education efforts (Kemper et al., 2005).

The International Society of Nurses in Genetics further supports the role of informed decision making with the following position recommendations (International Society of Nurses in Genetics, 2011):

- Nurses are responsible for alerting clients about their right for an informed decision-making process before genetic testing.
- Nurses need to advocate for client autonomy, privacy, and confidentiality in the informed decision-making process.
- Nurses need to ensure that the informed decision-making process includes discussion of benefits and risks including the potential psychological and societal injury by stigmatization, discrimination, and emotional stress, in addition to, if any, potential physical harm.
- Nurses need to be aware of the clinical and personal utility of genetic testing, such as positive predictive value, penetrance rates, background populations, and affected percentages, and advise clients of the meaning of the testing and results.
- Nurses need to advise clients on the difference between research versus clinical use of genetic testing, return of results, clinical utility, and advise clients of the status of a specific test.

- Nurses who have an established relationship and are providing ongoing care to a client contemplating genetic testing need to augment the informed decision-making process by assisting the client in the context of the client's specific circumstances of family, culture, and community life.
- Nurses need to integrate into their practice the guidelines for practice (e.g., privacy and confidentiality, truth telling and disclosure, and nondiscrimination) identified by the American Nurses Association.
- Nurses in preparation for providing genetic services need to receive appropriate education that includes knowledge of the implications and complexities of genetic testing; ability to interpret results; and knowledge of the ethical, legal, social, cultural, and psychological implications of genetic testing.
- Nurses need to be aware of genetic health professionals and services with whom they can collaborate to maximize the potential for the client to make an informed decision.

In summary, expanded NBS has highlighted significant gaps in the health care system, most notably lack of information for parents and valid, reliable evidence for the development of interventions and policy development (Botkin et al., 2011; Botkin et al., 2006; President's Council on Bioethics, 2008). Greater involvement by nurse scholars during the development, implementation, and assessment of prenatal and postnatal education about NBS can significantly improve these programs. As genomics continues to expand into the routine clinical care, ensuring patient understanding through education will be critical for informed decision-making and reduction of medical errors. Nurses will be a central component to any future genomic applications in clinical care.

REFERENCES

American Academy of Pediatrics. (2000). Serving the family from birth to the medical home. Newborn screening: A blueprint for the future—A call for a national agenda on state newborn screening programs. *Pediatrics*, *106*(2, Pt. 2), 389–422.

American Academy of Pediatrics. (2006). *AAP endores newborn screening report from the American College of Medical Genetics.* Retrieved from http://www.medicalhomeinfo.org/screening/Screen Materials/AAP Endorses ACMG 1.doc

American Academy of Pediatrics ad hoc task force on definition of the medical home: The medical home. (1992). *Pediatrics*, *90*(5), 774.

American College of Medical Genetics. (2005). Newborn screening: Toward a more uniform panel and system. Retrieved from www.acmg.net/resources/policies/NBS/NBS-sections.htm

American College of Obstetricians and Gynecologists Committee on Genetics. (2011). Committee Opinion No. 481: Newborn screening. *Obstetrics and Gynecology*, *117*(3), 762–765. http://dx.doi.org/10.1097/AOG.0b013e31821478a0

Arnold, C. L., Davis, T. C., Frempong, J. O., Humiston, S. G., Bocchini, A., Kennen, E. M., & Lloyd-Puryear, M. (2006). Assessment of newborn screening parent education materials. *Pediatrics, 117*(5, Pt. 2), S320–S325.

Balk, K. G. (2007). Recommended newborn screening policy change for the NICU infant. *Policy, Politics & Nursing Practice, 8*(3), 210–219. http://dx.doi.org/10.1177/1527154407309049

Bastable, S. B. (2006). *Essentials of patient education*. Sudbury, MA: Jones and Bartlett Publishers.

Berry, H. K., & Wright, S. (1967). Conference on treatment of phenylketonuria. *Journal of Pediatrics, 70*(1), 142–147.

Beucher, J., Leray, E., Deneuville, E., Roblin, M., Pin, I., Bremont, F., . . . Roussey, M. (2010). Psychological effects of false-positive results in cystic fibrosis newborn screening: A two-year follow-up. *Journal of Pediatrics, 156*(5), 771–776, 776.e1.

Bodegård, G., Fyrö, K., & Larsson, A. (1983). Psychological reactions in 102 families with a newborn who has a falsely positive screening tes for congential hypothyroidism. *Acta Paediatrica Scandinavica. Supplement*, 304, 1–21.

Bonhomme, N. (2009). *Integrating consumer perspectives into newborn screening*. Paper presented at the Genetic Alliance Annual Conference, Washington, DC.

Botkin, J., Anderson, R. A., & Rothwell, E. (2011). Expanded newborn screening: Contemporary challenges to the parens patriae doctrine and the use of public resources. In R. Rhodes, A. Silvers, & P. Battin (Eds.), *Medicine and social justice* (2nd ed.). New York, NY: Oxford University Press.

Botkin, J. R. (2005). Research for newborn screening: Developing a national framework. *Pediatrics, 116*(4), 862–871.

Botkin, J. R., Anderson, R., Staes, C., & Longo, N. (2009). Developing a National Registry for conditions identifiable through newborn screening. *Genetics in Medicine, 11*(3), 176–182.

Botkin, J. R., Clayton, E. W., Fost, N. C., Burke, W., Murray, T. H., Baily, M. A., . . . Ross, L. F. (2006). Newborn screening technology: Proceed with caution. *Pediatrics, 117*(5), 1793–1799.

Brase, T. (2009). *Newborn Genetic Screening; The New Eugenics?* Saint Paul, MN: Citizens' Council on Health Care.

Brennan, R. A. (2004). A nurse-managed universal newborn hearing screen program. *MCN. The American Journal of Maternal Child Nursing, 29*(5), 320–325.

Brosco, J. P. (2011). Hidden in the sixties: Newborn screening programs and state authority. *Archives of Pediatrics & Adolescent Medicine, 165*(7), 589–591.

Campbell, E. D., & Ross, L. F. (2004). Incorporating newborn screening into prenatal care. *American Journal Obstetrics and Gynocology, 190*(4), 876–877.

Centers for Disease Control and Prevention. (2011). Newborn screening. Retrieved from http://www.cdc.gov/newbornscreening/

Clemens, C. J., Davis, S. A., & Bailey, A. R. (2000). The false-positive in unversal newborn hearing screening. *Pediatrics*, 106(1), E7.

Davis, T. C., Humiston, S. G., Arnold, C. L., Bocchini, J. A., Jr., Bass, P. F., III, Kennen, E. M., . . . Lloyd-Puryear, M. (2006). Recommendations for effective newborn screening communication: Results of focus groups with parents, providers, and experts. *Pediatrics, 117*(5, Pt. 2), S326–S340.

Department of Health and Human Services. (2008). *Newborn screening contingency plan* (CONPLAN). Retrieved from http://www.cdc.gov/ncbddd/documents/NBS-CONPLAN.pdf

Diem, K. (2004). Newborn screening—Should it be part of prenatal care? *American Journal of Obstetrics and Gynecology, 190*(4), 874.

Douglas City v. Anaya, 269 Neb. 552 (2005)

Eunice Kennedy Shriver National Institute of Child Health and Human Development. (2009). *NIH newborn screening research program named in memory of Hunter Kelly.* Retrieved from http://www.nih.gov/news/health/oct2009/nichd-19.htm

Fan, J. Y., Chen, L. S., Lai, J. C., Chen, M. K., & Chen, H. C. (2010). A pre-paid newborn hearing screening programme: A community-based study. *B-ENT*, *6*(4), 265–269.T

Fyrö, K. (1988). Neonatal screening: Life-stress scores in families given a false-positive result. *Acta Paediatrica Scandinavica*, *77*(2), 232–238.

Fyrö, K., & Bodegård, G. (1987). Four-year follow-up of psychological reactions to false positive screening tests for congenital hypothyroidism. *Acta Paediatrica Scandinavica*, *76*(1), 107–114.

Fyrö, K., & Bodegård, G. (1988). Difficulties in psychological adjustment to a new neonatal screening programme. *Acta Paediatriaca Scandinavica*, *77*(2), 226–231.

Genetic Alliance. (n.d.) Retrieved from http://www.geneticalliance.org/

Gostin, L. O. (2008). *Public health law: Power, duty, restraint* (2nd ed.). London, England: The Regents of the University of California Press.

Grosse, S. D. (2005). Does newborn screening save money? The difference between cost-effective and cost-saving interventions. *The Journal of Pediatrics*, *146*(2), 168–170.

Gurian, E. A., Kinnamon, D. D., Henry, J. J., & Waisbren, S. E. (2006). Expanded newborn screening for biochemical disorders: The effect of a false-positive result. *Pediatrics*, *117*(6), 1915–1921.

Guthrie, R. (1996). *The origin of newborn screening*. Buffalo, NY: State University of New York at Buffalo.

Guthrie, R., & Susi, A. (1963). A simple phenylalanine method for detecting phenylketonuria in large populations of newborns infants. *Pediatrics*, *32*(32), 338–343.

Harrell, H. (2009). Currents in contemporary ethics: The role of parents in expanded newborn screening. *The journal of law, medicine & ethics: A journal of the American Society of Law, Medicine & Ethics*, *37*(4), 846–851.

Hayeems, R. Z., Miller, F. A., Little, J., Carroll, J. C., Allanson, J., Chakraborty, P., . . . Christensen, R. J. (2009). Informing parents about expanded newborn screening: Influences on provider involvement. *Pediatrics*, *124*(3), 950–958.

Health Resource Service Administration. Maternal and Child Health, Genetic Services. Retrieved from http://mchb.hrsa.gov/programs/geneticservices/index.html

Holtzman, N. A. (1997). Genetic screening and public health. *American Journal of Public Health*, *87*(8), 1275–1277.

Holtzman, N. A. (2003). Expanding newborn screening: How good is the evidence? *The Journal of the American Medical Association*, *290*(19), 2606–2608.

International Society of Nurses in Genetics. (n.d.). Retrieved from http://www.isong.org/

International Society of Nurses in Genetics. (2011). *Informed decision-making and consent: The role of nursing*. Retrieved from http://www.isong.org/ISONG_PS_informed_consent.php

Kemper, A. R., Fant, K. E., & Clark, S. J. (2005). Informing parents about newborn screening. *Public Health Nursing*, *22*(4), 332–338.

Kemper, A. R., Uren, R. L., Moseley, K. L., & Clark, S. J. (2006). Primary care physicians' attitudes regarding follow-up care for children with positive newborn screening results. *Pediatrics*, *118*(5), 1836–1841.

Kim, S., Lloyd-Puryear, M. A., & Tonniges, T. F. (2003). Examination of the communication practices between state newborn screening programs and the medical home. *Pediatrics*, *111*(2), E120–E126.

Kwon, C., & Farrell, P. M. (2000). The magnitude and challenge of false-positive newborn screening test results. *Archives of Pediatric & Adolescent Medicine*, *154*(7), 714–718.

Larsson, A., & Therrell, B. L. (2002). Newborn screening: The role of the obstetrician. *Clinical Obstetrics Gynecology*, *45*(3), 697–710, discussion 730–692.

Lewis, M. H., Goldenberg, A., Anderson, R., Rothwell, E., & Botkin, J. (2011). State laws regarding the retention and use of residual newborn screening blood samples. *Pediatrics*, *127*(4), 703–712.

Lilley, M., Christian, S., Hume, S., Scott, P., Montgomery, M., Semple, L., . . . Somerville, M. J. (2010). Newborn screening for cystic fibrosis in Alberta: Two years of experience. *Paediatrics & Child Health*, *15*(9), 590–594.

Lim, T. H., De Jesús, V. R., Meredith, N. K., Sternberg, M. R., Chace, D. H., Mei, J. V., & Hannon, W. H. (2011). Proficiency testing outcomes of 3-hydroxyisovalerylcarnitine measurements by tandem mass spectrometry in newborn screening. *Clinica Chimica Acta; International Journal of Clinical Chemistry*, *412*(7–8), 631–635.

Lord, J., Thomason, M. J., Littlejohns, P., Chalmers, R. A., Bain, M. D., Addison, G. M., . . . Seymour, C. A. (1999). Secondary analysis of economic data: A review of cost-benefit studies of neonatal screening for phenylketonuria. *Journal of Epidemiology and Community Health*, *53*(3), 179–186.

March of Dimes. (n.d.). *Bring baby home.* http://www.marchofdimes.com/baby/bringinghome_recommendedtests.html

Maschke, K. J. (2009). Disputes over research with residual newborn screening blood specimens. *Hastings Center Report*, *39*(4). Retrieved from http://www.thehastingscenter.org/Bioethicsforum/Post.aspx?id=3826

Mischler, E. H., Wilfond, B. S., Fost, N., Laxova, A., Reiser, C., Sauer, C. M., . . . Farrell, P. M. (1998). Cystic fibrosis newborn screening: Impact on reproductive behavior and implications for genetic counseling. *Pediatrics*, *102*(1, Pt. 1), 44–52.

Moran, J., Quirk, K., Duff, A. J., & Brownlee, K. G. (2007). Newborn screening for CF in a regional paediatric centre: The psychosocial effects of false-positive IRT results on parents. *Journal of Cystic Fibrosis*, *6*(3), 250–254.

Moyer, V. A., Calonge, N., Teutsch, S. M., & Botkin, J. R. (2008). Expanding newborn screening: Process, policy, and priorities. *The Hastings Center Report*, *38*(3), 32–39.

National Coordinating Center for the Regional Genetic and Newborn Screening Service Collaboratives. (n.d.). Retrieved from http://www.nccrcg.org

National Human Genome Research Institute. (n.d.a). Retrieved from http://www.genome.gov

National Human Genome Research Institute. (n.d.b). *The Ethical, Legal, and Social Implication (ELSI) Research Program.* Retrieved from http://www.genome.gov/10001618

National Human Genome Research Institute. (n.d.c). *Review of the Ethical, Legal, and Social Implication (ELSI) Research Program (1990–1995).* Retrieved from http://www.genome.gov/10001747

National Institute of Child Health and Human Development. (n.d.) Retrieved from http://www.nichd.nih.gov

National Institutes of Health. (n.d.). Retrieved from http://www.nih.gov

National Newborn Screening & Genetics Resource Center. (2011). *National newborn screening information system.* Retrieved from http://www2.uthscsa.edu/nnsis/

Natowicz, M. (2005). Newborn screening—Setting evidence-based policy for protection. *New England Journal of Medicine*, *353*(9), 867–870.

Newborn Screening Save Lives Act cccf, H.R. 1636, S.634 § 1109 of the Public Health Service Act (2008).

Newborn Screening Translational Network. (n.d.). Retrieved from http://www.nbstrn.org/about

New York State Task Force on Life and the Law. (2000). *Genetic testing and screening in the age of genomic medicine* (p. 135). Retrieved from http://www.health.ny.gov/regulations/task_force/reports_publications/#genetic_test

Olson, S., & Berger, A. C. (2010). *Challenges and opportunities in using residual newborn screening samples for translational research: A workshop summary.* Roundtable on Translating Genomic-Based Research for Health: Institute of Medicine. Retrieved from http://www.iom.edu/Reports/2010/Challenges-and-Opportunities-in-Using-Residual-Newborn-Screening-Samples-for-Translational-Research.aspx

Paul, D. B. (1998). *The politics of heredity: Essays on eugenics, biomedicine, and the nature-nurture debate*. Albany, New York: State University of New York Press.

Pollitt, R. J. (2004). Compliance with science: Consent or coercion in newborn screening. *European Journal of Pediatrics*, *163*(12), 757–758.

Powell, K., Van Naarden Braun, K., Singh, R., Shapira, S. K., Olney, R. S., & Yeargin-Allsopp, M. (2010). Prevalence of developmental disabilities and receipt of special education services among children with an inborn error of metabolism. *Journal Pediatrics*, *156*(3), 420–426.

President's Council on Bioethics. (2008). *The changing moral focus of newborn screening: An ethical analysis by the president's council on bioethics*. Washington, DC. Retrieved from http://bioethics.georgetown.edu/pcbe/reports/newborn_screening/

Prosser, L. A., Kong, C. Y., Rusinak, D., & Waisbren, S. L. (2010). Projected costs, risks, and benefits of expanded newborn screening for MCADD. Pediatrics, *125*(2), e286–e294. http://dx.doi.org/10.1542/peds.2009-0605

Rothenberg, M. B., & Sills, E. M. (1968). Iatrogenesis: The PKU anxiety syndrome. *Journal American Academy of Child Psychiatry*, *7*(4), 689–692.

Rothwell, E., Anderson, R., & Botkin, J. (2010). Policy issues and stakeholder concerns regarding the storage and use of residual newborn dried blood samples for research. *Policy, Politics & Nursing Practice*, *11*(1), 5–12.

Save Babies Through Screening Foundation. (n.d.). Retrieved from http://www.savebabies.org

Secretary's Advisory Commitee Secretary's Advisory Commitee on Heritable Disorders in Newborns and Children. (2010). *Letter to Sebelius*. Retrieved from http://www.hrsa.gov/advisorycommittees/mchbadvisory/heritabledisorders/recommendations/correspondence/uniformpanel022510.pdf

Secretary's Advisory Commitee Secretary's Advisory Commitee on Heritable Disorders in Newborns and Children. (2011). *Summary of 23rd meeting*. Retrieved from www.hrsa.gov/advisorycommittees/mchbadvisory/heritabledisorders/meetings/twentythird/minutes.pdf

Sobel, S. K., & Cowan, D. B. (2000). Impact of genetic testing for Huntington disease on the family system. *American Journal of Medical Genetics*, *90*(1), 49–59.

Sorenson, J. R., Levy, H., Mangione, T. W., & Sepe, S. J. (1984). Parental response to repeat testing of infants with false positive results in newborn screening program. *Pediatrics*, *73*(2), 183–187.

Southeast Regional NBS & Genetics Collaborative. (n.d.). *Judy Tuerck, MS, RN*. Retrieved from http://southeastgenetics.org/directory/user_details.php/217/Judy_Tuerck

Tarini, B. A. (2007). The current revolution in newborn screening: New technology, old controversies. *Archives of Pediatrics & Adolescent Medicine*, *161*(8), 767–772.

Tarini, B. A., Christakis, D. A., & Welch, H. G. (2006). State newborn screening in the tandem mass spectrometry era: More tests, more false-positive results. *Pediatrics*, *118*(2), 448–456.

Tarini, B. A., Clark, S. J., Pilli, S., Dombkowski, K. J., Korzeniewski, S. J., Gebremariam, A., . . . Grigorescu, V. (2011). False-positive newborn screening result and future health care use in a state Medicaid cohort. *Pediatrics*, *128*(4), 715–722.

Therrell, B. L., & Adams, J. (2007). Newborn screening in North America. *Journal of Inherited Metabolic Disease*, *30*(4), 447–465.

Therrell, B. L., Johnson, A., & Williams, D. (2006). Status of newborn screening programs in the United States. *Pediatrics*, *117*(5, Pt. 2), S212–S252.

Tluczek, A., Becker, T., Laxova, A., Grieve, A., Racine Gilles, C. N., Rock, M. J., . . . Farrell, P. M. (2010). Relationships among health-related quality of life, pulmonary health, and newborn screening for cystic fibrosis. *Chest*, *140*(1), 170–177.

Tluczek, A., Koscik, R. L., Farrell, P. M., & Rock, M. J. (2005). Psychosocial risk associated with newborn screening for cystic fibrosis: Parents' experince while awaiting the sweat-test appointment. *Pediatrics*, *115*(6), 1692–1703.

Tluczek, A., Mischler, E. H., Bowers, B., Peterson, N. M., Morris, M. E., Farrell, P. M., . . . Fost, N. (1991). Psychological impact of false-positive results when screening for cystic fibrosis. *Pediatric Pulmonology. Supplement*, *7*, 29–37.

Tluczek, A., Mischler, E. H., Farrell, P. M., Fost, N., Peterson, N. M., Carey, P., . . . McCarthy, C. (1992). Parents' knowledge of neonatal screening and response to false-positive cystic fibrosis testing. *Journal of Development of Behavioral Pediatrics*, *13*(3), 181–186.

United States General Accounting Office. (2003). Newborn screening: Characteristics of state programs. Retrieved from http://www.gao.gov/products/GAO-03-449

Waisbren, S. E. (2006). Newborn screening for metabolic disorders. *The Journal of the American Medical Association*, *296*(8), 993–995.

Waisbren, S. E., Albers, S., Amato, S., Ampola, M., Brewster, T. G., Demmer, L., . . . Levy, H. L. (2003). Effect of expanded newborn screening for biochemical genetic disorders on child outcomes and parental stress. *The Journal of the American Medical Association*, *290*(19), 2564–2572.

Wilcken, B., Wiley, V., Hammond, J., & Carpenter, K. (2003). Screening newborns for inborn errors of metabolism by tandem mass spectrometry. *New England Journal of Medicine*, *348*(23), 2304–2312.

Wilson, J. M., & Jungner, Y. G. (1968). [Principles and practice of mass screening for disease]. *Boletín de la Oficina Sanitaria Panamericana. Pan American Sanitary Bureau*, *65*(4), 281–393.

CHAPTER 7

Stop, Look, and Listen

Revisiting the Involvement of Children and Adolescents in Genomic Research

Martha Driessnack and Agatha M. Gallo

ABSTRACT

The intersection of the genomic era and information age has created novel ethical, legal, and social issues that may be beyond the reach of existing guidelines for children and adolescents in research. By taking the opportunity to *stop*, to *look*, and to *listen*, nurses are in an ideal situation to help children, adolescents, and families understand these emerging and important issues. This chapter thus reviews and highlights the issues and challenges that arise when children and adolescents are involved in genomic research. First, we *stop* and review existing guidelines for the protection of individual children and adolescents in research and existing gaps and inconsistencies in their implementation. Then we take a closer *look* at the unique features of genetic and genomic research that create particular ethical challenges for completing the informed consent process when the research participant is a child or adolescent. Finally, we challenge nurses to *listen* more intently to what children and their families need to know before they are included in genetic and genomic research. We emphasize the changing context of children's lives today and the emergence of their decision-making skills.

http://dx.doi.org/10.1891/0739-6686.29.133

INTRODUCTION

As children, we were often reminded to STOP! LOOK! and LISTEN! whenever we found ourselves at a busy or confusing intersection. The intersection of the genomic era and information age is not only busy and confusing, but it also appears to be a moving target; creating novel ethical, legal, and social issues that may be beyond the reach of existing guidelines for children and adolescents involved in research (Driessnack, 2009a; Hudson, 2011; Kohane, 2011). Before we take hold of young children's hands and/or direct older children and adolescents[1] across this busy intersection, nurse researchers need to (a) STOP!—to find a safe place to cross, (b) LOOK!—both to make sure we see what is coming and to make sure that what is coming sees the child or teen alongside us, and (c) LISTEN!—because sometimes we are able to hear what is coming long before we are able to see it.

Using this cautionary approach, the focus of this chapter is to review and highlight both issues and challenges that arise when children and adolescents are involved in genetic and/or genomic research. The chapter begins by taking a moment to *stop* and review existing guidelines for the protection of individual children and adolescents in research, as well as existing gaps and inconsistencies in their implementation. The chapter continues by taking a closer *look* at the unique features of genetic and genomic research that create particular ethical challenges for completing the informed consent process when the research participant is a child or adolescent. The chapter ends with a challenge to *listen* more intently for what nurse researchers need to know about and from children and their families before they are included in genetic and genomic research. Emphasis is placed on the changing context of children's lives in today's world and the emergence of their decision-making skills.

PART 1—STOP!

We begin by taking a moment to briefly review guidelines that are already in place for the protection of individual children and adolescents in research. We also highlight existing gaps and inconsistencies in the implementation of these guidelines that nurse researchers need to be aware of.

The guidelines and regulations for conducting research with children vary between countries and between states, creating challenges for nurse researchers in this new era of interdisciplinary and international research (Davidson & O'Brien, 2009). However, the underlying principles remain consistent (Friedman Ross,

[1] In this chapter, we use the term *children* to represent children and adolescents, unless the term *adolescent/adolescence* needs to be specified.

2008) and for the purposes of this chapter, we will focus on the United States (U.S.). One of the foundational documents guiding the ethical conduct of research in the United States is *The Belmont Report* published in 1979 by the National Commission for the Protection of Subjects of Biomedical and Behavioral Research (http://ohsr.od.nih.gov/guidelines/belmont.html). The three core principles outlined in the Belmont Report, which are well-known to nurse-researchers, include (1) respect for persons, (2) beneficence, and (3) justice. These core principles guide Institutional Review Boards (IRBs) and Ethics Committees, as well as inform the processes by which nursing research is conducted. In 1983, the Department of Health and Human Services (DHHS) published the first set of regulatory guidelines specifically focused on research involving children (http://www.hhs.gov/ohrp/humansubjects/guidance/45cfr46.htm). These regulations (Subpart D of 45 CFR 46) stipulate that parents provide permission before children can participate in research and, when appropriate, children should also be asked to assent to their own participation (Field & Berman, 2004). These early regulations also established the guidelines for when research involving children was appropriate, supporting the concept that although research studies should involve children, there should be no greater than minimal risk to the children who participate. If greater than minimal risk was involved, the regulations specifically outlined what justification was needed for such research to go forward. In combination, these guidelines set the foundation for ethical conduct of research with children.

More recently, amidst concerns about the adequacy of these guidelines along with the commitment to include children in research (Kohane, 2011), the Institute of Medicine (IOM) was charged with reviewing the state of regulations and preparing a report to make recommendations about best practices in the ethical conduct of research involving children. The report, requested in the 2002 Best Pharmaceuticals for Children Act (P.L.107–109), was rooted in claims that most of the medical treatments used with children had not actually been tested on children, even for basic safety and efficacy. The resultant report, entitled *The Ethical Conduct of Clinical Research Involving Children*, included three broad recommendations: (1) Well-designed and well-executed research involving children is essential to improve the health and health care of children; (2) a robust system for protecting human research participants is a necessary foundation for protecting child research participants; and (3) effective implementation of policies to protect child participants in research requires appropriate expertise in child health in the design, review, and conduct of the research involving children (Field & Berman, 2004). In short, the inclusion of children in research is a process that continues to be governed by ethical guidelines and procedures meant to safeguard children from harm and/or exploitation (Coyne, 2010; Friedman Ross, 2008).

Among populations identified as vulnerable, children remain a unique group in that they were previously considered to be entitled to protection (O'Lonergan & Milgrom, 2005). Children are afforded additional precautions above and beyond the basic protective guidelines established for adult research participants. Risk assessments are made using lower thresholds than those held for research on adults. Further, parents have far less discretion in what they can permit for their children than what they can consent to for themselves (O'Lonergan & Milgrom, 2005). Yet, a fine line exists between establishing guidelines to protect children and making guidelines so strict that they actually work to discourage and/or exclude children from participation in research (Kohane, 2011). The U.S. Congress, the Food and Drug Administration (FDA), and the National Institutes of Health (NIH) have each acted to expand research involving children (Field & Berman, 2004). Both the NIH and DHHS continue to emphasize the need to increase the participation of children in research through their *No More Hand-Me-Down Research* initiative and interactive website (http://www.nhlbi.nih.gov/childrenandclinicalstudies/index.php). These efforts speak to the IOM's recommendation about the importance of including children in research. But, what about the IOM's other two recommendations?

Although the IOM report acknowledged that a robust system for protecting human research participants is a necessary foundation for protecting child research participants, it also recognized that the effective implementation of additional policies to protect child participants would require expertise in child health in the design, review, and conduct of the research involving children. Ethics committees, peer review, and interdisciplinary efforts continue to fine-tune the foundational aspects of conducting sound scientific and ethical research. However, the infusion of child health expertise into the implementation of additional policies to protect child participants is less evident. In fact, the implementation of these protective policies by IRBs and researchers is often inconsistent, especially in terms of interpretation of what defines minimal and greater than minimal risk in children's everyday lives and implementation of assent procedures (Wendler & Varma, 2006). Of particular note was a study that uncovered, even though both the National Commission for the Protection of Subjects of Biomedical and Behavioral Research and the American Academy of Pediatrics (AAP) recommend that assent be obtained from children aged 7 and older, only 20% of the IRBs followed this recommendation (Whittle, Shah, Wilfond, Gensler, & Wendler, 2004). They also found that IRB practices varied widely for all pediatric considerations that do not have a specific analog in the adult regulations. More recently, the IOM report on *Child and Adolescent Health and Health Care Quality* (Committee on Pediatric Health and Health Care Quality Measures, 2011) highlighted that despite numerous data sets and measures, the United

States still does not have sufficient information about the health status and quality of care of children and adolescents, primarily because of inconsistencies in definitions and a lack of standardization in terms of methods and approaches to measurement. As more children participate in research, both researcher and IRB expertise about the needs of children and their rapidly changing social contexts increases (Field & Berman, 2004). Further, although the nursing literature is full of thoughtful articles spanning the last 20 years focused on involving children in research and, more specifically, in the assent process (Broome & Richards, 1998; Lindeke, Hauck, & Tanner, 2000; Neill, 2005), the call for nurse researchers to step forward by offering their expertise to IRBs, interdisciplinary research teams, and to each other remains.

There is long-standing consensus that children are considered minors by law (although the age varies between states) and therefore deemed legally incompetent to enter into any contractual agreement on their own behalf, including the provision of informed consent (Broome & Richards, 1998; Matutina, 2009). However, children's lack of legal authority to provide consent is too often equated with a corresponding lack of autonomy, self-determination, and cognitive abilities. In contrast, although children's legal status is fairly simple to determine, children's developing autonomy, self-determination, and cognitive abilities are a moving target (Metcalfe, Plumridge, Coad, Shanks, & Gill, 2011). Informed consent, which is an outgrowth of the core principle respect for persons, is considered to the cornerstone for ethical standards in research. The goal of informed consent is to assure that research participants are aware of the purpose of the study and the risks and benefits of the research before they enter into participation voluntarily (McGuire & Beskow, 2010). The informed consent process for minors must include parent(s) legally, either through parent(s) proxy consent on behalf of the young child or, for older children and teens, a combination of parental permission and child assent. Of note, however, is that child assent should not be confused with informed consent or with autonomous decision making (Miller & Nelson, 2006). The underlying value of assent is an expression of respect for the child as a person, but it does not acknowledge the child's individual autonomy, as assent is not sufficient to authorize inclusion in research. In fact, the requirement to obtain parental permission may actually restrict some children's ability to participate voluntarily in research (Coyne, 2010). This potential has sparked new debates about the need to re-evaluate the informed consent process with minors. Nursing, as well as other disciplines are beginning to challenge the blanket requirement for parental consent, as well as the current informed consent process' failure to recognize a child's advancing capabilities (Cavet & Sloper, 2005; Coyne, 2010). Years ago, Broome and Richards (1998) identified that children, as young as 6 years of age, already demonstrate a basic understanding of research, including its purpose, expectations for participation, and risks and benefits. Yet, despite continued reports of underestimation of children's capabilities, little has changed since the first guidelines for assent were established close to 30 years ago. Parental proxy and the acceptance of "substituted" autonomy diminish the already diminished legal autonomy of children in both society and research.

When children and teens are involved in research, there are other guidelines beyond the consent process that have been put in place to safeguard their participation. One guideline is the inclusion of a "best interest" analysis (Miller, 2010). Best interest of the child is a widely accepted legal standard when making proxy decisions for children (Beauchamp & Childress, 2001). Although the best interest standard is supposed to focus on the child, the child's best interest is linked to and/or dependent on the interests of parents and constructed via processes rooted in the broader context of the child's family and/or other family members (DeMarco, Powell, & Stewart, 2011; Geelen, Van Hoyweghen, Doevendans, Marcelis, & Horstman, 2011). Further, a child's "best interests," like so many other aspects about children, are not fixed in time (Geelen et al., 2011). Some have suggested that the best interest standard should be rejected in research because it ignores the special status of family as its own unique entity, both socially and legally, with its own interests, and offers no guidance or solution for those families in which there is more than one child, each with competing and conflicting interests (Ross, 1998; DeMarco et al., 2011). This is especially important as we consider genetic and genomic research, where one person's decision casts a wide net, with both immediate and far-reaching implications for the person and extended family (Petronio & Gaff, 2010).

Our goal in this first section was to review guidelines that are already in place for the protection of individual children and adolescents in research and highlight the current dilemmas. By highlighting potential dilemmas, nurse researchers can focus their attention on finding the mechanisms to resolve them. The dilemmas highlighted in this first section become increasingly relevant as we enter the intersection of genomic era and information age.

PART 2—LOOK!

With the call to increase research with children and adolescents, nowhere is this need for protection more confusing and challenging than in genetic and genomic research. Continuing with our cautionary approach, this next section takes a closer *look* at the unique features of genetic and genomic research that create some of the unique challenges at the busy intersection especially as they relate to the informed consent process with minors.

Advances in genetic and genomic research continue to challenge traditional conceptions of informed consent, especially with the emergence of large-scale population studies, the establishment of repositories or biobanks, the evolution of next-generation genetic and genomic technologies, and the availability of direct-to-consumer testing (Hens, Nys, Cassiman, & Dierickx, 2011; McGuire & Beskow, 2010; Patenaude, 2011; Samuël, Knoppers, & Avard, 2011; Tarini, Tercyak, & Wilfond, 2011; Wallace & Kent, 2011). Although guidelines for informed consent try to provide sufficient information to participants about the study before they volunteer, the completeness of this information is increasingly becoming a challenge for researchers who cannot foresee the future or predict what types of information will emerge over the course of the study or be applicable in the child's future (Wilfond & Carpenter, 2008). In the most recent *Annual Review of Genomics and Human Genetics*, the authors identify common features of genetic and genomic research and how each feature challenges established norms for informed consent (McGuire & Beskow, 2010). They take their discussion to the next level by proposing alternative models that have the potential to balance the obligation of researchers to respect and protect participants with the larger societal expectation for access to the promises of personalized genomic health care as quickly as possible. One of the proposed models, the partnership model, is applicable to the challenges faced in research with children. In the partnership model, the consent process is viewed as an open, bidirectional communication process that provides ongoing opportunities for the researcher and participant(s) to update the consent. One pediatric genetic research project that is using a partnership model is the Gene Partnership Project (GPP) based at the Children's Hospital Boston (http://www.genepartnership.org/). In this longitudinal study, children are assented at ages 7, 13, and then consented at age 18. The GPP also allows participants secure access to selected information about their own samples and keeps participants informed of emerging findings. However, it is not clear how the information is shared with children.

Genetic and genomic research involving children raises ethical and legal issues that do typically arise with adults. One of these issues indirectly involves the child's pre-existing family communication norms and patterns. When children are enrolled in genetic and genomic research, any findings, if provided, are most often provided to the parent(s), who in turn are tasked with providing the information to their child in a manner and at a time they decide to best suit their child. Yet, there is little evidence to suggest that parents actually perform such a function and even less evidence is available on the processes they might or do use (Gaff & Bylund, 2010). Further, in their meta-synthesis of the literature, Metcalfe and colleagues (2008) noted that most of the parents, across the studies included in their review, reported "a complete lack of support or advice from health professionals about discussing genetic conditions and inheritance with their children" (p. 5). This lack of support and advice is a wide-open area for intervention, development, and testing, and one that is ideally suited to nursing and its holistic approach to understanding patients and families. Nurses are perfectly positioned to partner with parents in the development of a staged plan for informing their children that is tailored, not only to the needs of the family and sensitive to the family's communication norms and patterns, but is also developmentally appropriate to the child (Hadley, Smith, Gallo, Angst, & Knafl, 2008; Sullivan & McConkie-Rosell, 2010). In their qualitative study with parents, children, and young people focused on family communication of genetic risk, Metcalfe and colleagues (2011) were able to identify patterns of understanding and emotional responses to that understanding as they developed across time, creating an innovative framework for others to work from.

Another issue relates to data collection and resultant findings that may reveal genetic information that can affect the child throughout her/his life. Although the stability of one's DNA is not unique to children, what is unique is that children mature and their capacity to make informed decisions increases over time. Therefore, the voice of the child in the decision-making process to participate in the study should carry more weight as the child matures. The unique challenge of including children in research is that children's progressive development, both physiologically and cognitively, raises *ongoing* issues with parental consent, child assent, the return of research results, and the confidentiality of the data collected as part of the research (Avard, Samuel, Black, Griener, & Knoppers, 2011). There are ongoing debates over (1) whether parents even have the right to enroll children in genetic and genomic research, and (2) if parents do have the right, at what point should the consent of the child be obtained in order to continue her/his participation and/or the use of her/his data and samples (Avard et al., 2011; Hens, Cassiman, Nys, & Dierickx, 2011). At the heart of this debate is the definition and assessment of risk.

Most people conceptualize and understand risk in their own terms within the context of their own experiences, including children. Current research has begun to provide insights into the different ways children perceive and communicate various risk experiences (Metcalfe et al., 2011; Peek, 2008; Tanner, 2010). To date, there is little guidance for how researchers and IRBs should interpret risk when assessing genetic and genomic research involving children. The concept of risk is already vague and, as stated earlier, it often has different contextual meanings. Federal regulations, noted earlier (Subpart D of 45 CFR 46) allow children to be enrolled in research that poses greater than "minimal risk" and offers no prospect for direct benefit when the risks do not exceed a "minor" increase over minimal risk. Further, "minimal risk," defined in the regulations as "the probability and magnitude of harm or discomfort anticipated in the research are not greater in and of themselves than those ordinarily encountered in daily life or during the performance of routine physical or psychological examinations or tests," does not specify what comprises the risks of daily life (Moreno & Kravitt, 2010). Therefore, much depends on how each IRB defines minimal risk because few systematic attempts to define what constitutes a minor increase over minimal risk are available (Moreno & Kravitt, 2010; Wendler, Belsky, Thompson, & Emmanuel, 2005; Wendler &, Emanuel, 2005; Wendler & Varma, 2007).

Attempts to quantify the risk of everyday life for children have been proposed. For example, it has been estimated that the risk of death from a car accident in the United States for children 14 years of age and younger is 0.06/million trips and 0.04/million trips for children 15–19 years of age (Davidson & O'Brien, 2009). In the same article, the authors also report attempts to quantify psychological risk using the Pediatric Quality of Life Inventory (PedsQL; http://www.pedsql.org/), stating 27% of children in the United States feel afraid sometimes. In their attempts to combine risk for all activity, the authors reported that the estimated risk of injury per day in a child is 1 in 250, and the estimated risk of death per day for that same child is 4 in a million (Davidson & O'Brien, 2009). Of note is that the authors could not find any systematic assessment, to date, of how IRBs tend to categorize genetic and genomic research involving children and/or whether the involvement of children in such research requires one or both parents' permission. There were some informal discussion points offered occasionally that suggest genetic and genomic research be considered as minimal risk for children in families where family health history is already being freely shared and family communication is more open. But again, without an agreed upon standardization of what minimal risk entails or any acknowledgement of a cumulative burden of research risk for children, we may actually be finding ourselves at a point in time when more ethical issues arise in genetic and genomic research with children than are able to be resolved.

The final issue is confidentiality as advances in information technology and bioinformatics continue to lead to the creation of increasingly interlinked databases (Davidson & O'Brien, 2009). There are some questions that individual genetic and genomic data cannot really be de-identified, especially when the individual has a very rare condition. Further, despite the passage of the Genetic Information Nondiscrimination Act (GINA; http://www.genome.gov/24519851), there are residual concerns about the ability of GINA to protect individuals against discrimination in obtaining health insurance and/or employment based on their genetic information. Children are not excluded from this risk as their research results will follow them for their entire life. Further, in research with children, the duty of confidentiality is not absolute because there are certain exceptions wherein researchers may have a legal obligation to disclose information that has been revealed to them in confidence. One well-known exception is the suspicion of child abuse and/or neglect, in which nurses, as mandatory responders, have a duty to disclose. It may not be such a stretch to imagine a situation in which parents' interests might be seen as overshadowing their child's best interests, interpreting it as potential abuse and/or neglect (http://www.childwelfare.gov/systemwide/laws_policies/statutes/manda.pdf).

In this section, we highlighted some of the challenges researchers face when conducting genetic and genomic research with children related to the informed consent process. In the next section, we move beyond *stop* and *look* and instead turn to listen to the children standing beside us.

PART 3—LISTEN!

We end our chapter by challenging nurse researchers to *listen* more intently to the children included in genetic and genomic research. The emphasis of the discussion is on what research to date says about the changing context of children's lives and children's ability to represent themselves, articulate their experiences, and engage in decision making on their own behalf. Two novel child-centered approaches for data collection with children are briefly reviewed.

The first challenge facing children involved in any research study is that there has been, and perhaps still is for many, a reluctance to take children seriously (Christensen & James, 2008). This reluctance persists despite 20 years of research to suggest otherwise and an international affirmation of children's right to self-determination, dignity, respect, noninterference, and the right to make informed decisions as suggested by the United Nations (U.N.) Convention of the Rights of the Children in 1989 (http://www.unicef.org/crc/). Further, despite increasing research highlighting the persistent underestimation of children's abilities to communicate on their own behalf, there has been a slow shift in health-related research, especially in the United States, from conducting research *on* children towards conducting research *with* them (Clavering & McLaughlin, 2010; Driessnack & Furukawa, 2011). This slow shift has persisted in health-related research, even with continued calls to increase the inclusion of children in research noted earlier in this chapter. Researchers in the past have "viewed children as passive and emphasized their potential as adults rather than their experience as children" (Fraser, Lewis, Ding, Kellet, & Robinson, 2004, p. 281). Others have summarized this first challenge, stating that "the predominant emphasis has been (1) on children as the objects of research rather than children as subjects, (2) on child-related outcomes rather than child-related processes, and (3) on child variables rather than children as persons" (Greene & Hogan, 2005, p. 1).

The second challenge facing children involved in research is that when researchers do recognize the value of engaging children and exploring children's experiences, the research team then needs to identify appropriate, or child-centered, ways to do so (Clavering & McLaughlin, 2010; Soderback, Coyne, & Harder, 2011). Within the past 10 years, there has been an increase in resources exploring novel child-centered approaches and methods, most of which are originate from the field of sociology and from outside of the United States (Christensen & James, 2008; Fraser et al., 2004; Freeman & Mathison, 2009; Greene & Hogan, 2005; Grieg, Taylor, & MacKay, 2007; Thomson, 2008). Of note, however, is that most of the approaches to research and data collection with children, including interviews, questionnaires, and surveys, remain adult-centered and adult-dominated, and children's perspectives and experiences consequently continue to be framed by an adult-centric worldview (Clavering & McLaughlin, 2010; Driessnack & Furukawa, 2011).

Both of these challenges carry over to genetic and genomic research with children and, in many ways, appear to be exaggerated. Genetics and genomic discovery is so fast-paced and steeped in technological advances that health care providers report concerns over their inability to comprehend and integrate this knowledge into care. Adult research participants and parents also report limited genetic literacy levels. It is no wonder that children's understanding of these complex concepts is often dismissed and communication with them postponed "until they are old enough to understand." Undoubtedly, adults have greater knowledge than children about many things, but they do not have greater knowledge about what children think about genetic and genomic concepts and/or research experiences or, more importantly, what it is like to be a child today and what children need or want to know (Kellett, Forrest, Dent, Ward, 2004). Today's children are growing up in a completely different world in which information is shared freely and instantly, both by the children themselves and by others who are invited, or may even be trespassing, in the child's rapidly expanding and evolving social world (Driessnack, 2009a). It is a remarkably different world from the one in which their parents and research teams emerged. In the last of their key messages, Cleverling and McLaughlin (2010) remind health researchers that,

> In a context where we are acutely aware of children's rights and policy agendas of inclusion in decision making, it is a requirement that research on children's experiences within health care asks in what way children can come to the fore as actors with something to say. (p. 609)

Children, by nature, are naturally equipped with the ability to acquire information about their worlds and represent and interpret that information in ways that make sense to them and, at the same time, equip them to solve the next set of often unpredictable problems that await them (Halford, 1993; Lucariello, Hudson, Fivush, & Bauer, 2004). As researchers, we are tasked not only with tapping that ability, but also with bringing it to the fore. Although some shy away from talking with children about genetic and genomic issues, citing potential psychological harm, the preponderance of evidence suggests that children who do receive genetic test results and family history information do not experience any significant changes in their psychosocial wellbeing (Wade, Wilfond, & McBride, 2010). There have been several classic studies over the past 20 years that have systematically described what children actually know about basic genetic and genomic concepts (Gelman & Kremer, 1991; Richards, 2000; Rowlands, 2001; Smith & Williams, 2007; Solomon, Johnson, Zaitchik, & Carey, 1996; Venville, Gribble, & Donovan, 2005). What they found were children, as young as 7 years of age and who had not yet received any formal education, were already assimilating information about these concepts from interactions with their family and peers, as well as from what children confirmed is a readily available and ever-expanding repertoire of information sources. More recently, one study (Smith & Williams, 2007) examining longitudinal changes in children's understanding of basic genetic concepts, found an increase in children's knowledge up to age 10, but few changes between 10 and 14 years of age and beyond.

The ability of nurses to translate basic and complex knowledge into practical education for children and families has played a significant role in promoting the health and welfare of children in clinical practice. This ability can be transferred to the research arena by leading the way in the development of child-centered approaches to data collection and engagement of children and their families in genetic and genomic research (Gallo, Angst, Knafl, Hadley, & Smith, 2005). It can also be transferred by advocating for true participation, rather than just inclusion, of children throughout the process. Shier (2001) describes progressive stages of participation that begins with listening to children, followed by supporting children in expressing their views, taking children's views into account, involving children in decision-making, and finally to the point where the power and responsibility for making a decision is fully shared. What is particularly attractive about this model is its description of how it is possible for children to participate at different levels, yet illustrates how each level is dependent of those beneath. For example, using this model, children cannot take part in decision-making processes without their views being taken into account (McLeod, 2008). However, as previously noted, there is a need for less adult-centric approaches for engaging children.

In their book, *Family Communication About Genetics*, Gaff and Bylund (2010) present a thorough overview of current theories and practice, including an entire chapter (Sullivan & McConkie-Rosell, 2010) on helping parents talk to their children. For the purposes of this chapter, we also want to highlight two specific child-centered approaches to data collection that have been developed or adapted by the authors for use with children in genetic and genomic research. The first approach, the Colored Eco-Genetic Relationship Map (CEGRM) was originally developed by Kenen and colleagues and then adapted for use with children (Driessnack, 2009b). The CEGRM uses arts-based inquiry to engage children in illustrating their social networks and communication patterns around genetic and genomic issues. The second approach is a specific application of the Draw-and-Tell Conversation (Driessnack, 2006) that engages children in discussion about what they know about genetic and genomic concepts using a series of children's drawings and specific inquiry probes (Driessnack & Gallo, in press). Both of these child-centered approaches to data collection have not only been used in research with children, but both have also given rise to unique insights into children's ideas and thought processes about disease causation, risk, and inheritance in the context of their families and social networks.

To end this third section about listening to children, we return to the informed consent process and decision making in genetic and genomic research. In research, we basically ask a few people to accept some burden, inconvenience, and, at times, some risk for the benefit of others. In short, we ask people, including children, to do something for someone else. Although they may gain some benefit, the goal in research is not to benefit or help the participants, the goal is to answer the research question, which in turn should move the science forward toward the benefit of all. Participants need to understand the question, weigh the risks and benefits, and then make their decision as to whether or not they want to volunteer. This is the decision all participants face when they enter into a research study. As was noted in the first section of this chapter, children are deemed legally incompetent to enter into any contractual agreement on their own behalf, including the provision of informed consent. This legal incompetence is linked to the assumption that children lack adequate decision-making skills to ensure competence. Decision-making skills are increasingly recognized to involve multi-faceted, complex, and context-specific processes. Although there are clearly nontrivial differences in the decision-making skills of children and adults, there is very little research that specifically identifies (1) which aspects of decision-making competence improve between childhood and adulthood and/or (2) which aspects of decision-making competence are needed to provide informed consent in research (Byrnes, 2002). Although few would argue that adults tend to have more knowledge and experience than children, to date this advantage has not been shown to systematically translate into an age difference in making better decisions, especially when the increased amount of knowledge is shown to be inaccurate (Byrnes, 2002). The use of age as a linear predictor for decision-making competence may be particularly ill-suited in genetic and genomic research and may need to be replaced with a more targeted and collaborative approach based on the child's current understanding of the information, pre-existing family communication norms and patterns, and specific aspects of decision-making competence (Gallo et al., 2005; Hadley et al., 2008; Metcalfe et al., 2011).

In summary, we have focused on a few of the ethical issues surrounding the inclusion of children in genetic and genomic research and highlighted areas that continue to need the attention of nurse researchers going forward. We recognize that we have not focused on all the issues or delved into parallel ethical issues with the increased use of genetic testing in pediatric clinical practice. Our cautionary approach, using—STOP! LOOK! and LISTEN!—reminds nurse-researchers that we are at a very busy intersection that includes children.

AUTHOR NOTE

For nurse-researchers who desire a more in-depth exploration of ethical issues surrounding the inclusion of children in research, we highly recommend Miller's book *Pediatric Bioethics* (2010).

ACKNOWLEDGMENTS

The authors would like to acknowledge Janet K. Williams, PhD, RN, FAAN, for her careful review and suggestions for this chapter.

REFERENCES

Avard, D., Samuël, J., Black, L., Griener, G., & Knoppers, B. M. (2011). *Best practices for health research involving children and adolescents: Genetic, pharmaceutical, longitudinal studies and palliative care research*. Centre of Genomics and Policy, McGill University Ethics Office, Canadian Institutes of Health Research. Retrieved from http://www.pediagen.org/ressources/BestPratices.pdf

Beauchamp, T. L., & Childress, J. F. (2001). *Principles of biomedical ethics*. New York, NY: Oxford University Press.

Broome, M., & Richards, D. (1998). Involving children in research. *Journal of Child and Family Nursing*, *1*(1), 3–7.

Byrnes, J. P. (2002). The development of decision-making. *Journal of Adolescent Health*, *31*(6, Suppl.), 208–215.

Cavet, J., & Sloper, P. (2005). *Children and young people's views on health and health services: A review of the evidence* (pp. 1–83). London, United Kingdom: National Children's Bureau.

Christensen, P., & James, A. (2008). *Research with children: Perspectives and practices* (2nd ed.). New York, NY: Routledge.

Clavering, E. K., & McLaughlin, J. (2010). Children's participation in health research: From objects to agents? *Child: Care, Health and Development*, 36(5), 603–611.

Committee on Pediatric Health and Health Care Quality Measures. (2011). *Child and Adolescent health and health care quality: Measuring what matters* [Consensus Report]. Washington, DC: National Academy of Sciences.

Coyne, I. (2010). Research with children and young people: The issue of parental (proxy) consent. *Children & Society*, *24*(3), 227–237.

Davidson, A. J., & O'Brien, M. (2009). Ethics and medical research in children. *Paediatric Anaesthesia*, *19*(10), 994–1004.

DeMarco, J. P., Powell, D. P., & Stewart, D. O. (2011). Best interest of the child: Surrogate decision making and the economics of externalities. *Journal of Bioethical Inquiry*, *8*(3), 289–298.

Driessnack M. (2006). Draw-and-tell conversations with children about fear. *Qualitative Health Research, 16*(10), 1414–1435.

Driessnack, M. (2009a). Growing up at the intersection of the genome era and information age. *Journal of Pediatric Nursing*, *24*(3), 189–193.

Driessnack, M. (2009b). Using the Colored Eco-Genetic Relationship Map (CEGRM) with children. *Nursing Research*, *58*(5), 304–311.

Driessnack, M., & Furukawa, R. (2012). Arts-based data collection techniques used in child research. *Journal for Specialists in Pediatric Nursing*, *17*(1), 3–9. http://dx.doi.org/10.1111/j.1744-6155.2011.00304.x

Driessnack, M., & Gallo, A. (in press). Children "draw-and-tell" their knowledge of genetics. *Pediatric Nursing*.

Field, M. J., & Berman, R. E. (Eds.). (2004). *The ethical conduct of clinical research involving children*. Washington, DC: National Academies Press.

Fraser, S., Lewis, V., Ding, S., Kellett, M., & Robinson, C. (Eds.). (2004). *Doing research with children and young people*. Thousand Oaks, CA: Sage.

Freeman, M., & Mathison, S. (2009). *Researching children's experiences*. New York, NY: Guilford Press.

Friedman Ross, L. (2008). *Children in medical research: Access versus protection. Issues in biomedical ethics*. New York, NY: Oxford University Press.

Gaff, C. L., & Bylund, C. L. (2010). *Family communication about genetics: Theory and practice*. New York, NY: Oxford University Press.

Gallo, A. M., Angst, D., Knafl, K. A., Hadley, E., & Smith, C. (2005). Parents sharing information with their children about genetic conditions. *Journal of Pediatric Health Care*, *19*(5), 267–275.

Geelen, E., Van Hoyweghen, I., Doevendans, P. A., Marcelis, C. L., & Horstman, K. (2011). Constructing "best interests": Genetic testing of children in families with hypertrophic cardiomyopathy. *American Journal of Medical Genetics*, *155A*(8), 1930–1938.

Gelman, S. A., & Kremer, K. E. (1991). Understanding natural cause: Children's explanations of how objects and their properties originate. *Child Development*, *62*(2), 396–414.

Greene, S., & Hogan, D. (Eds.). (2005). *Researching children's experience: Approaches and methods*. Thousand Oaks, CA: Sage.

Grieg, A., Taylor, J., & MacKay, T. (2007). *Doing research with children* (2nd ed.). Thousand Oaks, CA: Sage.

Hadley, E. K., Smith, C. A., Gallo, A. M., Angst, D. B., & Knafl, K. A. (2008). Parents' perspectives on having their children interviewed for research. *Research in Nursing & Health*, 31(1), 4–11.

Halford, G. S. (1993). *Children's understanding: The development of mental models*. Hillsdale, NJ: Lawrence Erlbaum Associates.

Hens, K., Cassiman, J. J., Nys, H., & Dierickx, K. (2011). Children, biobanks and the scope of parental consent. *European Journal of Human Genetics*, *19*(7), 735–739.

Hens, K., Nys, H., Cassiman, J. J., & Dierickx, K. (2011). Risks, benefits, solidarity: A framework for the participation of children in genetic biobank research. *The Journal of Pediatrics*, *158*(5), 842–848. http://dx.doi.org/10.1016/j.jpeds.2010.12.036

Hudson, K. L. (2011). Genomics, health care, and society. *New England Journal of Medicine*, *365*(11), 1033–1041.

Kellett, M., Forrest, R., Dent, N., & Ward, S. (2004). "Just teach us the skills please, we'll do the rest": Empowering ten-year-olds as active researchers. *Children & Society*, *18*(5), 329–343.

Kohane, I. S. (2011). No small matter: Qualitatively distinct challenges of pediatric genomic studies. *Genome Medicine*, *3*(9), 62. http://dx.doi.org/10.1186/gm278

Lindeke, L. L., Hauck, M. R., & Tanner, M. (2000). Practical issues in obtaining child assent for research. *Journal of Pediatric Nursing*, *15*, 99–103.

Lucariello, J. M., Hudson, J. A., Fivush, R., & Bauer, P. J. (Eds.). (2004). *The development of the mediated mind: Sociocultural context and cognitive development*. Mahwah, NJ: Lawrence Earlbaum Associates.

Matutina, R. E. (2009). Ethical issues in research with children and young people. *Pediatric Nursing*, *21*(8), 38–44.

McGuire, A.L., & Beskow, L.M. (2010). Informed consent in genomics and genetic research. *Annual Review of Genomics and Human Genetics, 11*, 361–381

McLeod, A. (2008). *Listening to children: A practitioner's guide*. Philadelphia, PA: Jessica Kingsley Publishers.

Metcalfe, A., Coad, J., Plumridge, G. M., Gill, P., & Farndon, P. (2008). Family communication between children and their parents about inherited genetic conditions: A meta-synthesis of the research. *European Journal of Human Genetics*, *16*(10), 1193–1200.

Metcalfe, A., Plumridge, G., Coad, J., Shanks, A., & Gill, P. (2011). Parents' and children's communication about genetic risk: A qualitative study, learning from families' experiences. *European Journal of Human Genetics*, *19*(6), 640–646.

Miller, G. (2010). *Pediatric bioethics*. New York, NY: Cambridge University Press.

Miller, V. A., & Nelson, R. M. (2006). A developmental approach to child assent for nontherapeutic research. *The Journal of Pediatrics, 149*(Suppl. 1), S25–S30.

Moreno, J. D., & Kravitt, A. (2010). The ethics of pediatric research. In G. Miller (Ed.), *Pediatric bioethics* (pp. 54–72). New York, NY: Cambridge University Press.

Neill, S. J. (2005). Research with children: A critical review of the guidelines. *Journal of Child Health Care*, *9*(1), 2005, 46–58.

O'Lonergan, T. A., & Milgrom, H. (2005). Ethical considerations in research involving children. *Current Allergy and Asthma Reports*, *5*(6), 451–458.

Patenaude, A. F. (2011). Save the children: Direct-to-consumer testing of children is premature, even for research. *Journal of Pediatric Psychology, 36*(10), 1122–1127. http://dx.doi.org/10.1093/jpepsy/jsr068

Peek, L. (2008). Children and disasters: Understanding vulnerability, developing capacities, and promoting resilience—An introduction. *Children, Youth and Environments*, *18*(1), 1–29.

Petronio, S., & Gaff, C. L. (2010). Managing privacy ownership and disclosure. In C. L. Gaff and C. L. Bylund (Eds.), *Family communication about genetics: Theory and practice* (pp. 120–135). New York, NY: Oxford University Press.

Richards, M. (2000). Children's understanding of inheritance and family. *Child Psychology and Psychiatry Review*, *5*, 2–8.

Ross, L. F. (1998). *Children, families, and health care decision making*. New York, NY: Oxford University Press.

Rowlands, M. (2001). The development of children's biological understanding. *Journal of Biological Education*, *35*(2), 66–68.

Samuël, J., Knoppers, B. M., & Avard, D. (2011). Paediatric biobanks: What makes them so unique? *Journal of Paediatrics and Child Health*, *48*(2), E1–E3. http://dx.doi.org/10.1111/j.1440-1754.2011.02072.x

Shier, H. (2001). Pathways to participation: Openings, opportunities and obligations. *Children and Society*, *15*(2), 107–117.

Smith, L. A., & Williams, J. M. (2007). "It's the X and Y thing": Cross-sectional and longitudinal changes in children's understanding of genes. *Research in Science Education*, *37*(4), 407–422.

Soderback, M., Coyne, I., & Harder, M. (2011). The importance of including both a child perspective and the child's perspective within health care settings to provide truly child-centred care. *Journal of Child Health Care*, *15*, 99–106.

Solomon, G. E., Johnson, S. C., Zaitchik, D., & Carey, S. (1996). Like father, like son: Children's understanding of how and why offspring resemble their parents. *Child Development*, *67*(1), 151–171.

Sullivan, J., & McConkie-Rosell, A. (2010). Helping parents talk to their children. In C. L. Gaff & C. L. Bylund (Eds.), *Family communication about genetics: Theory and practice* (pp. 227–242). New York, NY: Oxford University Press.

Tanner, T. (2010). Shifting the narrative: Child-led responses to climate change and disasters in El Salvador and the Philippines. *Children & Society*, *24*(4), 339–351.

Tarini, B., Tercyak, K. P., & Wilfond, B. S. (2011). Commentary: Children and predictive genomic testing: Disease prevention, research protection, and our future. *Journal of Pediatric Psychology*, *36*(10), 1113–1121. http://dx.doi.org/10.1093/jpepsy/jsr040

Thomson, P. (Ed.). (2008). *Doing visual research with children and young people*. New York, NY: Routledge.

Venville, G., Gribble, S. J., & Donovan, J. (2005). An exploration of young children's understanding of genetics concepts from ontological and epistemological perspectives. *Science Education*, *89*(4), 614–633.

Wade, C. H., Wilfond, B. S., & McBride, C. M. (2010). Effects of genetic risk information on children's psychosocial wellbeing: A systematic review of the literature. *Genetics in Medicine*, *12*(6), 317–326.

Wallace, S. E., & Kent, A. (2011). Population biobanks and returning individual research results: Mission impossible or new directions? *Human Genetics*, *130*(3), 393–401.

Wendler, D., Belsky, L., Thompson, K. M., & Emanuel, E. J. (2005). Quantifying the federal minimal risk standard: Implications for pediatric research without a prospect of direct benefit. *The Journal of the American Medical Association*, *294*(7), 826–832.

Wendler, D., & Emanuel, E. J. (2005). What is a "minor" increase over minimal risk? *The Journal of Pediatrics*, *147*(5), 575–578.

Wendler, D., & Varma, S. (2006). Minimal risk in pediatric research. *The Journal of Pediatrics*, *149*(6), 855–61.

Whittle, A., Shah, S., Wilfond, B., Gensler, G., & Wendler, D. (2004). International review board practices regarding assent in pediatric research. *Pediatrics*, *113*(6), 1747–1752.

Wilfond, B. S., & Carpenter, K. J. (2008). Incidental findings in pediatric research. *Journal of Law, Medicine & Ethics*, *36*(2), 332–340.

CHAPTER 8

Genomics Education in Nursing in the United States

Kathleen A. Calzone and Jean Jenkins

ABSTRACT

Discovery of the genetics/genomics underpinnings of health, risk for disease, sickness, and treatment response have the prospects of improving recognition and management of at risk individuals; improving screening, prognostics, and therapeutic decision-making; expanding targeted therapies; and improving the accuracy of medication dosing and selection based on drug metabolism genetic variation. Thus, genetics/genomics science, information, and technologies influence the entire health care continuum and are fundamental to the nursing profession. Translating the benefits of genetics and genomics into health care requires that nurses are knowledgeable about and able to integrate this information and technology into their practice. This chapter explores the development of essential nursing competences in genetics and genomics and outcome indicators. Included is an overview of projects aimed at measuring and/or supporting adoption and integration of such competencies. Included as well is an update reviewing current evidence of the state of genomics nursing education in the United States and recommendations for next steps.

INTRODUCTION

The entire health care continuum is being impacted by research developments in genetics and genomics that are rapidly being integrated into practice. This

http://dx.doi.org/10.1891/0739-6686.29.151

rapid translation has moved genetics from specialty services into mainstream health care as the clinical utility of genetics and genomics expands. No longer are genetic applications limited to the 10–20 million Americans estimated to have or have been diagnosed with a genetic condition (National Institutes of Health, 2010). Current applications include identifying people at risk of disease; using genomic information and technology to screen, diagnose, inform prognosis, and guide treatment; informing drug selection; and expanding knowledge of the biologic underpinnings of disease leading to new treatments. The area of pharmacogenomics, or knowledge of variation in drug response and toxicity because of genetic variation, has experienced exceptional growth and alludes to the massive extent of potential genomic applications. The U.S. population stands at approximately 307,000,000 with each individual having a realistic probability of being prescribed a medication at some point in their lifespan (United States Census Bureau, 2009). Medications are now on the market whose dose, inhibitors, and/or inducers already are or may be optimized in the future based on genetic and genomic information.

A look back at the field of infectious disease provides a model of the process for current genetic/genomic information translation. At one time, the specialty of infectious disease was commonplace. Antimicrobials were first developed and introduced into health care by infectious disease specialists. But the reduction in morbidity and mortality of infections occurred as all health care providers were trained in the prevention and therapy of infections (Kass, 1987). This transition into routine health care has not eliminated the need for infectious disease specialists. Complex infections, drug resistance, and new infectious diseases have all illustrated the continuing need of the specialty. The field of genetics and genomics is striving for a similar conclusion, an improvement in health outcomes through the integration of genomic information and technology into mainstream health care. That successful conclusion is hinged on a genomically competent health care workforce and served to inspire the projects reviewed in this chapter.

Until recently, genetic services have been predominately consultation services for prenatal, dysmorphology, congenital, and/or single gene disorders services. In certain conditions, long-term care may be provided such as the case of metabolic genetic syndromes. Genetic health care professionals have had specific training in this field and can be nurses, physicians, or genetic counselors. As the genetic and genomic contribution to common diseases has been illuminated, the genetic specialists may have begun to practice closely with other health care providers, including working in general health care services to assist in integration of this knowledge (Drury, Bethea, Guilbert, & Qureshi, 2007). In addition, there has been continued expansion of genetic services' scope of practice in the laboratory, diagnostic, counseling, case management, and treatment services for a wide range of conditions including adult onset disorders such as cancer and cardiovascular diseases.

As the clinical utility of genomic information and technology continues to increase, there are insufficient numbers of genetic health care professionals to meet the demand. In addition, as genomic applications continue to expand, many genetic providers lack the specialty training needed to use the new approaches. Cancer care and genomic tumor profiling to inform therapeutic decisions represents an application where this is clearly demonstrated. The therapeutic decisions associated with genomic prognostic indicators such as tumor profiles falls into the domain of the oncologist and further patient education about the prescribed therapies is in the oncology nursing domain. This illustrates the need for genomic competency of all health professionals including nurses.

An influencing factor of genomic applications in care is the insurance coverage where genetic and/or genomic services and tests coverage differ significantly. One study conducted in Illinois found that the extent of coverage varied, testing criteria differed depending on the insurance carrier, and different provider types had varied levels of reimbursement for their services (Latchaw, Ormond, Smith, Richardson, & Wicklund, 2010). Nurses serve as advocates for insurance coverage and therefore are central to assuring that genetic and genomic translation does not further expand health care inequality (Calzone et al., 2010).

BACKGROUND

Genetic/Genomic Nursing Integration Efforts

Nurses are the largest segment of the health care provider community in the United States with more than 3 million licensed registered nurses of which more than 2.5 million are actively practicing (U.S. Department of Health and Human Services, 2010). Nurses practice in every health care setting and continue to hold the distinction of being trusted by the public as the most honest and ethical health care providers. This credibility positions nurses to be effective change agents, as illustrated by the success of the End-of-Life Nursing Education Consortium (ELNEC; Ferrell, Virani, & Malloy, 2006). This initiative has sponsored train-the-trainer seminars that have reached more than 4,500 nurses across the country. The results have included content integration into the academic curricula, clinical care translation including expansion of palliative care, professional nursing organization dissemination, and establishment of ELNEC efforts internationally (Malloy, Paice, Virani, Ferrell, & Bednash, 2008). Such an educational model is valuable when contemplating strategies to educate such a large and diverse nursing workforce in genetics and genomics.

As the largest health care profession, nurses are vital to the successful translation of genetics/genomics into health care and should not serve as a translation barrier (Calzone et al., 2010). Yet, nurses have limited competency in this foundational science that is hindering their ability to integrate this into their practice and academic programs preparing the next generation of the profession. This insufficient competency can result in decreased capability to provide education in genetics/genomics to patients and families; perform assessments that identify people at risk for diseases; facilitate referrals to genetic specialists; know how to interpret genetic test results; understand the potential toxicities of genetically targeted therapies; recognize inhibitors or inducers of medications based on genetic drug metabolism; handle the ethical, legal, and social components of genetic/genomic information and technology; and provide sufficient information consent for research that includes genomics (Calzone et al., 2010; Feetham, Thomson, & Hinshaw, 2005).

For more than 4 decades, there have been consistent calls for more genetics and, recently, genomics integration into the nursing curricula (Brantl & Esslinger, 1962). Despite these appeals, Anderson (1996) found in a review of the literature spanning 1983–1995 that genetic integration into academic nursing education remained scarce. Scanlon and Fibison (1995) further documented this finding in their national survey of 1,000 practicing nurses, in which 91% reported no training in genetics or the consequences of genetic information.

This issue is not unique to the United States but represents a universal issue for nursing around the globe. Recently, studies have begun to assess the health care impact associated with a genetically illiterate workforce. One study examined the non-genetic specialist health care encounter experiences of consumers and their family members who are affected with a genetic condition (Harvey et al., 2007). The *Genetic Alliance*, an advocacy organization of more than 600 genetic advocacy groups, served as the source for participant recruitment. Sixty-four percent of the 5,915 respondents indicated they had received no genetic education materials from their health care provider, which resulted in their pursuit of information and resources unaided. Respondents also rated their primary care physician's ability to establish a treatment plan in collaboration with others, identify necessary services, and understand the medical and psychosocial implications of their genetic condition as fair (Harvey et al., 2007). These data provide insight into the health care implications of a workforce not yet competent in genetics.

There are many reasons that health care providers, including nurses, have genetic/genomic competency deficits. These include

- limited comprehension of genetic/genomic relevancy to practice,
- insufficient knowledge to understand emerging literature,

- faculty are inadequately prepared to teach genetic/genomic content,
- existing genetic/genomic competencies were long and complex,
- changes in the field are occurring rapidly making it difficult to remain current, and
- there were no regulations requiring integration of genetics and genomics (i.e., accreditation of academic programs or hospitals, and licensure examinations; Burke & Kirk, 2006; Calzone et al., 2010; Prows, Calzone, & Jenkins, 2006).

Focused Genetic/Genomic Nursing Competency Efforts

To address these competency deficits, the Genetic/Genomic Nursing Competency Initiative (GGNCI) was established by Calzone and Jenkins in 1995. GGNCI activities were focused at both organizational issues in addition to government and regulatory policies. Figure 8.1 summarizes the timeline associated with genetic integration activities. The GGNCI launched their initiative in September 1995 by hosting the Genetics Education in Nursing Workshop. Representatives from 23 professional nursing specialty organizations with a constituency of more than 400,000 members sent representatives to discuss the state of nursing knowledge in genetics and brainstorm strategies to address competency deficits. Consensus from the workshop was to develop a genetic core curriculum for practicing nurses.

Following this workshop, the Oncology Nursing Society (ONS) began to establish an action plan for their organization. A 2-day Think Tank on Cancer Genetics was convened in November, 1996. Participants in the Think Tank established levels for the roles of oncology nurses in genetics and confirmed the nursing role in genetic service delivery. ONS followed with Position Statements documenting Think Tank outcomes, published initially in 1997 and maintained and updated since that time (Oncology Nursing Society, 2009a, 2009b).

At the same time, a Task Force on Cancer Genetic Education was established by the American Society of Clinical Oncology (ASCO). ASCO efforts focused on a multi-dimensional education initiative that began with the development of the Cancer Genetics and Cancer Predisposition Testing curriculum resource consisting of prepared slides and accompanying support materials (ASCO, 1998). Other efforts included establishing a detailed genetic curriculum as well as developing a position statement on inherited cancer predisposition genetic testing that has continued to be updated (American Society of Clinical Oncology, 1996, 2003; Robson, Storm, Weitzel, Wollins, & Offit, 2010).

In an effort to expand beyond oncology health care professionals and address genetic/genomic education issues for all health care disciplines, not specific to

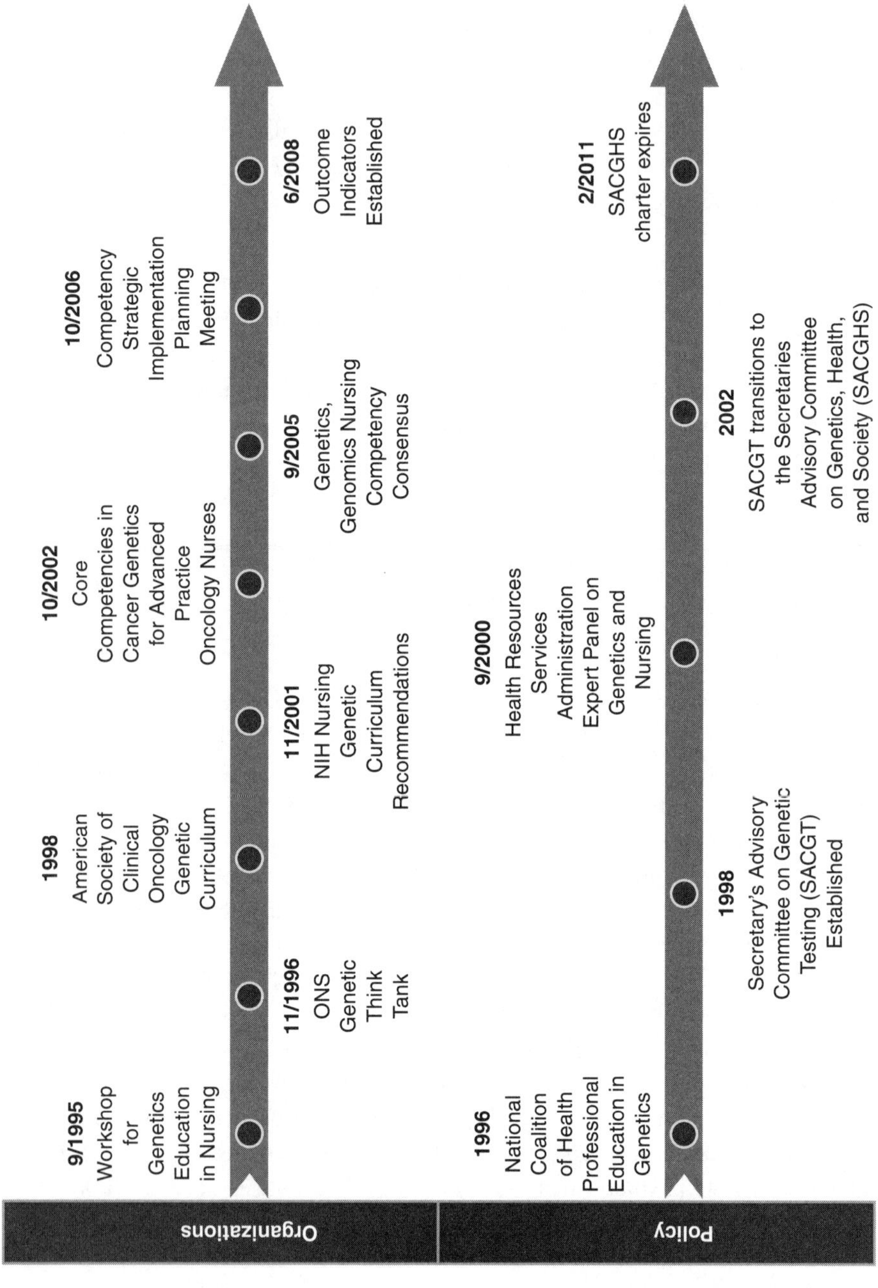

Organizations

9/1995 Workshop for Genetics Education in Nursing

11/1996 ONS Genetic Think Tank

1998 American Society of Clinical Oncology Genetic Curriculum

11/2001 NIH Nursing Genetic Curriculum Recommendations

10/2002 Core Competencies in Cancer Genetics for Advanced Practice Oncology Nurses

9/2005 Genetics, Genomics Nursing Competency Consensus

10/2006 Competency Strategic Implementation Planning Meeting

6/2008 Outcome Indicators Established

Policy

1996 National Coalition of Health Professional Education in Genetics

1998 Secretary's Advisory Committee on Genetic Testing (SACGT) Established

9/2000 Health Resources Services Administration Expert Panel on Genetics and Nursing

2002 SACGT transitions to the Secretaries Advisory Committee on Genetics, Health, and Society (SACGHS)

2/2011 SACGHS charter expires

FIGURE 8.1 Timeline of U.S. genetic/genomic organizational and policy integration efforts.

just nursing as in the GGNCI, the National Coalition for Health Professional Education in Genetics (NCHPEG) was established in 1996. A collaborative effort of the American Nurses Association, the American Medical Association, and the National Human Genome Research Institute, NCHPEG membership consists of more than 50 professional specialty health care provider organizations working on education in genetics and genomics. An early NCHPEG effort was to establish genetic core competencies applicable to all health care disciplines (Core Competency Working Group of the National Coalition for Health Professional Education in Genetics, 2001). Initially approved in 2000 and published in 2001, these competencies were not widely implemented as they were extensive (44 competencies), not health care discipline specific, and addressed genetics not genetics and genomics that applies more broadly to the entire nursing profession (Jenkins & Calzone, 2007).

Building on the NCHPEG efforts and responding to the primary Genetics Education in Nursing Workshop recommendations, Jenkins and colleagues (2001) conducted a project that established genetic curriculum content recommendations for nursing education. Using this core curriculum and the oncology specialty recommendations that genetic nursing practice occur at the advance practice level, Calzone and colleagues (2002) applied the Delphi technique to nurse expert responses to establish core competencies in cancer genetics for Advanced Practice Oncology Nurses that applied to any oncology nurse with a graduate degree. The limitations to this effort included application only to one specialty (i.e., oncology), advanced practice nurses, the large number of competencies identified (52), and the continuing emphasis on genetics and not genomics (Note: Genetics was the primary scientific focus at that time).

Policy Efforts

Policy recommendations and guidance are important when providing the justification for efforts to address health care professional competency in genetics. Such guidance was provided by an expert panel on Genetics Nursing September 2000, when the U.S. Department of Health and Human Service, Health Resources and Services Administration, Bureau of Health Professions convened a group of experts in nursing, medicine, genetics, and nursing education. This panel recommended federal funding for an interdisciplinary resource clearinghouse, creation of interdisciplinary programs and collaborative partnerships, regulatory environment enhancements, and attention to workforce issues (Expert Panel Report on Genetics and Nursing, 2000).

Additionally, policy efforts at the federal level highlighted ethical, legal, and social issues that require an educated workforce to be able to identify genetic testing associated policy issues and make recommendations to address those issues. The U.S. Department of Health and Human Services (DHHS) established the Secretary's Advisory Committee on Genetic Testing (SACGT, 2006) in response to recommendations from two advisory groups (Task Force on Genetic Testing and the Joint National Institutes of Health (NIH)/ Department of Energy Committee to Evaluate the Ethical, Legal, and Social Implications Program of the Human Genome Project) that evolved to a new committee in 2002, the Secretary's Advisory Committee on Genetics, Health, and Society (SACGHS). Both committees advised the DHHS secretary on genetic testing plus human health and societal issues associated with genetic technologies (Secretary's Advisory Committee on Genetics, 2010). Nurses have been active participants in these policy initiatives adding a biobehavioral, prevention, health promotion, and holistic perspective to the discussions. An informed and competent nurse can advocate and inform the developing genetic/genomic policies, standards, and practices influencing standards of care (Calzone et al., 2010).

Outcome Evaluation

Despite the potential benefits achieved by an informed health care workforce, evidence indicates that the progress of U.S. nursing competency in genetics and genomics remains limited (Edwards, Maradiegue, Seibert, Macri, & Sitzer, 2006; Hetteberg, Prows, Deets, Monsen, & Kenner, 1999; Prows et al., 2006).

Surveys done in 1996 and 2005 documented minimal change in the amount of genetic and genomic content in the curricula of basic preparatory nursing academic programs (Hetteberg et al., 1999; Prows et al., 2006) with only 30% of academic nursing programs reporting a curriculum thread in genetics/ genomics (Prows et al., 2006). Similarly, a survey of advanced practice nurses documented comparable finding with the majority reporting minimal training and knowledge in genetics (Maradiegue, Edwards, Seibert, Macri, & Sitzer, 2005). Genetic and genomic content is nominally found within National Council Licensure Examination (NCLEX®) or certification examinations. Continuing education resources for practicing nurses are available but with only 30% of professional nursing organizations offering some form of genetic content (Monsen & Anderson, 1999). In summary, existing reviews indicate that the nursing profession continues to have deficits in understanding and using genetic and genomic information in health care despite existing resources, established competencies in genetics/genomics for all health professionals, and published model curricula (Prows, Glass, Nicol, Skirton, & Williams, 2005). These issues are not unique to the United States, but represent a problem shared globally (Iino et al., 2002; Kirk, Calzone, Arimori, & Tonkin, 2011; Kirk, Lea, & Skirton, 2008; Kirk, Tonkin, & Burke, 2008; Nicol, 2002).

There are many reasons for the lack of progress in nursing genetics/genomics education that influence the capacity of this diverse discipline to acknowledge the value of this information to care including:

1. The relevance of genetics/genomics to nursing practice is not fully appreciated. Most nurses view genetics as being relevant only to a few specialty nurses as opposed to the entire profession.
2. Competency expectations appear daunting, long, and complex challenging educators given packed curricula and practicing professionals with limited time and resources for continuing education.
3. Most nurses, including faculty, have not had a foundation in understanding the science of genetics resulting in insufficient numbers of faculty prepared to teach this content.
4. State Boards of Nursing currently do not require competency in genetics/genomics as part of licensure or re-licensure (National Council of State Boards of Nursing, 2010; Prows et al., 2005)

These influencing factors represent challenges and opportunities. Challenges associated with the nursing workforce size and diversity in educational preparation requires a centralized approach with the creation of new resources and educational models for learning. Opportunities because nurses are central to health care delivery with the potential to be contributing change agents as illustrated by the End of Life Nursing Education Consortium and Geriatric initiatives (Kelly, Ersek, Virani, Malloy, & Ferrel, 2008; Sherman, Matzo, Rogers, McLaughlin, & Virani, 2002). Nurses, therefore, are the cornerstone for a focused effort to prepare the health care workforce in using genetics and genomics information in quality care.

Overcoming Barriers to Change

Success in a focused educational effort requires an understanding of best methods and strategies to overcome barriers and hindrances of change. The use of a strong theoretical underpinning such as a tested change theory (i.e., Rogers Diffusion of Innovations) was valuable to informing the process, to identify the reasons for lack of progress, and present mechanisms for moving forward that could garner greater success (Rogers, 2003). The literature delineates factors that can serve as successful facilitators for change such as a strong research evidence base. Five research phases have been identified in the translation of research into application. The phases are:

1. Basic research studies that explore different phenomena.
2. Methods development that involves the development of technology, instruments, or equipment needed for basic research.
3. Efficacy trials that assess whether there is any value to a given intervention.

4. Effectiveness trials that assess the extent of usefulness associated with one or more interventions.
5. Dissemination trials which explore the propagation and uptake of a given intervention (Sussman, Valente, Rohrbach, Skara, & Pentz, 2006).

The phases are not necessarily a linear progression and attention must be paid to the translation end-point from the onset of any research agenda to assure translation of research innovation to application (Glasgow, Lichtenstein, & Marcus, 2003). Furthermore, critical to any translational research are the validity criteria used including adequate sample representation, program implementation issues, decision-making considerations in light of existing guidelines, complexity of the innovation, and long-term sustainability evaluations (Glasgow & Emmons, 2007). Attention to data emerging from each of these phases informs and improves the quality of research discoveries when applied in health care and are appropriate for consideration when assessing the translation of genetics/genomics research into application.

The U.S. Institute of Medicine, an independent, non-governmental body aimed at providing unbiased advice to the public and policy makers, reported that several elements need to be in place for successful translation of discoveries that can improve health care and health outcomes (Glasgow, Lichtenstein, & Marcus, 2003):

1. Organizational support for change (i.e., hospitals, managed care organizations, medical groups, multi-specialty clinics, integrated delivery systems).
2. Evidence reports and practical implementation into clinical practice guidelines that improve application of evidence to health care delivery.
3. Use of information technology such as clinical decision support tools.
4. Alignment of payment policies with quality improvement (Institute of Medicine, 2001).

Key individuals are important variables in this change process including involvement of:

1. Key interdisciplinary stakeholders who are the end-users of the innovation such as health care providers and health care consumers.
2. Context adapters such as administrators or executives who can facilitate implementation of the innovation within the existing health care system.
3. Policy makers who establish policies, regulations, and the needed oversight for the protection of the public's welfare (Sussman, 2006).

The engagement of such key individuals can be accelerated using opinion leaders, individuals who have both the respect and social ties within the health care environment who can help with the transmission of the information within their social network (Valente & Fasados, 2006). All of this is hinged on an educated health care workforce and an educated public who need a sufficient command of the scientific underpinnings of the innovation to understand the literature, the research, the potential application to practice, and the policy implications to contribute to the innovation dissemination and adoption (Institute of Medicine, 2001). Thus, it was identified that to have successful adoption of quality-based care that integrated genetic/genomic research into health care, a competency initiative for nursing workforce education was important to moving this identified goal forward.

ESTABLISHING THE ESSENTIAL GENETIC AND GENOMIC COMPETENCIES FOR NURSING

The competency initiative (GGNCI) began with the identification of Essential Nursing Competencies and Curricula Guidelines for Genetics and Genomics to guide academic curriculum content/learning activities and continuing education for practicing registered nurses. The Competencies were defined by a competency consensus panel and were framed by the domains of professional responsibilities and professional practice that includes assessment, identification, referral, and provision of education, care, and support. The second edition document includes competency specific outcome indicators consisting of specific areas of knowledge and clinical performance indicators. The consensus process used for competency development and endorsement was guided by the Diffusion of Innovations theory (Rogers, 2003).

Methods

There were two phases to the competency development process including establishing the essential competencies and identifying what to do with the document through the creation of a strategic implementation plan. Faculty response motivated the development of the corresponding competency outcome indicators. Figure 8.2 provides a flow diagram of the competency development process.

Phase 1

The essential genetic and genomic competencies for all U.S. registered nurses were defined in phase 1. The objectives for this phase were:

1. Create a steering committee made up of identified stakeholders including U.S. federal government, academic, and clinical nursing leaders. These leaders were instrumental in guiding the competency development and consensus process.
2. Funding for bringing together resources necessary for competency development, dissemination and implementation was obtained.
3. A consensus panel (see Table 8.1) representing academic, research, clinical, government agencies, and international representation was established to review and approve the draft competencies.

TABLE 8.1
Consensus Panel Participant Organizations

Nursing Organization Participants:
Academy of Neonatal Nursing
American Academy of Ambulatory Care Nursing
American Academy of Nurse Practitioners
American Academy of Nursing
American Association of Colleges of Nursing
American Association of Nurse Anesthetists
American Nurses Association
American Psychiatric Nurses Association
American Society for Pain Management Nursing
Asian American/Pacific Islander Nurses Association
Association of Women's Health, Obstetric and Neonatal Nurses
Dermatology Nurses' Association
Developmental Disabilities Nurses Association
National League for Nursing
National Nursing Staff Development Organization
National Organization of Nurse Practitioner Faculties
Oncology Nursing Society
Pediatric Endocrinology Nursing Society
Philippine Nurses Association of America
Sigma Theta Tau International
Society of Pediatric Nurses
The International Society of Psychiatric-Mental Health Nurses
National Alaska Native American Indian Nurses Association
National Association of Clinical Nurse Specialists
National Association of Hispanic Nurses
National Association of Neonatal Nurses
National Association of Pediatric Nurse Practitioners
Medical Centers:
Cincinnati Children's Hospital Medical Center

(*Continued*)

TABLE 8.1
Consensus Panel Participant Organizations (Continued)

Academic Programs:
Johns Hopkins University School of Nursing
Oregon Health and Science University
University of Glamorgan, Wales
University of Iowa
University of Tennessee Health Science Center
Virginia Commonwealth University
Governmental Agencies:
Health Resources and Services Administration
National Cancer Institute, NIH
National Human Genome Research Institute, NIH
National Institute of Nursing Research, NIH
United Kingdom National Health Service, National Genetics Education and Development Centre
Regulatory Bodies:
American Nurses Credentialing Center
National Council of State Boards of Nursing
National League for Nursing Accrediting Commission

4. A consensus panel meeting brought together representatives of nursing organizations for final modification and sanction of the essential competencies.
5. Plans for dissemination of the essential competencies for endorsement by U.S. academic and clinical nursing organizations were made.
6. Publication of the final document in monograph and internet format facilitated wider distribution to the nursing community.

Steering Committee

The development of the competencies began with the assembly of a steering committee that included varied stakeholder representation that was an extremely lengthy process. To assure that this steering committee included adequate representation of the diverse nursing profession, a process of social network building, engaging leaders, disseminating knowledge to persuade them to adopt genetics and genomics and engage in this GGNCI effort was necessary. A snowballing effort to work through initial contacts to gain access to other key nursing leaders was repeated. Throughout this process, what was learned from these well-positioned leaders were key characteristics about the receiver, antecedent, and nursing social system characteristics all of which informed the optimal ways to engage and influence this massive and diverse discipline. After almost a year's worth of effort, the final steering committee was compiled consisting of 15 key representatives from federal agencies and key nursing organizations, recognized academic educators, and genetic nurse experts. The committee's responsibilities were to help guide the consensus panel, advise on the methods for establishing consensus, and provide expert review of each competency document draft as illustrated in Figure 8.2.

The consensus process for the Competencies involved soliciting feedback from a number of groups and consisted of the following:

1. Review and revision of the competency draft by the steering committee.
2. Review of the revised document by nursing representatives at a 2005 National Coalition for Health Professional Education in Genetics meeting.
3. Public comment from the nursing community at large was solicited using the American Nurses Association website http://nursingworld.org in August 2005.
4. Consensus panel (Table 8.1) convened on September 24–25, 2005, which was comprised of key stakeholders from the U.S. nursing community including academic, research, clinical, as well as minority nursing representation, and crucial regulatory bodies including certification and accrediting organizations, and the National Council of State Boards of Nursing.

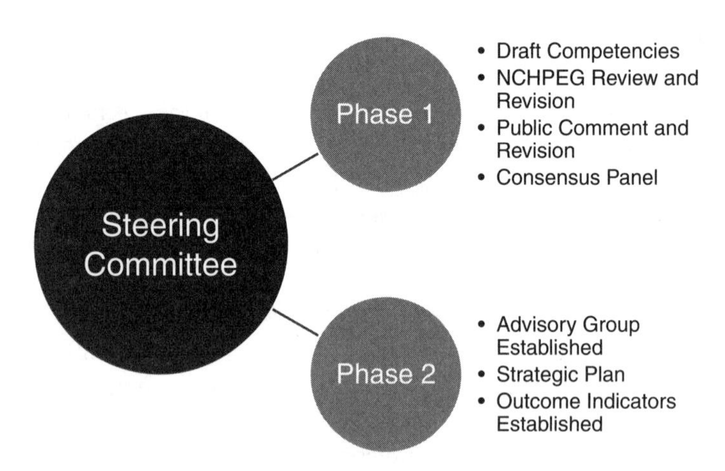

FIGURE 8.2 Competency development Flow Diagram.

Endorsement

All nursing professional organizations registered as a member of the Nursing Organizations Alliance were offered the opportunity to review, comment, and endorse the competencies. Each was contacted with first, a letter including the competencies, followed by a personal contact using standardized talking points established by members of the consensus panel and the steering committee. Fifty organizations have now endorsed the competencies. The GGNCI co-chairs received unsolicited endorsements from two schools of nursing: Johns Hopkins University School of the Nursing and Nell Hodgson Woodruff School of Nursing at Emory University. Both these schools indicated that, although they were not formally asked to endorse, they felt strongly enough about this effort to provide their endorsement. The Genetic Alliance, an advocacy genetic organization, also endorsed the competencies. A copy of the final Essential Nursing Competencies and Curricula Guidelines for Genetics and Genomics monograph can be found at http://www.genome.gov/Pages/Careers/HealthProfessionalEducation/geneticscompetency.pdf

Phase 2

The implementation of the competencies began with the assembly of an essentials advisory group that consisted of representation from each endorsing organization as well as clinical, research, academic, and government settings. The responsibility of the advisory group was to help develop a plan to guide implementation activities.

Strategic Plan

The first advisory group activity was to establish a 5-year Strategic Implementation Plan for the Essential Nursing Competencies and Curricula Guidelines for Genetics and Genomics. This plan was defined at a meeting held October 22–24, 2006. The final strategic plan focuses on four areas including recommended infrastructure, focus on practicing nurses, initiative focused on academics, and regulatory/quality control efforts (http://www.genome.gov/Pages/Health/HealthCareProvidersInfo/CompetencyStrategicPlan3-2-07.pdf). One of the priority recommendations was to establish outcome indicators for each of the essential competencies to facilitate faculty use of the competencies.

Outcome Indicator Development

As a separate second consensus initiative, a writing team consisting of members of the essentials advisory group drafted the outcome indicators that consist of two components: specific areas of knowledge and clinical performance indicators (Calzone, Jenkins, Prows, & Masny, 2011). The consensus process for approval of the outcome indicators involved soliciting feedback from a number of groups:

1. Creation, review, and revision of the document by the essentials advisory group.
2. Review by representatives at another convened group meeting for development of a Genetics/Genomics Toolkit for Faculty, with written comments accepted following the meeting and revisions incorporated.
3. Review by workshop attendees at American Association of Colleges of Nursing Baccalaureate and Master's Education Conferences with revisions incorporated, and
4. Review and final approval by the essentials advisory group.

The outcome indicators were published in the second edition of the essential competencies that define specific knowledge areas and suggesting clinical performance indicators for each competency. Both the first and second editions of the essential competencies do not represent the opinions or position of the genetics specialty nursing community, a single nursing organization or government body. Instead, the competencies and outcome indicators were both developed through separate consecutive consensus building efforts using independent advisory panels of nurse leaders from organizations, clinical, research, and academic settings in collaboration with genetic nurse experts.

STATUS OF GENOMIC NURSING COMPETENCY

The competency document has been used to lobby for and guide the integration of genetics and genomics into the revision of the American Association Colleges of Nursing (AACN) Essentials of Baccalaureate Education for Professional Nursing Practice (American Association of Colleges of Nursing, 2008). The AACN Essentials are used by the Commission on Collegiate Nursing Education (CCNE) to establish nursing program accreditation standards, and since 2010, nursing programs undergoing CCNE accreditation review are evaluated to determine that efforts are ongoing for genetics/genomics to be included in the Baccalaureate curriculum.

A survey to assess the impact of this regulatory development was conducted in 2008, using the stages of change framework to assess the intention of nursing faculty to integrate genetic and genomic curriculum content into entry level nursing education (Prochaska, Redding, & Evers, 2002). Most faculty participants were found to be in the *contemplation stage* with 35% planning to adopt curriculum changes within the next 6 months. Four percent were in the *preparation* stage and had plans to make changes within the next 30 days. Some faculty were in the *action stage* with 9% already reporting curriculum changes that included genetics/genomics for less than 6 months and 10% in *maintenance* stage with curriculum changes that included genetics/genomics for greater than 6 months (Jenkins & Calzone, 2012).

These data indicate that the CCNE policy change resulted in a rapid change in faculty attitudes and appreciation of genetics/genomics for nursing education and practice. More than half the surveyed faculty were already in some form of action to be able to meet the new CCNE accreditation standards. Clearly, the ability of the competencies to influence policy has resulted in immediate impact stimulating GGNCI to prioritize efforts aimed at influencing policy to stimulate change.

Genetics and Genomics in Nursing Practice Survey

Next steps were to establish a benchmark of genomic nursing competency. To address this issue, the GGNCI developed the Genetics and Genomics in Nursing Practice Survey (GGNPS) instrument in 2007. The GGNPS is a web-based survey instrument designed to evaluate the integration of genetics and genomics in nursing practice using family history as the critical benchmark for assessment of attitudes, practices, receptivity, confidence, and competency. Questions assessed the value nurses placed on family history and their understanding of the role of genetics in common diseases such as cancer, heart disease, diabetes, and psychiatric illnesses. Use of family history within the past 3 months was the measure used to assess current practice. Receptivity to genetics/genomics application was measured though evaluation of barriers to family history utilization and difficulties encountered in practice as a result of family history assessments or patient questions on genetics. Nurse confidence was evaluated using self-assessment exercises on use of family history and referral patterns. Competency in family history utilization was measured through questions regarding the extent of family history that should be collected and other knowledge questions. The survey instrument was tailored from a validated instrument used to assess Family Physicians (FP) competency in genomics (Jenkins, Woolford, Stevens, Kahn, & McBride, 2010). The FP instrument was developed by a multidisciplinary team consisting of family physicians, a behavioral scientist, online survey designers, and genetic/genomic experts as an online tool to assess FP practice. The domains for measurement were selected from the constructs of Rogers Diffusion of Innovations theory and considered FP use of family history as a benchmark for genetic/genomic competency. Each item was reviewed for content validity by outside content experts including a convenience sample of family physicians. A second phase of item analysis occurred through a pilot online survey with a different convenience sample of family physicians and factor analysis was then used to revise the instrument. The final FP instrument included assessments of genetic/genomic knowledge/competency, persuasion, and decision domains from the Diffusion of Innovations theory (Jenkins et al., 2010). Most of these questions were equally appropriate for nurses, thus they were replicated in the GGNPS nursing instrument using the same questions except leveled for scope of practice when applicable.

Two studies have now been completed using this instrument. The pilot study done in collaboration with the National Institutes of Health (including the Clinical Center and the National Cancer Institute) and the National Nursing Workforce study done in collaboration with the American Nurses Association (ANA). In total, 859 licensed registered nurses have completed the survey instrument. There are 239 in the pilot study consisting of nurses from a large single research institution and 620 from the National Nursing Workforce study done in collaboration with ANA. Findings from both the pilot (2008) and National Nursing Workforce studies (2011) were congruent revealing that most nurses completing the survey thought genomics was important but felt inadequately prepared to incorporate genomics into practice. In addition, knowledge gaps were found in all nurses regardless of education level indicating that all nurses would benefit from a broad-scale education intervention to assure genetic/genomic competency.

Next Steps

At this juncture, despite a burgeoning body of evidence regarding the genetics/genomics influences on health and illness, genomic competency of the nursing workforce remains limited and a national multifaceted education effort is indicated. In addition, the evidence specific to outcomes of genomically competent nursing practice and the impact on the public's health is extremely limited, if not entirely absent. This paucity of outcome data is further hindering efforts to translate genetics/genomics discoveries into patient care as well as efforts to influence incorporation of genetics/genomics information into academic curricula, licensure examinations, schools of nursing, and health care organization accreditation.

Although some regulatory bodies have embraced genomics, this is not universal. For instance, academic accrediting bodies that have not yet integrated genomics (National League for Nursing Accrediting Commission), the National Council of State Boards of Nursing (NCSBN) that develops the National Council Licensure Examination for Registered Nurses (NCLEX®) licensing examination, and the accrediting bodies for health care facilities such as Joint Commission, are clear that modifications to existing requirements must be based on evidence that nursing care that incorporates genetic and genomic principles, information, and technology improves the public's health. However, this kind of evidence-base remains sparse. The GGNCI is currently engaged in developing a state of the science effort to examine and evaluate the current state of the science regarding nursing care that incorporates genetic and genomic principles, information, and technology. GGNCI will then use that evidence base to identify the gaps and establish through consensus research priorities that will serve as the underpinning for the development of research that produces the essential outcome data needed to define outcomes associated with a genetic/genomically competent nurse, and ultimately how that influences the quality of the public's health.

The generation of this research base may be a vehicle for influencing the regulatory environment including NCLEX® and the Joint Commission. Coupled with a nationwide genomic competency education initiative, this could serve as a model for other health care disciplines to replicate in advancing genomic competency to optimize health care outcomes.

CONCLUSION

The translation of genetics and genomics into the clinical arena is progressing rapidly and has implications for the entire nursing profession. Genetics and genomics when integrated competently into health care have the potential to improve health outcomes. Yet, despite efforts in the United States and elsewhere around the world, genetic/genomic nursing competency continues to be limited. Failure of the nursing profession to understand the relevancy of genetics/genomics for health care, to have a sufficient scientific foundation in genetics to comprehend the literature, and to have the capacity to teach this material is contributing to this limited progress. This was a driving force behind the GGNCI whose primary aim is to prepare the nursing workforce to be competent in genetics and genomics. Additional evidence is needed on health outcomes associated with genomically competent nursing practice. Nurse scientists conducting translation research will document the outcomes associated with competency applying this new knowledge into health care. Progress in implementing the competencies and associated outcomes will have a global effect as we work together to optimize health care both nationally and internationally.

REFERENCES

American Association of Colleges of Nursing. (2008). *The essentials of baccalaureate education for professional nursing practice*. Retrieved from http://www.aacn.nche.edu/Education/pdf/BaccEssentials08.pdf

American Society of Clinical Oncology. (1996). Statement of the American Society of Clinical Oncology: Genetic testing for cancer susceptibility. *Journal of Clinical Oncology*, *14*(5), 1730–1736.

American Society of Clinical Oncology. (1998). *Cancer Genetics and Cancer Predisposition Testing: An ASCO Curriculum*. Washington, DC: American Society of Clinical Oncology.

American Society of Clinical Oncology. (2003). American Society of Clinical Oncology policy statement update: Genetic testing for cancer susceptibility. *Journal of Clinical Oncology*, *21*(12), 2397–2406.

Anderson, G. W. (1996). The evolution and status of genetics education in nursing in the United States 1983–1995. *Image—Journal of Nursing Scholarship*, *28*(2), 101–106.

Brantl, V. M., & Esslinger, P. N. (1962). Genetic implications for the nursing curriculum. *Nursing Forum*, *1*(2), 90–100.

Burke, S., & Kirk, M. (2006). Genetics education in the nursing profession: Literature review. *Journal of Advanced Nursing*, *54*(2), 228–237.

Calzone, K., Jenkins, J., & Masny, A. (2002). Core competencies in cancer genetics for the advanced practice oncology nurses. *Oncology Nursing Forum*, *29*(4), 1327–1333.

Calzone, K. A., Cashion, A., Feetham, S., Jenkins, J., Prows, C. A., Williams, J. K., & Wung, S. F. (2010). Nurses transforming health care using genetics and genomics. *Nursing Outlook*, *58*(1), 26–35.

Calzone, K. A., Jenkins, J., Prows,C. A., & Masny, A. (2011). Establishing the outcome indicators for the essential nursing competencies and curricula guidelines for genetics and genomics. *Journal of Professional Nursing*, *27*(3), 179–191.

Core Competency Working Group of the National Coalition for Health Professional Education in Genetics. (2001). Recommendations of core competencies in genetics essential for all health professionals. *Genetics in Medicine*, *3*(2), 155–159.

Drury, N., Bethea, J., Guilbert, P., & Qureshi, N. (2007). Genetics support to primary care practitioners—A demonstration project. *Journal of Genetic Counseling*, *16*(5), 583–591.

Edwards, Q. T., Maradiegue, A., Seibert, D., Macri, C., & Sitzer, L. (2006). Faculty members' perceptions of medical genetics and its integration into nurse practitioner curricula. *Journal of Nursing Education*, *45*(3), 124–130.

Expert Panel Report on Genetics and Nursing. (2000). *Report of the expert panel on genetics and nursing: Implications for education and practice* (HRS00296 HRSA Publication Catalog). Washington, DC: U.S. Department of Health and Human Services, Human Resources and Services Administration.

Feetham, S., Thomson, E. J., & Hinshaw, A. S. (2005). Nursing leadership in genomics for health and society. *Journal of Nursing Scholarship*, *37*(2), 102–110.

Ferrell, B. R., Virani, R., & Malloy, P. (2006). Evaluation of the end-of-life nursing education consortium project in the USA. *International Journal of Palliative Nursing*, *12*(6), 269–276.

Glasgow, R. E., & Emmons, K. M. (2007). How can we increase translation of research into practice? Types of evidence needed. *Annual Review of Public Health*, *38*, 413–433.

Glasgow, R. E., Lichtenstein, E., & Marcus, A. C. (2003). Why don't we see more translation of health promotion research to practice? Rethinking the efficacy-to-effectiveness transition. *American Journal of Public Health*, *93*(8), 1261–1267.

Harvey, E. K., Fogel, C. E., Peyrot, M., Christensen, K. D., Terry, S. F., & McInerney, J. D. (2007). Providers' knowledge of genetics: A survey of 5915 individuals and families with genetic conditions. *Genetics in Medicine*, *9*(5), 259–267.

Hetteberg, C. G., Prows, C. A., Deets, C., Monsen, R. B., & Kenner, C. A. (1999). National survey of genetics content in basic nursing preparatory programs in the United States. *Nursing Outlook*, *47*(4), 168–180.

Iino, H., Tsukahara, M., Murakami, K., Lambert, V. A., Lambert, C. E., & Tsujino, K. (2002). Genetic education in baccalaureate and associate degree nursing programs in Japan. *Nursing & Health Sciences*, *4*(4), 173–180.

Institute of Medicine. (2001). *Crossing the quality chasm: A new health system for the 21st century*. Washington, DC: National Academy Press.

Jenkins, J., Woolford, S., Stevens, N., Kahn, N., & McBride, C. M. (2010). Family physicians' likely adoption of genomic-related innovations. *Case Studies in Business, Industry and Government Statistics*, *3*(2). Retrieved from http://legacy.bentley.edu/csbigs/documents/jenkins.pdf

Jenkins, J. F., & Calzone, K. A. (2007). Establishing the essential nursing competencies for genetics and genomics. *Journal of Nursing Scholarship*, *39*(1), 10–16.

Jenkins, J. F., & Calzone, K. A. (2012). Are nursing faculty ready to integrate genomic content into curricula? *Nurse Educator*, *37*(1), 25–29.

Jenkins, J. F., Prows, C., Dimond, E., Monsen, R., & Williams, J. (2001). Recommendations for educating nurses in genetics. *Journal of Professional Nursing*, *17*(6), 283–290.

Kass, E. H. (1987). History of the specialty of infectious diseases in the United States. *Annals of Internal Medicine*, *106*(5), 745–756.

Kelly, K., Ersek, M., Virani, R., Malloy, P., & Ferrell, B. (2008). End-of-life nursing education consortium. Geriatric training program: Improving palliative care in community geriatric care settings. *Journal of Gerontological Nursing*, *34*(5), 28–35.

Kirk, M., Calzone, K., Arimori, N., & Tonkin, E. (2011). Genetics-genomics competencies and nursing regulation. *Journal of Nursing Scholarship*, *43*(2), 107–116.

Kirk, M., Lea, D., & Skirton, H. (2008). Genomic health care: Is the future now? *Nursing & Health Sciences*, *10*(2), 85–92.

Kirk, M., Tonkin, E., & Burke, S. (2008). Engaging nurses in genetics: The strategic approach of the NHS National Genetics Education and Development Centre. *Journal of Genetic Counseling*, *17*(2), 180–188.

Latchaw, M., Ormond, K., Smith, M., Richardson, J., & Wicklund, C. (2010). Health insurance coverage of genetic services in Illinois. *Genetics in Medicine*, *12*(8), 525–531.

Malloy, P., Paice, J., Virani, R., Ferrell, B. R., & Bednash, G. P. (2008). End-of-life nursing education consortium: 5 years of educating graduate nursing faculty in excellent palliative care. *Journal of Professional Nursing*, *24*(6), 352–357.

Maradiegue, A., Edwards, Q. T., Seibert, D., Macri, C., & Sitzer, L. (2005). Knowledge, perceptions, and attitudes of advanced practice nursing students regarding medical genetics. *Journal of the American Academy of Nurse Practitioners*, *17*(11), 472–479.

Monsen, R. B., & Anderson, G. (1999). Continuing education for nurses that incorporates genetics. *Journal of Continuing Education in Nursing*, *30*(1), 20–24.

National Council of State Boards of Nursing. (2010). *2010 NCLEX examination candidate bulletin*. Retrieved from https://www.ncsbn.org/2010_NCLEX_Candidate_Bulletin.pdf

National Institutes of Health. (2010). *NIH announces genetic testing registry*. Retrieved from http://www.nih.gov/news/health/mar2010/od-18.htm

Nicol, M. J. (2002). The teaching of genetics in New Zealand undergraduate nursing programmes. *Nurse Education Today*, *22*(5), 401–408.

Oncology Nursing Society. (2009a). *Cancer predisposition genetic testing and risk assessment counseling*. Retrieved from http://www.ons.org/Publications/Positions/Predisposition

Oncology Nursing Society. (2009b). *The role of the oncology nurse in cancer genetic counseling*. Retrieved from http://www.ons.org/Publications/Positions/GeneticCounseling

Prochaska, J. O., Redding, C. A., & Evers, K. E. (2002). The transtheoretical model and stages of change. In K. Glanz, B. Rimer, & F. M. Lewis (Eds.), *Health behavior and health education: Theory, research, and practice* (3rd ed., pp. 97–122). San Francisco, CA: Josey-Boss.

Prows, C., Calzone, K., Jenkins, J. (2006). *Genetics content in nursing curriculum*. Paper presented at the meeting of the National Coalition Health Professional Education in Genetics, Bethesda, MD.

Prows, C. A., Glass, M., Nicol, M. J., Skirton, H., & Williams, J. (2005). Genomics in nursing education. *Journal of Nursing Scholarship*, *37*(3), 196–202.

Robson, M. E., Storm, C. D., Weitzel, J., Wollins, D. S., & Offit, K. (2010). American Society of Clinical Oncology policy statement update: Genetic and genomic testing for cancer susceptibility. *Journal of Clinical Oncology*, *28*(5), 893–901.

Rogers, E. (2003). *Diffusion of Innovations* (5th ed.). New York, NY: Free Press.

Scanlon, C., & Fibison, W. (1995). *Managing genetic information: Implications for nursing practice*. Washington, DC: American Nurses Association.

Secretary's Advisory Committee on Genetics, Health, and Society. (2006). *Coverage and reimbursement of genetic tests and services: Report of the Secretary's Advisory Committee on Genetics, Health, and Society*. Washington, DC: Department of Health and Human Services. Retrieved from http://oba.od.nih.gov/oba/sacghs/reports/CR_report.pdf

Secretary's Advisory Committee on Genetics, Health, and Society. (2010). *Secretary's Advisory Committee on Genetics, Health, and Society*. Retrieved from http://oba.od.nih.gov/SACGHS/sacghs_home.html

Sherman, D. W., Matzo, M. L., Rogers, S., McLaughlin, M., & Virani, R. (2002). Achieving quality care at the end of life: A focus of the End-of-Life Nursing Education Consortium (ELNEC) curriculum. *Journal of Professional Nursing*, *18*(5), 255–262.

Sussman, S., Valente, T. W., Rohrbach, L. A., Skara, S., & Pentz, M. A. (2006). Translation in the health professions: Converting science into action. *Evaluation and the Health Professions*, *29*(1), 7–32.

United States Census Bureau. (2009). *National and state population estimates: Annual population estimates 2000 to 2009*. Retrieved from http://www.census.gov/newsroom/releases/archives/population/cb10-81.html

U.S. Department of Health and Human Services, Health Resources Services Administration. (2010). *The registered nurse population: Findings from the 2008 national sample survey of registered nurses*. Washington, DC: Retrieved from http://bhpr.hrsa.gov/healthworkforce/rnsurveys/rnsurveyfinal.pdf

Valente, T. W., & Fosados, R. (2006). Diffusion of innovations and network segmentation: The part played by people in the promoting health. *Sexually Transmitted Diseases*, *33*(7, Suppl), S23–S31.

CHAPTER 9

Development of the Essential Genetic and Genomic Competencies for Nurses With Graduate Degrees

Karen E. Greco, Susan Tinley, and Diane Seibert

ABSTRACT

Scientific advances in genetics and genomics are rapidly redefining our understanding of health and illness and creating a significant shift in practice for all health care disciplines. Nurses educated at the graduate level are well-prepared to assume clinical and leadership roles in health care systems and must also be prepared to assume similar roles related to genetic/genomic health care. This chapter describes the processes used to create a consensus document identifying the genetic/genomic competencies essential for nurses prepared at the graduate level. Three groups were involved in the competency development; a steering committee provided leadership and used qualitative methods to review and analyze pertinent source documents and create an initial competency draft; an advisory board evaluated and revised the draft, and a consensus panel refined and validated the final set of competencies. The concensus process resulted in 38 competencies organized under the following categories: Risk Assessment and Interpretation; Genetic Education, Counseling, Testing and Results Interpretation; Clinical Management; Ethical, Legal, and Social Implications; Professional Role; Leadership, and Research. These competencies apply to all individuals functioning

http://dx.doi.org/10.1891/0739-6686.29.173

at the graduate level in nursing, including but not limited to advanced practice registered nurses, clinical nurse leaders, nurse educators, nurse administrators, and nurse scientists and are intended to inform and guide their practice.

INTRODUCTION

In less than a decade, genetic and genomic clinical concepts have become inextricably woven into the fabric of health care as advances in genomic science and information technologies transform our understanding of health and illness. Educators across virtually every health care discipline struggle to incorporate genetic/genomic content into already crowded curricula, and clinicians who graduated before this content became a core requirement often struggle to obtain and maintain basic genetic/genomic competencies. As scientists continue to explore the relationships between genes, proteins, receptors, and biochemical pathways, new explanations for common diseases such as diabetes, cancer, human immunodeficiency virus (HIV) infection, psychiatric illnesses, Parkinson's disease, Alzheimer's disease, and so forth are being recognized. The result is that many clinicians are being challenged to reconsider how they manage patients with common diseases, and they lack many of the genomic tools needed to make sound clinical decisions. This chapter describes the development and consensus process used to create a set of essential genetic and genomic competencies that will guide the preparation and practice of nurses prepared at the graduate level.

In 2005, a consensus panel was formed to develop essential genetic/genomic competencies for all nurses (Jenkins & Calzone, 2007). The goal of this effort was to provide support to nursing educators as they integrate this new genetic/genomic content into undergraduate nursing curricula and ensure that nurses entering the workforce receive the basic genomic knowledge and skills they needed to practice. These competencies were revised in 2009 and now include curricular guidelines as well as outcome indicators, providing nurses and nurse educators with a powerful tool to guide nursing practice and education (Consensus Panel, 2009). There is growing evidence of the application of genetics and genomics across various nursing practice arenas including mental health, (Clark, 2007; Elder & Mosack, 2011), public health (Holtzclaw-Williams, 2008), neonatal (Thorngate & Rios, 2008), gastroenterology (Kelly, 2008), oncology (Jenkins, 2011), and clinical nurse specialists (Kelly, 2009). For additional background on the essential genetic/genomic competencies applicable to all nurses, the reader is referred to the chapter by Calzone and Jenkins in this *Annual Review of Nursing Research* volume.

While appreciating the value of the existing competencies (Consensus Panel, 2009), graduate faculty recognized a need for a set of genetic/genomic competencies specific to the expanded roles of nurses prepared at the graduate level. All nurses with advanced degrees are expected to apply advanced genetic/genomic competencies in their roles as educators, administrators, researchers, or clinicians. Advanced practice registered nurses (APRNs) are expected to translate genetics and genomics directly to patient care and/or health care systems. Genetic/genomic competencies that better inform the education and practice of nurses with graduate degrees were needed.

In 2009, the American Association of Colleges of Nursing (AACN) invited the co-chairs of the original competency document to provide recommendations for critical genetic/genomic competencies needed for masters prepared advanced practice nurses to function in these roles in preparation for the upcoming revision of the *AACN Masters Essentials* (AACN, 1996). The first author was invited to provide input into these AACN competency recommendations. This stimulated a discussion regarding a need for developing genetic/genomic competencies for nurses practicing in advanced practice roles, which would facilitate the integration of genetics/genomics into the revised Masters Essentials. The first author was also invited to help lead the effort to develop graduate level competencies and subsequently formed a Steering Committee to provide leadership in the development of essential graduate level genetic/genomic nursing competencies.

Key assumptions guiding the creation of the new graduate level competencies were that nurses with graduate degrees would have achieved the essential genetic/genomic competencies for all nurses prior to entering graduate nursing programs, and that the new competencies would apply to anyone with an advanced degree in nursing including clinical nurse leaders, nurse educators, nurse administrators, and nurse scientists. Because the scope of practice is different for nurses functioning in APRN roles, some of the graduate level competencies were applicable only to nurses working in those specialty areas.

COMPETENCY DEVELOPMENT

In the spring of 2009, a three-person Steering Committee was formed to develop a set of new genetic/genomic competencies for nurses with graduate degrees. A highly methodical approach was used to develop these new competencies (see Figure 9.1). The Steering Committee conducted a structured, systematic literature review of articles and other documents published in the United States between May 1997 and May 2009. Searches were performed in MEDLINE, CINAHL, Web of Science, and Google and Google Scholar. In addition, health professional websites and references from relevant articles were examined using the search terms genetic(s) competencies, genomic(s) competencies, genetic(s) curriculum, genomic(s) curriculum, genetic(s) education, genomic(s) education, nursing, and advanced practice nursing/nurses. Key documents were identified, reviewed, and analyzed using qualitative methods (see Table 9.1).

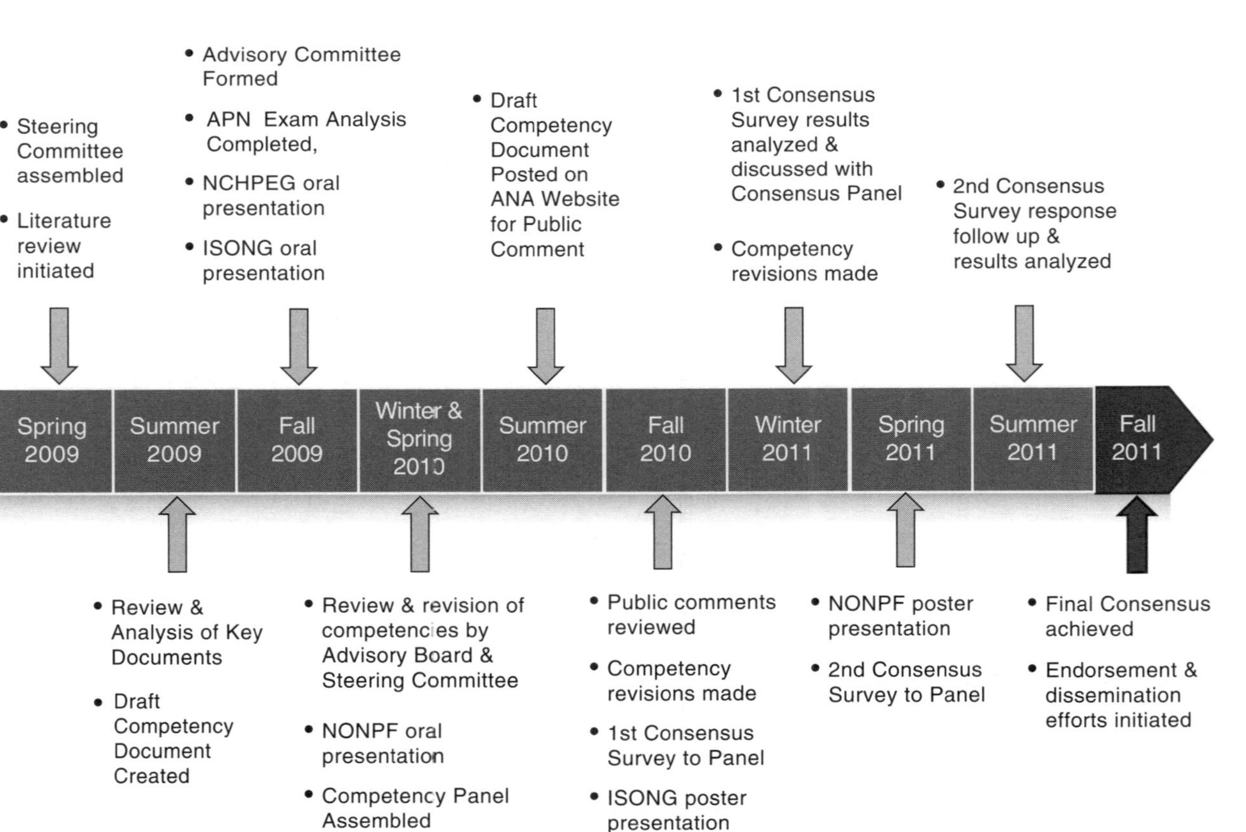

FIGURE 9.1 Graduate-Level Genetic Competencies Development Timeline.

TABLE 9.1

Key Documents Used in the Competency Development Process

Published Genetic/Genomic Competencies Applicable to Graduate Level Health Professionals
Competencies that include but are not specific to graduate level nurses • Genomic Competencies for All Health Professionals (CDC, 2001) • Core Competencies in Genetics Essential for All Health Care Professionals (NCHPEG, 2001, 2007) • Essentials of Genetic and Genomic Nursing: Competencies, Curricula Guidelines, and Outcome Indicators 2nd Edition (Consensus Panel, 2009)
Competencies for graduate level health professionals from other professional organizations • American Academy of Physician's Assistants (Rackover et al., 2007) • American Academy of Family Physicians (2008) • Association of Professors of Human and Medical Genetics/American Society of Human Genetics (2001) • Association of Schools & Colleges of Optometry (2008) • American Speech-Language-Hearing Association (ASLA; 2005) • National Association of Social Workers (NASW; 2003)
Key Findings from Published Nursing Articles Addressing What Nurses With Graduate Degrees Should Know About Genetics/Genomics • Advanced Practice Registered Nurse (APRN) students' knowledge and attitudes toward genetics (Maradiegue, Edwards, Seibert, Macri, & Sitzer, 2005). • Description of a genetics course for APRNs (Horner, 2004) • Faculty perception of genetics and level of integration in nurse practitioner (NP) curricula (Edwards, Maradiegue, Seibert, Macri, Sitzer, 2006). • Perceived genetics competencies for advanced practice oncology nurses and genetics content contained in licensing and certification examinations (Lea, Jenkins, & Monsen, 1999) • Overview of the integration of genetics/genomics into advanced nursing practice, education, and research (Lea, Feetham, & Monsen, 2002; Berry & Hern, 2004). • Core competencies in cancer genetics for advanced practice nurses (APNs) in oncology (Calzone, Jenkins, & Masny, 2002).

Advanced Practice Nursing Certification Exams Evaluated for Genetic/Genomic Content
• American Nurses Credentialing Commission (ANCC)
    ◦ Family NP
    ◦ Adult NP
    ◦ Pediatric NP
    ◦ Gerontological NP
    ◦ CNS "Core" exam

(*Continued*)

TABLE 9.1

Key Documents Used in the Competency Development Process (Continued)

Published Genetic/Genomic Competencies Applicable to Graduate Level Health Professionals

• National Certification Corporation (NCC)
  ◦ Women's Health Care NP
  ◦ Neonatal NP
• American Academy of Nurse Practitioners (AANP)
  ◦ Family NP
• Pediatric Nursing Certification Board (PNCB)
• American Midwifery Certification Board (AMCB)
• Council of Certification of Nurse Anesthetists (CCNA)

Other Key Nursing Documents • American Academy of Nursing (2009). Nurses Transforming Health Care Using Genetics and Genomics. • American Association of Colleges of Nursing (2006). *The Essentials of Doctoral Education for Advanced Nursing Practice.* • American Association of Colleges of Nursing (1996, 2011). *The Essentials of Masters Education in Nursing Practice.* • International Society of Nurses in Genetics/American Nurses Association (2007). Genetics and genomics nursing: Scope and standards of practice. Washington DC: American Nurses Publishing. • National Organization of Nurse Practitioner Faculties in partnership with the American Association of Colleges of Nursing (2002). Nurse Practitioner Primary Care Competencies in Specialty Areas: Adult, Family, Gerontological, and Women's Health, published by HRSA, division of nursing.

Source: Greco, K. E., Tinley, S. & Seibert, D. (2012). *Essential Genetic and Genomic Competencies for Nurses with Graduate Degrees.* Silver Spring, MD: American Nurses Association and International Society of Nurses in Genetics. ISBN-13: 978-1-55810-434-1. Available at http://nursingworld.org/MainMenuCategories/EthicsStandards/Genetics-1

Genetic/Genomic Competencies Applicable to Health Professionals

The steering committee examined several different sets of existing genetics/genomics competencies. Some included but were not specific to graduate level nurses, whereas others were developed for other types of health professionals prepared at the graduate level (see Table 9.1).

The Centers for Disease Control and Prevention (CDC) genomic competencies were among the first genomic competencies published, and they target broad categories of public health professionals who have not received formal training in genetics, including nurses (CDC, 2001). The CDC competencies were later endorsed by the Institute of Medicine (Institute of Medicine, 2005). The National Coalition for Health Professional Education in Genetics (NCHPEG) core competencies in genetics were also published very early (NCHPEG, 2001, 2007). Like the CDC's competencies, NCHPEG competencies were also intended to guide the education and practice of all health professions, and are the most widely cited of all the genetics competency documents. Social workers and speech and language pathologists both developed competencies based on the 2001 NCHPEG competencies (American Speech-Language-Hearing Association, 2005; National Association of Social Workers, 2003). In 2007, the original 44 NCHPEG genetics competencies published in 2001 were rewritten and condensed into 18 competencies (NCHPEG, 2007).

A literature review identified published genetic/genomic competencies for the following graduate level health professional groups: family practice physicians, physician assistants, physicians, optometrists, social workers, and speech-language pathologists and audiologists (American Academy of Family Physicians, 2008; Rackover et al., 2007; Association of Professors of Human and Medical Genetics/American Society of Human Genetics, 2001; Association of Schools & Colleges of Optometry, 2008; National Association of Social Workers, 2003; American Speech-Language-Hearing Association, 2005). Because no genetic/genomic competencies could be found for chiropractors, dentists, naturopaths, nutritionists, physical therapists, podiatrists, psychologists, and occupational therapists, most do acknowledge the importance of genetics and/or genomics in their publications.

Qualitative methods were used to analyze these competency documents to identify and compare common competency themes and identify competency gaps. Each document was analyzed by individual members of the steering committee and results were compared, discussed, and evaluated in regard to their relevance to nurses functioning at the graduate level.

What Nurses With Graduate Degrees Should Know About Genetics/Genomics

The steering committee then conducted a careful review of the nursing literature related to what nurses with graduate degrees should know about genetics/genomics. Seven relevant articles were identified (see Table 9.1) and critically appraised both individually and collectively. The seven articles offered unique perspectives with very little overlap (see Table 9.1), but a common theme among all the articles was that genetic/genomic knowledge and skills needed to be integral to nursing practice at the graduate level regardless of academic preparation, practice setting, role, or specialty.

Advanced Practice Nursing Certification Exam Guidelines

National certifying body examination guidelines were examined to see what nursing specialty organizations expected new clinicians to know about genetics and genomics. Although actual exam content is closely protected, the authors were able to access 11 different examination guidelines published by six major nursing certification bodies (see Table 9.1). Inclusion of genetic/genomic content in exams, based on exam guidelines, appears to vary considerably across specialty organizations. Much of the genetic/genomic content was implied versus specific, and content was more often related to individual disorders than global genetic/genomic concepts.

Other Key Nursing Documents

Finally, key professional nursing documents were examined to identify what the nursing profession is formally expecting graduate nurses to know about genetics/genomics. These documents included *The Essentials of Masters Education for Advanced Nursing Practice* (1996), *The Essentials of Doctoral Education for Advanced Nursing Practice* (2006), and the American Academy of Nursing's 2009 document on *Nurses Transforming Health Care Using Genetics and Genomics* (see Table 9.1). During the development of these graduate level genetic/genomic nursing competencies, AACN was in the process of revising their 1996 *Masters Essentials*, which did not contain genetic/genomics content. The authors participated in the development and submission of recommendations from the International Society of Nurses in Genetics (ISONG) to the AACN *Masters Essentials* development team. These recommendations contained key elements based on the draft of these graduate level essential genetic/genomic nursing competencies. The revised *Masters Essentials* published in March 2011 now includes genetic/genomics content in three of the nine essentials (American Association of Colleges of Nursing, 2011). The Commission on Collegiate Nursing Education requires that graduate nursing programs seeking accreditation demonstrate the incorporation of the AACN essentials (Commission on Collegiate Nursing Education, 2009), which will have a significant impact on the integration of genetics/genomics into graduate nursing programs.

RESULTS

As a result of the exhaustive review, it was clear to the steering committee that there was a need for a new, more advanced set of genetic/genomic competencies for nurses prepared at the graduate level.

Eight overarching themes with 51 nested competencies were identified in this first draft:

1. Risk assessment and interpretation
2. Genetic education, counseling, testing, and results interpretation
3. Clinical management

4. Collaboration
5. Ethical, Legal, and Social Implications (ELSI)
6. Professional role
7. Leadership
8. Research

A 12-member Advisory Board, representing a wide variety of nursing leaders and genetics experts, was recruited to provide comment and review on the initial competency draft (see Table 9.2). Over the next 6 months, the Steering Committee met twice with the advisory board and numerous times as a small group to discuss the draft competencies. During this iterative process, several revisions were proposed and adopted.

CONSENSUS PROCESS

The second draft of the competencies, containing 51 items, was adopted by the steering committee and advisory board. The two groups then identified individuals to form a Consensus Panel. Individuals from key nursing organizations, including advanced practice organizations, were contacted and ultimately 31 representatives from stakeholder organizations comprised the final Consensus Panel (see Table 9.2). The Consensus Panel reviewed the relevance and comprehensiveness of each draft competency, providing input via numerous e-mails and conference calls. The draft competencies were posted for 45 days (from August 6 to September 20 of 2010), for public comment on the American Nurses Association (ANA) website. During the review period, a total of 15 public comments were received; seven were related to content, three were related to grammar, and five supported the document without changes. The comments were reviewed and summarized by the steering committee, discussed at length with the consensus panel, and further adjustments to the draft competencies were made.

Once the consensus panel was satisfied with the revised draft, a formal survey was sent out to all consensus panel members via Survey Monkey. Each competency was individually evaluated by every Consensus Panel member for uniqueness, clarity, and appropriateness to the roles of nurses with graduate degrees. While the survey was deployed, the Steering Committee held regular meetings to monitor comments and feedback as they were provided. Follow-up reminders were sent until responses had been received from all consensus panel members.

All comments from the consensus survey were reviewed, discussed, and summarized by the Steering Committee and circulated back out to the Consensus Panel. During a consensus panel conference call in January of 2011, each competency item that failed to receive 100% agreement on all 3 elements was revised or deleted as appropriate.

TABLE 9.2

Steering Committee, Advisory Committee, and Consensus Panel

Steering Committee Members	Affiliations
Karen Greco, PhD, RN, ANP-BC, FAAN	National Cancer Institute, Genetics Branch
Susan Tinley, PhD, RN, CGC	Creighton University
Diane Seibert, PhD, CRNP, FAANP	Uniformed Services University
Advisory Committee Members	**Affiliations**
Kathleen Calzone, PhD, RN, APNG, FAAN	National Cancer Institute, Genetics Branch
Jean F. Jenkins PhD, RN, FAAN	National Human Genome Research Institute
Carol J. Bickford, PhD, RN-BC, CPHIMS	American Nurses Association
Quannetta Edwards, PhD, FNP, WHCNP, FAANP	American Academy of Nurse Practitioners
Suzanne Feetham, PhD, RN, FAAN	Children's National Medical Center & University of Wisconsin, Milwaukee
Tracy Klein, PhD, RN, FNP-BC, FAANP	Oregon State Board of Nursing
Kathy McGuinn, MSN, RN, CPHQ	American Association of Colleges of Nursing
Karen Pehrson, MS, PMHCNS-BC	Sigma Theta Tau International
Cynthia A. Prows, MSN, CNS, FAAN	Cincinnati Children's Hospital Medical Center
Jo Ellen Rust, MSN, RN, CNS	National Association of Clinical Nurse Specialists
Elizabeth Thomson, DNSc, MSN, RN, CGC, FAAN	National Human Genome Research Institute
Janet K. Williams, PhD, RN, PNP, FAAN	University of Iowa
***Consensus Panel Members**	**Affiliations**
Michelle Beauchesne, DNSc, RN, CPNP, FNAP, FAANP	National Association of Pediatric Nurse Practitioners
Linda Callahan, PhD, CRNA, PMHNP	American Association of Nurse Anesthetists
Duck-Hee Kang, PhD, RN, FAAN	Asian American/Pacific Islander Nurses Association
Carole Kenner, PhD, NNP, RNC-NIC,FAAN	Council of International Neonatal Nurses
Elizabeth A. Kostas-Polston, PhD, APRN, WHNP-BC	National Association of Nurse Practitioners in Women's Health

TABLE 9.2

Steering Committee, Advisory Committee, and Consensus Panel (Continued)

Consensus Panel Members[a]	Affiliations
Ann Maradiegue PhD, FNP-BC, FAANP	National Organization of Nurse Practitioner Faculties
Carrie Merkle, PhD, RN, FAAN	American Academy of Nursing, Genetic Health Care Expert Panel
Carmen T. Paniagua, EdD, RN, CPC, ACNP-BC, APNG, FAANP	National Association of Hispanic Nurses
Nancy Roehnelt, PhD, NP-BC	Oncology Nursing Society
Barbara M. Raudonis, PhD, RN, FNGNA, FPCN	National Gerontological Nursing Association
Lynneece Rooney, MSN, CNM, RNC-OB	American College of Nurse-Midwives
Catherine Ruhl, MS, CNM	Association of Women's Health, Obstetric and Neonatal Nurses
Kathleen Sparbel, PhD, RN, FNP-BC	International Society of Nurses in Genetics
Ida Johnson-Spruill, PhD, RN, LISW, FAAN	National Black Nurses Association
Lois A. Tully, PhD	National Institute of Nursing Research
Mary Weber, PhD, APRN, PMHNP-BC	International Society of Psychiatric-Mental Health Nurses

Note. From Greco, K. E., Tinley, S., & Seibert, D. (2012). *Essential genetic and genomic competencies for nurses with graduate degrees.* Silver Spring, MD: American Nurses Association and International Society of Nurses in Genetics.

[a] The Consensus Panel also includes the Steering Committee and Advisory Board.

As a result of the survey comments and group discussion, eight competencies were retained as initially written, 12 were consolidated or deleted, 12 underwent significant revision, and the remaining 19 received minor modifications. As a result of the revision process, the "collaboration" category was deleted. Those competencies receiving minor modifications were sent out to the consensus panel for vote via e-mail. A second Survey Monkey questionnaire was created to capture votes and comments on the 24 competencies that were deleted or substantively revised to allow for more detailed analysis. To streamline the decision making process on this second survey, prior to sending out the second questionnaire, the consensus panel established a "90%" decision rule; any consensus item receiving a score of 90% or greater was considered to have achieved consensus and was identified as "final." Thirty-eight revised and validated competencies remained after the entire consensus process had been completed (see Table 9.3).

TABLE 9.3

The Essential Genetic and Genomic Competencies for Nurses With Graduate Degrees

Professional Practice Domain
Risk Assessment and Interpretation The nurse with a graduate degree engages in a more active role in risk assessment and interpretation than the registered nurse without a graduate degree. All nurses with graduate degrees in nursing 1. Identify individuals with inherited predispositions to diseases as appropriate to the nurse's practice setting. Nurses with graduate degrees functioning in APRN roles also perform a more detailed evaluation, gather an expanded history, assess for modifiers of risk, confirm reported family health histories, ensure that histories are updated, integrate psychosocial aspects of the family history, and assess for other complex variables (e.g., consanguinity within a family pedigree). 2. Analyze a pedigree to identify potential inherited predisposition to disease. 3. Estimate risks for Mendelian and multifactorial disorders in affected families as appropriate. 4. Use family history and pedigree information to plan and conduct a targeted physical assessment. 5. Interpret the findings from the physical assessment, family history, laboratory findings, diagnostic tests, and/or radiology results that may indicate genetic/genomic disease, disease risk, or the need for a genetic/genomics referral. 6. Refer at-risk family members for assessment of an inherited predisposition to disease.
Genetic Education, Counseling, Testing, and Results Interpretation Nurses with graduate degrees provide genetic/genomic education, counseling and testing and client support throughout the lifespan within their licensure, scope of practice and clinical setting, and seek consultation as appropriate. All nurses with graduate degrees in nursing 1. Incorporate knowledge of clients' attitudes, values, and beliefs rooted in varying ethnic, cultural, social, and religious backgrounds when communicating genetic/genomic information. 2. Provide genetic/genomic information that is appropriate to client's level of health literacy and numeracy. 3. Educate clients about possible risks, benefits, and limitations of genetic testing and/or therapy. 4. Provide anticipatory guidance to assist clients in the decision-making process related to genetics/genomics. 5. Obtain informed consent for genetic testing and/or therapy. 6. Assess the influence of genetic/genomic risk and disease on family communication and functioning.

TABLE 9.3

The Essential Genetic and Genomic Competencies for Nurses With Graduate Degrees *(Continued)*

Professional Practice Domain

7. Assess the clinical and psychosocial outcomes including benefits, limitations, and risks of genetic/genomic information and/or therapies for clients.
8. Support client coping and client use of genetic/genomic information in promoting health, reducing risk, managing symptoms, and/or preventing illness.

Nurses with graduate degrees functioning in APRN roles also

9. Provide genetic/genomic education and counseling appropriate to practice setting.
10. Select appropriate genetic/genomic tests and/or studies.
11. Communicate results of genetic/genomic screening and/or testing at a level that clients can understand.

Clinical Management

Nurses with graduate degrees need to be able to provide personalized care and/or care coordination that incorporates genetic- and genomic-based technology into client care.

All nurses with graduate degrees in nursing

1. Apply knowledge about the interaction of genetic/genomic and environmental factors to the care of clients.
2. Make appropriate referrals to genetic professionals or other health care resources.
3. Evaluate effectiveness of prevention, risk reduction, health promotion, and disease management interventions related to genetics/genomics.

Nurses with graduate degrees functioning in APRN roles also

4. Manage care of clients incorporating genetic/genomic information and technology (e.g., risk-based genetic screening and testing, prescription of pharmacogenomic-based drugs, gene targeted therapy, and use of genetic/genomic information in symptom management).
5. Collaborate with genetic specialists, health professionals, and those in relevant disciplines to develop a comprehensive plan to evaluate and manage clients with genetic/genomic disease or risk.

Ethical, Legal, and Social Implications (ELSI)

Nurses with graduate degrees need to recognize the significance of ethical, legal, and social implications in genetics and genomics. While ethical, legal, and social implications apply across all areas of practice, genetic testing is a component of health care where ethical issues may be most apparent.

All nurses with graduate degrees in nursing

1. Facilitate ethical decision-making related to genetics/genomics congruent with the client's values and beliefs.
2. Inform health care and research policy related to ELSI issues in genetics/genomics.
3. Implement effective strategies to resolve ELSI issues related to genetics/genomics.
4. Apply ethical principles when making decisions regarding management of genetic/genomic information identified through clinical or research technologies.

(Continued)

TABLE 9.3

The Essential Genetic and Genomic Competencies for Nurses With Graduate Degrees (*Continued*)

Professional Responsibilities Domain
Professional Role Nurses with graduate degrees need to maintain a solid foundation in genetics/genomics to provide safe and competent care to clients. All nurses with graduate degrees in nursing 1. Integrate best genetic/genomic evidence into practice that incorporates client values and clinical judgment. 2. Mentor other nurses in the application of genetics/genomics to nursing care within their practice setting. 3. Identify genetic/genomic learning needs of other health professionals and disciplines. 4. Conduct educational interventions to address the genetic/genomic learning needs of health professionals and clients. 5. Participate in the development of professional practice guidelines related to genetics/genomics.
Leadership Nurses with graduate degrees assume an active role in genetic/genomic policies at the local, state, national, and international levels in nursing and other health care organizations. All nurses with graduate degrees in nursing 1. Contribute a nursing perspective to genetic/genomic clinical and policy discussions. 2. Facilitate an organizational climate that is responsive to genetic/genomic discoveries. 3. Use care delivery strategies that incorporate genetics/genomics. 4. Influence health policy at the local, state, national, and international levels related to genetics/genomics.
Research Nurses with graduate degrees must understand how genetic/genomic research can provide insight into human biology and disease pathogenesis, leading to improved health outcomes. Nurses prepared at the doctoral level are expected to provide leadership in the conduct of research and translation of genetic/genomic findings into practice. All nurses with graduate degrees in nursing 1. Participate in the application and translation of genetic/genomic research in nursing practice and/or education. 2. Identify genetic/genomic health care methods and outcomes that can be influenced by nursing. 3. Collaborate with researchers in relevant disciplines in the conduct, dissemination and/or translation of genomic inquiry and research.

Note. From Greco, K. E., Tinley, S., & Seibert, D. (2012). *Essential genetic and genomic competencies for nurses with graduate degrees.* Silver Spring, MD: American Nurses Association and International Society of Nurses in Genetics.

NEXT STEPS

Achieving consensus on the "Essential Genetic and Genomic Competencies for Nurses with Graduate Degrees" is the first step toward assuring that nurses with graduate degrees are prepared to deliver competent genomic care. Endorsement, dissemination, and implementation of these competencies are critical additional steps needed to ensure that fundamental genomic knowledge and skills needed by graduate level nurses is integrated into practice, education, and research. Endorsement from a wide variety of nursing organizations is currently being sought. At the time of publication of this document, 19 nursing organizations have endorsed these competencies. Dissemination is being conducted through publication on the ANA website (Greco, Tinley, & Seibert, 2012) and via this chapter. The next major effort will be to develop performance indicators and identify educational resources to assist educators in teaching competency concepts. Some high-quality genetics resources already exist (see Table 9.4 for examples) but more will be needed as knowledge continues to expand. Performance indicators will be developed using a process similar to the one used to develop outcome indicators for the *Essentials of Genetic and Genomic Nursing: Competencies, Curricula Guidelines, and Outcome Indicators* (Calzone, Jenkins, Prows, & Masny, 2011; Consensus Panel, 2009) and educational resources will be made available through the Genetics/Genomics Competency Center (G2C2: Genetics/Genomics Competency Center for Education, 2011), an open access educational resource for health care providers and educators that provide curricula materials and resources to teach genomics content.

CONCLUSION

The "Essential Genetic and Genomic Competencies for Nurses with Graduate Degrees" establishes baseline genetic/genomic competencies for all nurses functioning at the graduate level in nursing. Developed using a highly structured, methodical approach, the competencies were validated using a consensus model by a diverse panel of 31 nursing leaders and genetics experts representing professional nursing organizations, academic institutions, regulatory bodies, and government agencies. The importance of these competencies to the provision of health care delivery by nurses prepared at the graduate level was recently underscored by the American Association of Colleges of Nursing (AACN) with the publication of the 2011 AACN Essentials of Masters Education for Advanced Nursing Practice. The previous edition, published in 1996, did not include any mention of "genetics" or "genomics," and the 2011 edition contains seven references to genetics/genomics distributed across three discreet essentials.

All health care professionals should be knowledgeable about, comfortable with, and competent in the provision of genetic/genomic health care. Nurses with advanced degrees are at the interface of translating genetic and genomic advances

TABLE 9.4
Genetic/Genomic Online Resources

Genetics/Genomics Competency Center for Education Searchable database with genetic/genomic resources and learning activities for educators of nurses, genetic counselors, and physician assistants http://www.g-2-c-2.org
Gene Tests/Gene Reviews Expert-authored disease reviews for clinicians. Site includes a database of clinical and research genetic testing laboratories and links to genetic services & providers http://www.genetests.org/
Genes and Disease Articles on genetic disorders, arranged by body systems http://www.ncbi.nlm.nih.gov/books/nbk22183
American Medical Association Family History Resources Family history forms for prenatal, pediatric, and adult populations http://www.ama-assn.org/ama/pub/physician-resources/medical-science/genetics-molecular-medicine/family-history.page?
Surgeon General's Family Health History Initiative Online and paper-based family history collection tool offered in both English and Spanish languages www.hhs.gov/familyhistory
Online Mendelian Inheritance in Man Comprehensive database about genetic conditions that have been reported in the professional literature http://www.ncbi.nlm.nih.gov/entrez/query.fcgi?db=OMIM
The Future of Medicine, Pharmacogenomics: An Online Course Options to select an overview sheet or complete in-depth modules on 13 different pharmacogenomic topics http://www.lithiumstudios.com/fda/Sample_Home.htm
Public Health Genomics Application of genomics to Public Health http://www.cdc.gov/genomics/default.htm

into the care of individuals, families, and communities. Nurses prepared at the graduate level are functioning in key leadership roles in health care and must be equipped with the knowledge and skills they need to integrate genetics and genomics into health care systems. Achievement of these vital roles in the translation of genomic discoveries into practice requires the integration of the appropriate genetic/genomic content into nursing curricula to provide the knowledge and skills needed for graduate level nursing practice, education, and research. Graduate nursing faculty and practicing clinicians needed an expanded set of genetic/genomic competencies to meet these health care challenges. The "Essential Genetic and Genomic Competencies for Nurses with Graduate Degrees" provides this essential foundation for the genomic competence of graduate nursing practice.

REFERENCES

American Academy of Family Physicians. (2008). *Recommended curriculum guidelines for family medicine residents*. Retrieved from http://www.aafp.org/online/etc/medialib/aafp_org/documents/about/rap/curriculum/medical_genetics.Par.0001.File.tmp/medicalgenetics.pdf

American Academy of Nursing. (2009). *Nurses transforming health care using genetics and genomics*. Retrieved from http://www.aannet.org/files/public/Genetic_White_Paper_1.22.09_FINAL2.pdf

American Association of Colleges of Nursing. (1996). *The essentials of master's education for advanced nursing practice*. Retrieved from http://www.aacn.nche.edu/Education/pdf/MasEssentials96.pdf

American Association of Colleges of Nursing. (2006). *The essentials of doctoral education for advanced nursing practice*. Retrieved from http://www.aacn.nche.edu/DNP/pdf/Essentials.pdf

American Association of Colleges of Nursing. (2011). *The essentials of master's education in nursing*. Retrieved from http://www.aacn.nche.edu/education-resources/MastersEssentials11.pdf

American Speech-Language-Hearing Association. (2005). *What does the speech-language pathologist or audiologist need to know about genetics when conducting assessments?* Retrieved from http://www.asha.org/academic/questions/Genetics-Education.htm

Association of Professors of Human and Medical Genetics/American Society of Human Genetics. (2001). *Medical school core curriculum in genetics*. Retrieved from http://www.meddean.luc.edu/lumen/MedEd/genetics/core_curric.htm

Association of Schools & Colleges of Optometry. (2008). *Core competencies in genetics*. Retrieved from http://www.opted.org/i4a/pages/index.cfm?pageid=3366

Beery, T. A., & Hern, M. J. (2004). Genetic practice, education, and research: An overview for advanced practice nurses. *Clinical Nurse Specialist*, *18*(3), 126–134.

Calzone, K. A., Jenkins, J., & Masny, A. (2002). Core competencies in cancer genetics for advanced practice oncology nurses. *Oncology Nursing Forum*, *29*(9), 1327–1333.

Calzone, K. A., Jenkins, J., Prows, C. A., & Masny, A. (2011). Establishing the outcome indicators for the essential nursing competencies and curricula guidelines for genetics and genomics. *Journal of Professional Nursing*, *27*(3), 179–191.

Centers for Disease Control and Prevention. (2001). *Genomic competencies for all public health professionals*. Retrieved from http://www.cdc.gov/genomics/translation/competencies/index.htm

Clark, W. G. (2007). Schizophrenia and genomics: Linking research to practice. *Journal of Psychosocial Nursing and Mental Health Services*, *45*(6), 24–28.

Commission on Collegiate Nursing Education. (2009). *Standards for accreditation of baccalaureate and graduate degree nursing programs*. Retrieved from http://www.aacn.nche.edu/ccne-accreditation/standards09.pdf

Consensus Panel Genetic/Genomic Nursing Competencies. (2006). *Essential Nursing Competencies and Curricula Guidelines for Genetics and Genomics*. Silver Spring, MD: American Nurses Association.

Consensus Panel Genetic/Genomic Nursing Competencies. (2009). *Essentials of genetic and genomic nursing: Competencies, curricula guidelines, and outcome indicators* (2nd ed.). Silver Spring, MD: American Nurses Association.

Edwards, Q. T., Maradiegue, A., Seibert, D., Macri, C., & Sitzer, L. (2006). Faculty members' perceptions of medical genetics and its integration into nurse practitioner curricula. *Journal of Nursing Education*, *45*(3), 124–130.

Elder, B. L., & Mosack, V. (2011). Genetics of depression: An overview of the current science. *Issues in Mental Health Nursing, 32*(4), 192–202.

G2C2: Genetics/Genomics Competency Center for Education. (2011). Retrieved from http://www.g-2-c-2.org/

Greco, K. E., Tinley, S., & Seibert, D. (2012). *Essential genetic and genomic competencies for nurses with graduate degrees*. Silver Spring, MD: American Nurses Association and International Society of Nurses in Genetics. Available at http://nursingworld.org/MainMenuCategories/EthicsStandards/Genetics-1.

Holtzclaw-Williams, P. S. (2008). Genetic and genomic public health strategies: Imperatives for neonatal nursing genetic competency. *Newborn and Infant Nursing Reviews, 8*(1), 43–50.

Horner, S. D. (2004). A genetics course for advanced clinical nursing practice. *Clinical Nurse Specialist, 18*(4), 194–199.

Institute of Medicine. (2005). *Implications of genomics for public health: Workshop summary*. Washington, DC: The National Academies Press.

International Society of Nurses in Genetics/American Nurses Association. (2006). *Genetics and genomics nursing: Scope and standards of practice*. Washington, DC: American Nurses Association.

Jenkins, J. (2011). Essential genetic and genomic nursing competencies for the oncology nurse. *Seminars in Oncology Nursing, 27*(1), 64–71.

Jenkins, J., & Calzone, K. A. (2007). Establishing the essential nursing competencies for genetics and genomics. *Journal of Nursing Scholarship, 39*(1), 10–16.

Kelly, P. (2008). Understanding genomics: No longer optional for gastroenterology nurses. *Gastroenterology Nursing, 31*(1), 45–54.

Kelly, P. (2009). The clinical nurse specialist and essential genomic competencies: Charting the course. *Clinical Nurse Specialist, 23*(3), 145–150.

Lea, D. H., Feetham, S. L., & Monsen, R. B. (2002). Genomic-based health care in nursing: A bidirectional approach to bringing genetics into nursing's body of knowledge. *Journal of Professional Nursing, 18*(3), 120–129.

Lea, D. H., Jenkins, J., & Monsen, R. B. (1999). Incorporating genetics into nursing practice. *Nurse Educator, 24*(5), 4–5.

Maradiegue, A., Edwards, Q. T., Seibert, D., Macri, C., & Sitzer, L. (2005). Knowledge, perceptions, and attitudes of advanced practice nursing students regarding medical genetics. *Journal of the American Academy of Nurse Practitioners, 17*(11), 472–479.

National Association of Social Workers. (2003). *NASW standards for integrating genetics into social work practice*. Retrieved from http://www.naswdc.org/practice/standards/GeneticsStdFinal4112003.pdf

National Coalition for Health Professional Education in Genetics. (2001). Recommendations of core competencies in genetics essential for all health professionals. *Genetics in Medicine, 3*(2), 155–159.

National Coalition for Health Professional Education in Genetics. (2007). *Core competencies in genetics essential for all health care professionals*. Retrieved from http://www.nchpeg.org/index.php?option=com_content&view=article&id=237&Itemid=84

National Organization of Nurse Practitioner Faculty in partnership with the American Association of Colleges of Nursing. (2002). *Nurse practitioner primary care competencies in specialty areas: Adult, family, gerontological, and women's health, published by HRSA, division of nursing*. Retrieved from http://www.aanp.org/NR/rdonlyres/E1B37354-2195-401D-8027-A5A2763E3006/0/finalaug2002.pdf

Rackover, M., Goldgar, C., Wolpert, C., Healy, K., Feiger, J., & Jenkins, J. (2007). Establishing essential physician assistant clinical competencies guidelines for genetics and genomics. *Journal of Physician Assistant Education, 18*(2), 47–48. Retrieved from http://www.paeaonline.org/index.php?ht=a/GetDocumentAction/i/25416

Thorngate, L., & Rios, C. (2008). Clinical care at the genomic interface: Current genetic issues in neonatal nursing. *Newborn and Infant Nursing Reviews, 8*(1), 36–42.

CHAPTER 10

Skeletal Muscle and Genetics

Christine E. Kasper

ABSTRACT

The regulation of hypertrophy or atrophy of skeletal muscle is highly regulated by genetic signals closely tied to function. This chapter focuses on the genetic alteration of structural and cytoskeletal proteins that influence exercise capacity, self-care, and activities of daily living by modulation of the physiologic cross-sectional area of skeletal muscle. In addition to a discussion of genetic mechanisms of atrophy and sarcopenia, the muscular dystrophies along with the laminopathies, both diseases of cytoskeletal proteins will be reviewed.

INTRODUCTION

Key patient care concepts in nursing and nursing science are those that address issues of functional capacity, mobility, self-care, and activities of daily living. All of these require mobility, which in mammalian systems is derived from the musculoskeletal organ system. The striated muscle tissues of mammalian systems are a marvel of architectural engineering. Each of its proteins and anatomical structures are adapted to bear structural loads, high levels of shear stresses, as well as sustained or sudden extreme contractile forces. Some muscles such as the gastrocnemious can generate forces in excess of 250 N. Each skeletal muscle is designed for a specific anatomical site and function; and, will adapt each and every protein to optimally contract following sustained altered use. Skeletal muscle as an organ system personifies the classic physiologic phrase "structure

http://dx.doi.org/10.1891/0739-6686.29.191

recapitulates function." It is fundamentally the structural integrity of hundreds of proteins wrapped in multiple fascial sheaths and contained by tendons, which permit the generation of forces and organismal movement. Structural failures of either the contractile proteins or the surrounding connective tissues form the fundamental basis of the genetic defects of striated and skeletal muscle.

This review of the genetics involved in skeletal muscle structure and function attempts only to point to a few areas of interest, which can be later explored at length by the interested investigator or clinician as a thorough review would fill volumes on multiple library shelves. A search of the terms "skeletal muscle disease" in the Online Mendelian Inheritance in Man (OMIM) yields 280 entries; whereas the same search in the Online Mendelian Inheritance in Animals (OMIA) lists another 79 diseases of genetic origin (Nichols, 2011; "OMIM: Online Mendilian Inheritance in Man," 2011). These genetic diseases of skeletal muscle represent either deformation or absence of key proteins, which result in a disruption of structural integrity with loss of contractile force. On the organismic or human level, loss of contractile force often translates to a limitation in the ability to move or breathe.

The ability to conduct one's activities of daily living is dependent on skeletal muscle that is of a sufficient size to generate the forces required for movement. Simply stated, muscle force is proportional to physiologic cross-sectional area (PCSA) and muscle velocity is proportional to muscle fiber length (Lieber & Ward, 2011). If there is insufficient PCSA due to atrophy, aging, or disease, then movement will be impaired. PCSA can be increased by either adding muscle fibers in parallel or by increasing the size of existing myofibers (Lieber & Ward, 2011). Loss of PCSA, also known as atrophy, occurs by a decrease in the number of fibers in parallel or a loss of size of the individual cells or both (Kasper, 2001). How cell size is controlled in striated muscle systems remains somewhat obscure; however, recent studies using a multitude of genetic methods have begun to describe the mechanisms involved in the regulation of myofiber volume (White, Biérinx, Gnocchi, & Zammit, 2010).

SKELETAL MUSCLE ATROPHY

Atrophy of skeletal muscle occurs rapidly whenever an individual's pattern of exercise and activity is significantly decreased. These relatively sedentary periods are known to result in disuse atrophy of skeletal muscle and a loss of strength as well as endurance. Hospitalization, injury, or other illnesses are common occurrences and result in a loss of exercise capacity, increased fatigue, and muscle atrophy. One week of bed rest has been demonstrated to cause a loss of about 20% of the total muscle mass of a young adult male (18–30 years old); and, severity of atrophy due to disuse increases with increasing age of the individual (Kasper, Talbot, & Gaines,

2002). In addition, repeated bouts of atrophy coupled with aging may lead to a permanent inability to recover exercise capacity and appropriate mobility; thus, leading to loss of deployable forces. Later in life, these accumulated events may impair recovery in the elderly leading to severe losses in muscle and the potential of confinement to a wheelchair (Elkina, von Haehling, Anker, & Springer, 2011).

The study of genetic control in relation to activity was first studied by Booth and colleagues (Watson, Stein, & Booth, 1984). In this study, they demonstrated a pre-translational decrement in (Williams, 1986) skeletal α-actin mRNA, following 7 days of decreased contractile activity due to fixing the soleus muscle in a shortened position. A series of studies followed, which together demonstrated that exercise and inactivity interact with genes to produce adaptive changes in skeletal muscle proteins, ultimately altering function. Stimulation of skeletal muscle for 21 days, as a proxy for increased exercise, was shown to increase cytochrome b mRNA fivefold (Williams, Salmons, Newsholme, Kaufman, & Mellor, 1986). Later, Spangenburg and colleagues (2003) found significant differences in global profiles of mRNA at the 10th day of soleus muscle immobilization. This study demonstrated that the significantly large sets of changes in mRNA levels are a reflection of the number of genes responsive to physical inactivity (Spangenburg et al., 2003).

DISEASE AND DISUSE

Comparison of the process of atrophy across various diseases and disuse revealed a common set of genes, which were up or down-regulated and are now known as atrophy-related genes or atrogenes (Sacheck et al., 2007). Two novel skeletal muscle-specific ubiquitin ligases, atrogin-1/MAFbx and MuRF1, are the most frequently upregulated genes. Both are upregulated in various models of muscle atrophy and are responsible for the increased protein degradation through proteasome system (Bodine et al., 2001; Eddins et al., 2011; Jagoe & Goldberg, 2001; Pattison, Folk, Madsen, Childs, & Booth, 2003). At this time, both of these genes, regarded as the best markers or indicators of muscle wasting, are the master genes controlling atrophy (Sandri, 2008). Additional genes, which play a role in atrophy, remain of potential interest. These include genes coding for lysosomal protease, transcription factors, protein synthesis regulation, and enzymes of metabolic pathways, but their particular role in muscle atrophy has yet to be clarified.

SARCOPENIA

The loss of skeletal muscle mass and strength due to aging is known as *sarcopenia*. Sarcopenia is a normal progressive process of mammalian aging, which is a major contributor to increased frailty, limitations of mobility and strength, difficulty adapting to stress, and eventual mortality (Doherty, 2003). In addition to loss of muscle tissue, there are associated losses of alpha motor neurons and the ability to sustain contractile forces (Bae et al., 2008; Fling, Knight, & Kamen, 2009; Ling, Conwit, Ferrucci, & Metter, 2009; Suetta et al., 2009). Research of the past decade has focused on increasing strength training and activity to stabilize loss and promote activities of daily living. Chronic low-grade inflammation is also a known contributor to atrophy due to aging (Beyer et al., 2011). Sarcopenia has also been associated with a number of other health problems, which include airway diseases (Sermet-Gaudelus et al., 2003), metabolic syndrome, cardiovascular disease (Janssen, 2006; Payne, 2006), osteoporosis and fractures (Szulc, Beck, Marchand, & Delmas, 2005), and sarcopenic obesity (Roubenoff, 2000, 2004).

Numerous studies in the past 2 decades have focused on mitigating the effects of sarcopenia through the maintenance of physical activity and proper nutrition in the elderly. However, it was consistently observed that there was a large variation in the rate and magnitude of atrophy along with the loss of strength in geriatric populations. Although some of these differences could be accounted for by environment, nutritional status, and regular exercise, the remaining variance could not. Recent studies demonstrate that there is a strong genetic association with progressive atrophy during aging (Garatachea & Lucía, 2011; Hai et al., 2011; Tan, Liu, Lei, Papasian, & Deng, 2012). Using genome-wide association study (GWAS) of copy number variants (CNVs), Hai and colleagues (2011) identified CNV2073 that locates at chromosome 15q13.3. CNV2073 has been implicated as a candidate region for lean body mass by this group's previous linkage studies. The nearest gene, gremlin1, has a pivotal role in the regulation of skeletal muscle formation and repair (Hai et al., 2011). Tan and colleagues (2012) list 35 known markers with their chromosome loci for the various skeletal muscle sarcopenia phenotypes. Few linkage studies have been completed and the findings are inconsistent due to the extensive genetic variation in the sarcopenia phenotype; however, five of the 37 chromosome loci have been reported in more than one study. Given the complexity of skeletal muscle contractile proteins, it is expected that extensive further studies will need to be conducted prior to clearly defining the genetic origins of sarcopenia.

STRUCTURAL PROTEINS DYSTROPHIES

The muscular dystrophies (MD) as a group are well-known to the lay public due to extensive fund raising activities over the last few decades. OMIM lists 84 various dystrophies ("OMIM: Online Mendilian Inheritance in Man," 2011). These diseases weaken contractile proteins by the mutation or loss of various structural proteins (Dahiya et al., 2011; Flanigan et al., 2011; Hoffman et al., 2011; Punnoose, Burke, & Golub, 2011). Unfortunately, most of these MD are progressive over time resulting in death after extensive tissue loss. The most common of these dystrophies is Duchenne muscular dystrophy (DMD). The incidence of DMD is 1 in 3,000 boys and is a recessive X-linked form of muscular dystrophy. Girls are rarely affected and are usually carriers. The disease is the result of a mutation in the muscle structural protein dystrophin, which is a part of the crucial dystroglycan complex (DGC) in the muscle sarcolemmal membrane. The altered structure of dystrophin renders the DGC unstable under contractile loading resulting in ripping of the membrane and subsequent cell death. The mutation has been tracked in humans to the X chromosome (Xp21; "OMIM: Online Mendilian Inheritance in Man," 2011).

LAMINOPATHIES

In the past decade, a new set of dystrophies, the laminopathies, have been identified (Wilson et al., 2005; Wilson, Zastrow, & Lee, 2001). In general, these mutations affect proteins residing in the nuclear envelope and significantly impact basic nuclear functions. In skeletal muscle, myonuclear architecture and position change significantly during normal neonatal development, during compensatory hypertrophy-induced adaptation from fast to slow fiber type, and upon loss of weight-bearing (Kasper & Xun, 1995; Tseng, Kasper, & Edgerton, 1994). These changes include alterations in nuclear volume, cytoplasmic volume per nucleus, nuclear shape, and nuclear positioning within the fiber. Nuclear architecture is linked to both the normal and pathologic control of gene expression, through mechanisms that are not yet understood. The emerging concept is that fundamental activities including DNA synthesis, chromatin organization, and gene expression all depend on the interior ultrastructure of the nucleus, particularly the nuclear lamina and a growing number of lamin-associated proteins. Furthermore, mutations in nuclear lamins and lamin-binding proteins cause inherited diseases that affect both skeletal and cardiac muscle. The lamins and lamin-associated proteins have also been demonstrated to have a role in skeletal muscle aging and disease (Hyder, Isoniemi, Torvaldson, & Eriksson, 2011; Maraldi et al., 2007; Reddy & Comai, 2012; Rodríguez & Eriksson, 2010; Vlcek & Foisner, 2007). The A-, B-, and C-type lamins intermediate filament proteins are found enmeshed in the nuclear lamina on the interior face of the nuclear membrane (Sinensky et al., 1994). In mammalian cells, there are two forms of B-type lamins that are required during embryonic development (Vlcek & Foisner, 2007). The LMNA gene encodes both lamin A and C in differentiated cells and these have an important role in cell stability and homeostasis (Fong et al., 2006; see Table 10.1).

TABLE 10.1
Laminopathies

Disease	Gene	OMIM Reference
AD- & AR-Emery-Dreifuss muscular dystrophy	EDMD2, EDMD3	#181350 & #604929
Limb-girdle muscular dystrophy type 1B	LGMD1B	#159001
Dilated cardiomyopathy with conduction defect	DCM-CD/CMD1A	#115200
Autosomal recessive Charcot-Marie-Tooth-disease	CMT2B1	#605588
Familial partial lipodystrophy type Dunnigan	FPLD2	#151660
MandibuloAcral Dysplasia	MADA	#248370
Hutchinson-Gilford Progeria syndrome	HGPS	#176670
Restrictive Dermopathy	DR	#275210
Atypical Werner Syndrome	WRN-like	#277700
Other Premature Aging syndromes, Generalized Lipoatrophy, Insulin-resistant diabetes, disseminated Leuko-melanodermic papules, Liversteatosis, and Cardiomyopathy	LIRLLC/LDHCP	#608056

LAMINS

Lamins are nuclear-specific intermediate filament proteins, which form dimers and tetramers that polymerize into filaments (Stuurman, Heins, & Aebi, 1998). There are two major types of lamins known as A-type and B-type. A-type includes lamins A, AD10, C, and C2, which are encoded by alternative splicing of the LMNA gene (Capell & Collins, 2006; Lin & Worman, 1993; Stuurman et al., 1998). Two other genes encode the major B-type lamins, B_1 and B_2. Two minor lamins, C2 and B3, are expressed uniquely in germ cells (Furukawa, Fritze, & Gerace, 1998; Nakajima & Abe, 1995). The expression of A-type lamins is highly regulated during development (Moir et al., 2000; Stuurman et al., 1998). B-type lamins appear to be less specialized because every mammalian cell expresses at least one B-type lamin. The single B-type lamin in *Caenorhabditis elegans* is essential for cell viability (Liu et al., 2000). In contrast, A-type lamins are not essential, in the broad sense of the word, because LMNA-knockout mice are born normal and have no detectable phenotype until 2–3 weeks after birth. After 3 weeks, the LMNA-knockout mice show gait defects and then rapidly develop a severe form of muscular dystrophy, which leads to death by 8 weeks (Sullivan et al., 1999).

Lamins and Muscular Dystrophy

Mutations in A-type lamins cause the autosomal dominant form of Emery–Dreifuss muscular dystrophy (EDMD; Bonne et al., 1999). EDMD is characterized by slowly progressive skeletal muscle weakening, contractures of major tendons (Achilles, elbow, neck), and cardiac conduction system defects that can cause sudden cardiac arrest in the absence of a pacemaker (Bonne et al., 2000; Nagano & Arahata, 2000). Importantly, EDMD is also caused by recessive loss of emerin (Bione et al., 1994; Raffaele Di Barletta et al., 2000), a nuclear membrane protein that binds directly to A-type lamins (Clements, Manilal, Love, & Morris, 2000) and to a DNA-bridging protein named BAF (Barrier-to-autointegration factor; Furukawa, 1999; Shumaker, Lee, Tanhehco, Craigie, & Wilson, 2001; Zheng et al., 2000). The situation with A-type lamins recently became more complicated by the discoveries that different mutations in LMNA cause either EDMD, dilated cardiomyopathy with conduction system disease, limb-girdle muscular dystrophy, or Dunnigan-type familial partial lipodystrophy (Bonne, et al., 2000; Holt, Clements, Manilal, Brown, & Morris, 2001; Wilson et al., 2001). So far, many tissues that are affected (dense connective tissue/tendons, adipose) appear to derive from the mesenchymal stem cell lineage, suggesting that the differentiation or stability of connective tissue lineages are particularly sensitive to the correct expression of A-type lamins (Wilson, Benavente, et al., 2001; Wilson, Zastrow, et al., 2001). Given that A-type lamins appeared late in evolution and are only expressed in higher eukaryotes (Cohen, Lee, Wilson, & Gruenbaum, 2001), it seems likely that A-type lamins have specialized functions.

Despite a plethora of models ranging from gene expression to mechanical instability, the molecular mechanism(s) of these disorders are not yet completely understood (Dechat, Gesson, & Foisner, 2010; Prokocimer et al., 2009; Worman, Ostlund, & Wang, 2010). Thus, the roles of lamins and lamin-binding proteins during muscle development and muscle adaptation represent an important open and exciting question in cell biology, with important implications for a growing number of inherited diseases.

Nuclear architecture is dramatically altered in LMNA knockout cells. Nuclei from wild-type epithelial cells were ovoid, whereas those from the LMNA knockout cells were highly elongated or irregular, and lacked lamin B staining at one end of the nucleus (Sullivan et al., 1999). Thus, the nuclear envelope in LMNA cells is structurally compromised, meaning that A-type lamins maintain nuclear integrity, assist in nuclear envelope formation after mitosis, and have also been linked to nuclear activities including DNA replication and transcription (Marji et al., 2010; Prokocimer et al., 2009; Taimen et al., 2009). Dystrophic LMNA muscle cells also showed an increase in the number of nuclei and have occasional centrally-located nuclei.

Lamins and muscle differentiation: Muscle differentiation in vitro has been used to model the relationship between lamins and gene expression. Chaly and colleagues (1996) used L6E9 cells to show that as myoblasts fused into multinucleated myotubes, nuclear architecture changed: the centrosomes moved to the nuclear periphery and condensed chromatin began to aggregate at the nuclear periphery. One possible mechanism for such changes involves the altered expression or activity of nuclear lamina proteins or chromatin modifying machinery, or both. Consistent with this idea, there are characteristic changes in lamin isoform expression during differentiation in rat and chicken embryonic development (Lehner, Stick, Eppenberger, & Nigg, 1987; Lourim & Lin, 1989a; Wedrychowski, Bhorjee, & Briggs, 1989). In L6E9 cells, nuclear shape also changed during myoblast fusion, from ovoid to increasingly elongated after cell fusion (Chaly & Munro, 1996).

Chaly and Munro (1996) proposed that the nuclear periphery reorganizes during myogenesis with increased expression of lamins A, B, and C. Interestingly, lamin A was expressed earlier than "classic" muscle genes such as MHC, MLC, tropomyosin, troponin C, and desmin (Lourim & Lin, 1989b). Furthermore, lamin A expression was reversibly blocked when differentiation and fusion were inhibited, suggesting that A-type lamins are integral to muscle fiber formation.

The nuclear lamins maintain structural integrity of the nucleus much in the same manner that the various cytoskeletal proteins and connective tissues maintain the structural integrity of myofibers and muscle during contraction. As seen in the muscle pathologies listed in Table 10.1, mutations in LMNA result in some magnitude of myonuclear fragility. Mutations in emerin, one of the lamina-associated proteins, are believed to affect the function of mechanosensitive genes in response to contractile forces (Lammerding, Kamm, & Lee, 2004; Lammerding et al., 2004). Altered nuclear shape and structure is a prominent feature of the premature aging syndromes, such as Hutchinson–Gilford Progeria Syndrome (Gonzalez, Pla, Perez-Sala, & Andres, 2011; Reddy & Comai, 2012).

In all, the research into the various aspects of function, aging, sarcopenia, and the laminopathies have exponentially increased in the past decade. Previously, only the traditional dystrophies such as DMD were studied in depth.

The next 10 years should shed significant light on the mechanisms of skeletal muscle structure as related to function at the level of the gene. These amazing advances are certain to provide multifaceted interventions to maintain patient mobility and activities of daily living leading to significant savings in the health care budgets everywhere.

REFERENCES

Bae, J. S., Sawai, S., Misawa, S., Kanai, K., Isose, S., Shibuya, K., & Kuwabara, S. (2008). Effects of age on excitability properties in human motor axons. *Clinical Neurophysiology*, *119*(10), 2282–2286.

Beyer, I., Njemini, R., Bautmans, I., Demanet, C., Bergmann, P., & Mets, T. (2012). Inflammation-related muscle weakness and fatigue in geriatric patients. *Experimental Gerontology*, *47*(1), 52–59.

Bione, S., Maestrini, E., Rivella, S., Mancini, M., Regis, S., Romeo, G., & Toniolo, D. (1994). Identification of a novel X-linked gene responsible for Emery-Dreifuss muscular dystrophy. *Nature Genetics*, *8*(4), 323–327.

Bodine, S. C., Latres, E., Baumhueter, S., Lai, V. K., Nunez, L., Clarke, B. A., . . . Glass, D. J. (2001). Identification of ubiquitin ligases required for skeletal muscle atrophy. *Science*, *294*(5547), 1704–1708.

Bonne, G., Di Barletta, M. R., Varnous, S., Bécane, H. M., Hammouda, E. H., Merlini, L., . . . Schwartz, K. (1999). Mutations in the gene encoding lamin A/C cause autosomal dominant Emery-Dreifuss muscular dystrophy. *Nature Genetics*, *21*(3), 285–288.

Bonne, G., Mercuri, E., Muchir, A., Urtizberea, A., Bécane, H. M., Recan, D., . . . Muntoni, F. (2000). Clinical and molecular genetic spectrum of autosomal dominant Emery-Dreifuss muscular dystrophy due to mutations of the lamin A/C gene. *Annals of Neurology*, *48*(2), 170–180.

Capell, B. C., & Collins, F. S. (2006). Human laminopathies: Nuclei gone genetically awry. *Nature Reviews. Genetics*, *7*(12), 940–952.

Chaly, N., & Munro, S. B. (1996). Centromeres reposition to the nuclear periphery during L6E9 myogenesis in vitro. *Experimental Cell Research*, *223*(2), 274–278.

Chaly, N., Munro, S. B., & Swallow, M. A. (1996). Remodelling of the nuclear periphery during muscle cell differentiation in vitro. *Journal of Cellular Biochemistry*, *62*(1), 76–89.

Clements, L., Manilal, S., Love, D. R., & Morris, G. E. (2000). Direct interaction between emerin and lamin A. *Biochemical and Biophyical Research Communications*, *267*(3), 709–714.

Cohen, M., Lee, K. K., Wilson, K. L., & Gruenbaum, Y. (2001). Transcriptional repression, apoptosis, human disease and the functional evolution of the nuclear lamina. *Trends in Biochemical Science*, *26*(1), 41–47.

Dahiya, S., Givvimani, S., Bhatnagar, S., Qipshidze, N., Tyagi, S. C., & Kumar, A. (2011). Osteopontin-stimulated expression of matrix metalloproteinase-9 causes cardiomyopathy in the mdx model of Duchenne muscular dystrophy. *Journal of Immunology*, *187*(5), 2723–2731.

Dechat, T., Gesson, K., & Foisner, R. (2010). Lamina-independent lamins in the nuclear interior serve important functions. *Cold Spring Harbor Symposia on Quantitative Biology*, *75*, 533–543.

Doherty, T. J. (2003). Invited review: Aging and sarcopenia. *Journal of Applied Physiology*, *95*(4), 1717–1727.

Eddins, M. J., Marblestone, J. G., Suresh Kumar, K. G., Leach, C. A., Sterner, D. E., Mattern, M. R., & Nicholson, B. (2011). Targeting the ubiquitin E3 ligase MuRF1 to inhibit muscle atrophy. *Cell Biochemistry and Biophysics*, *60*(1–2), 113–118.

Elkina, Y., von Haehling, S., Anker, S. D., & Springer, J. (2011). The role of myostatin in muscle wasting: An overview. *Journal of Cachexia, Sarcopenia and Muscle*, *2*(3), 143–151.

Flanigan, K. M., Dunn, D., Larsen, C. A., Medne, L., Bönnemann, C. B., & Weiss, R. B. (2011). Becker muscular dystrophy due to an inversion of exons 23 and 24 of the DMD gene. *Muscle & Nerve*, *44*(5), 822–825.

Fling, B. W., Knight, C. A., & Kamen, G. (2009). Relationships between motor unit size and recruitment threshold in older adults: Implications for size principle. *Experimental Brain Research*, *197*(2), 125–133.

Fong, L. G., Ng, J. K., Lammerding, J., Vickers, T. A., Meta, M., Coté, N., . . . Young, S. G. (2006). Prelamin A and lamin A appear to be dispensable in the nuclear lamina. *The Journal of Clinical Investigation*, *116*(3), 743–752.

Furukawa, K. (1999). LAP2 binding protein 1 (L2BP1/BAF) is a candidate mediator of LAP2-chromatin interaction. *Journal of Cell Science*, *112*(Pt. 15), 2485–2492.

Furukawa, K., Fritze, C. E., & Gerace, L. (1998). The major nuclear envelope targeting domain of LAP2 coincides with its lamin binding region but is distinct from its chromatin interaction domain. *Journal of Biological Chemistry*, *273*(7), 4213–4219.

Garatachea, N., & Lucía, A. (2011). Genes and the ageing muscle: A review on genetic association studies. *Age*. http://dx.doi.org/10.1007/s11357-011-9327-0

Gonzalez, J. M., Pla, D., Perez-Sala, D., & Andres, V. (2011). A-type lamins and Hutchinson-Gilford progeria syndrome: Pathogenesis and therapy. *Frontiers in Bioscience*, *3*, 1133–1146.

Hai, R., Pei, Y. F., Shen, H., Zhang, L., Liu, X. G., Lin, Y., . . . Deng, H. W. (2011). Genome-wide association study of copy number variation identified gremlin1 as a candidate gene for lean body mass. *Journal of Human Genetics*, *57*(1), 33–37.

Hoffman, E. P., Bronson, A., Levin, A. A., Takeda, S., Yokota, T., Baudy, A. R., & Connor, E. M. (2011). Restoring dystrophin expression in duchenne muscular dystrophy muscle progress in exon skipping and stop codon read through. *The American Journal of Pathology*, *179*(1), 12–22.

Holt, I., Clements, L., Manilal, S., Brown, S. C., & Morris, G. E. (2001). The R482Q lamin A/C mutation that causes lipodystrophy does not prevent nuclear targeting of lamin A in adipocytes or its interaction with emerin. *European Journal of Human Genetics*, *9*(3), 204–208.

Hyder, C. L., Isoniemi, K. O., Torvaldson, E. S., & Eriksson, J. E. (2011). Insights into intermediate filament regulation from development to ageing. *Journal of Cell Science*, *124*(Pt. 9), 1363–1372.

Jagoe, R. T., & Goldberg, A. L. (2001). What do we really know about the ubiquitin-proteasome pathway in muscle atrophy? *Current Opinion in Clinical Nutrition and Metabolic Care*, *4*(3), 183–190.

Janssen, I. (2006). Influence of sarcopenia on the development of physical disability: The Cardiovascular Health Study. *Journal of the American Geriatrics Society*, *54*(1), 56–62.

Kasper, C. E. (2001). Functional loss: Aging muscle during rehabilitation. *Orthopaedic Physical Therapy Clinics of North America*, *10*(2), 291–302.

Kasper, C. E., Talbot, L. A., & Gaines, J. M. (2002). Skeletal muscle damage and recovery. *AACN Clinical Issues*, *13*(2), 237–247.

Kasper, C. E., & Xun, L. (1995). Cytoplasm-to-myonucleus ratios in atrophic skeletal muscle. *Medicine Science Sport & Exercise*, *27*(5), S121.

Lammerding, J., Kamm, R. D., & Lee, R. T. (2004). Mechanotransduction in cardiac myocytes. *Annals of the New York Academy of Science*, *1015*, 53–70.

Lammerding, J., Schulze, P. C., Takahashi, T., Kozlov, S., Sullivan, T., Kamm, R. D., . . . Lee, R. T. (2004). Lamin A/C deficiency causes defective nuclear mechanics and mechanotransduction. *The Journal of Clinical Investigation*, *113*(3), 370–378.

Lehner, C. F., Stick, R., Eppenberger, H. M., & Nigg, E. A. (1987). Differential expression of nuclear lamin proteins during chicken development. *Journal of Cell Biology*, *105*(1), 577–587.

Lieber, R. L., & Ward, S. R. (2011). Skeletal muscle design to meet functional demands. *Philosophical Transactions of the Royal Society of London. Series B, Biological Sciences*, *366*(1570), 1466–1476.

Lin, F., & Worman, H. J. (1993). Structural organization of the human gene encoding nuclear lamin A and nuclear lamin C. *Journal of Biological Chemistry*, *268*(22), 16321–16326.

Ling, S. M., Conwit, R. A., Ferrucci, L., & Metter, E. J. (2009). Age-associated changes in motor unit physiology: Observations from the Baltimore Longitudinal Study of Aging. *Archives of Physical Medicine and Rehabilitation*, *90*(7), 1237–1240.

Liu, J., Rolef Ben-Shahar, T. R., Riemer, D., Treinin, M., Spann, P., Weber, K., . . . Gruenbaum, Y. (2000). Essential roles for Caenorhabditis elegans lamin gene in nuclear organization, cell cycle progression, and spatial organization of nuclear pore complexes. *Molecular Biology of the Cell*, *11*(11), 3937–3947.

Lourim, D., & Lin, J. J. (1989a). Expression of nuclear lamin A and muscle-specific proteins in differentiating muscle cells in ovo and in vitro. *The Journal of Cell Biology*, *109*(2), 495–504.

Lourim, D., & Lin, J. J. (1989b). Expression of nuclear lamin A and muscle-specific proteins in differentiating muscle cells in ovo and in vitro. *The Journal of Cell Biology*, *109*(2), 495–504.

Maraldi, N. M., Mazzotti, G., Rana, R., Antonucci, A., Di Primio, R., & Guidotti, L. (2007). The nuclear envelope, human genetic diseases and ageing. *European Journal of Histochemistry*, *51*(Suppl. 1), 117–124.

Marji, J., O'Donoghue, S. I., McClintock, D., Satagopam, V. P., Schneider, R., Ratner, D., . . . Djabali, K. (2010). Defective lamin A-Rb signaling in Hutchinson-Gilford Progeria Syndrome and reversal by farnesyltransferase inhibition. *PLoS One*, *5*(6), e11132.

Moir, R. D., Spann, T. P., Lopez-Soler, R. I., Yoon, M., Goldman, A. E., Khuon, S., & Goldman, R. D. (2000). Review: The dynamics of the nuclear lamins during the cell cycle—Relationship between structure and function. *Journal of Structural Biology*, *129*(2–3), 324–334.

Nagano, A., & Arahata, K. (2000). Nuclear envelope proteins and associated diseases. *Current Opinion in Neurology*, *13*(5), 533–539.

Nakajima, N., & Abe, K. (1995). Genomic structure of the mouse A-type lamin gene locus encoding somatic and germ cell-specific lamins. *FEBS Letters*, *365*(2–3), 108–114.

Nichols, F. (2011). *Online Mendelian Inheritance in Animals (OMIA)*. Retrieved from http://www.ncbi.nlm.nih.gov/sites/entrez?db=omia&tool=toolbar

OMIM: Online Mendilian Inheritance in Man. (2011). Retrieved from http://www.omim.org/

Pattison, J. S., Folk, L. C., Madsen, R. W., Childs, T. E., & Booth, F. W. (2003). Transcriptional profiling identifies extensive downregulation of extracellular matrix gene expression in sarcopenic rat soleus muscle. *Physiological Genomics*, *15*(1), 34–43.

Payne, G. W. (2006). Effect of inflammation on the aging microcirculation: Impact on skeletal muscle blood flow control. *Microcirculation*, *13*(4), 343–352.

Prokocimer, M., Davidovich, M., Nissim-Rafinia, M., Wiesel-Motiuk, N., Bar, D. Z., Barkan, R., . . . Gruenbaum, Y. (2009). Nuclear lamins: Key regulators of nuclear structure and activities. *Journal of Cellular and Molecular Medicine*, *13*(6), 1059–1085.

Punnoose, A. R., Burke, A. E., & Golub, R. M. (2011). JAMA patient page. Muscular dystrophy. *JAMA: The Journal of the American Medical Association*, *306*(22), 2526.

Raffaele Di Barletta, M., Ricci, E., Galluzzi, G., Tonali, P., Mora, M., Morandi, L., . . . Toniolo, D. (2000). Different mutations in the LMNA gene cause autosomal dominant and autosomal recessive Emery-Dreifuss muscular dystrophy. *American Journal of Human Genetics*, *66*(4), 1407–1412.

Reddy, S., & Comai, L. (2012). Lamin A, farnesylation and aging. *Experimental Cell Research*, *318*(1), 1–7.

Rodríguez, S., & Eriksson, M. (2010). Evidence for the involvement of lamins in aging. *Current Aging Science*, *3*(2), 81–89.

Roubenoff, R. (2000). Sarcopenic obesity: Does muscle loss cause fat gain? Lessons from rheumatoid arthritis and osteoarthritis. *Annals of the New York Academy of Science*, *904*, 553–557.

Roubenoff, R. (2004). Sarcopenic obesity: The confluence of two epidemics. *Obesity Research*, *12*(6), 887–888.

Sandri, M. (2008). Signaling in muscle atrophy and hypertrophy. *Physiology (Bethesda), 23*, 160–170.

Sacheck, J. M., Hyatt, J. P., Raffaello, A., Jagoe, R. T., Roy, R. R., Edgerton, V. R., . . . Goldberg, A. L. (2007). Rapid disuse and denervation atrophy involve transcriptional changes similar to those of muscle wasting during systemic diseases. *FASEB Journal, 21*(1), 140–155.

Sermet-Gaudelus, I., Souberbielle, J. C., Azhar, I., Ruiz, J. C., Magnine, P., Colomb, V., . . . Lenoir, G. (2003). Insulin-like growth factor I correlates with lean body mass in cystic fibrosis patients. *Archives of Disease in Childhood, 88*(11), 956–961.

Shumaker, D. K., Lee, K. K., Tanhehco, Y. C., Craigie, R., & Wilson, K. L. (2001). LAP2 binds to BAF.DNA complexes: Requirement for the LEM domain and modulation by variable regions. *EMBO Journal, 20*(7), 1754–1764.

Sinensky, M., Fantle, K., Trujillo, M., McLain, T., Kupfer, A., & Dalton, M. (1994). The processing pathway of prelamin A. *Journal of Cell Science, 107*(Pt. 1), 61–67.

Spangenburg, E. E., Abraha, T., Childs, T. E., Pattison, J. S., & Booth, F. W. (2003). Skeletal muscle IGF-binding protein-3 and -5 expressions are age, muscle, and load dependent. *American Journal of Physiology. Endocrinology and Metabolism, 284*(2), E340–E350.

Stuurman, N., Heins, S., & Aebi, U. (1998). Nuclear lamins: Their structure, assembly, and interactions. *Journal of Structural Biology, 122*(1–2), 42–66.

Suetta, C., Hvid, L. G., Justesen, L., Christensen, U., Neergaard, K., Simonsen, L., . . . Aagaard, P. (2009). Effects of aging on human skeletal muscle after immobilization and retraining. *Journal of Applied Physiology, 107*(4), 1172–1180.

Sullivan, T., Escalante-Alcalde, D., Bhatt, H., Anver, M., Bhat, N., Nagashima, K., . . . Burke, B. (1999). Loss of A-type lamin expression compromises nuclear envelope integrity leading to muscular dystrophy. *Journal of Cell Biology, 147*(5), 913–920.

Szulc, P., Beck, T. J., Marchand, F., & Delmas, P. D. (2005). Low skeletal muscle mass is associated with poor structural parameters of bone and impaired balance in elderly men—the MINOS study. *Journal of Bone and Mineral Research: The Official Journal of the American Society for Bone and Mineral Research, 20*(5), 721–729.

Taimen, P., Pfleghaar, K., Shimi, T., Möller, D., Ben-Harush, K., Erdos, M. R., . . . Goldman, R. D. (2009). A progeria mutation reveals functions for lamin A in nuclear assembly, architecture, and chromosome organization. *Proceedings of the National Academy of Sciences of the United States of America, 106*(49), 20788–20793.

Tan, L. J., Liu, S. L., Lei, S. F., Papasian, C. J., & Deng, H. W. (2012). Molecular genetic studies of gene identification for sarcopenia. *Human Genetics, 131*(1), 1–31. http://dx.doi.org/10.1007/s00439-011-1040-7

Tseng, B. S., Kasper, C. E., & Edgerton, V. R. (1994). Cytoplasm-to-myonucleus ratios and succinate dehydrogenase activities in adult rat slow and fast muscle fibers. *Cell and Tissue Research, 275*(1), 39–49.

Vlcek, S., & Foisner, R. (2007). Lamins and lamin-associated proteins in aging and disease. *Current Opinion in Cell Biology, 19*(3), 298–304.

Watson, P. A., Stein, J. P., & Booth, F. W. (1984). Changes in actin synthesis and alpha-actin-mRNA content in rat muscle during immobilization. *The American Journal of Physiology, 247*(1, Pt. 1), C39–C44.

Wedrychowski, A., Bhorjee, J. S., & Briggs, R. C. (1989). In vivo crosslinking of nuclear proteins to DNA by cis-diamminedichloroplatinum (II) in differentiating rat myoblasts. *Experimental Cell Research, 183*(2), 376–387.

White, R. B., Biérinx, A. S., Gnocchi, V. F., & Zammit, P. S. (2010). Dynamics of muscle fibre growth during postnatal mouse development. *BMC Developmental Biology, 10*, 21.

Williams, R. S. (1986). Mitochondrial gene expression in mammalian striated muscle. Evidence that variation in gene dosage is the major regulatory event. *The Journal of Biological Chemistry*, *261*(26), 12390–12394.

Williams, R. S., Salmons, S., Newsholme, E. A., Kaufman, R. E., & Mellor, J. (1986). Regulation of nuclear and mitochondrial gene expression by contractile activity in skeletal muscle. *The Journal of Biological Chemistry*, *261*(1), 376–380.

Wilson, K. L., Benavente, R., Burke, B., Craigie, R., Foisner, R., Furukawa, K., . . . Worman, H. J. (2001). Problems with LAP nomenclature. *Nature Cell Biology*, *3*(4), E90.

Wilson, K. L., Holaska, J. M., Montes de Oca, R., Tifft, K., Zastrow, M., Segura-Totten, M., . . . Bengtsson, L. (2005). Nuclear membrane protein emerin: Roles in gene regulation, actin dynamics and human disease. *Novartis Found Symposium*, *264*, 51–58, discussion 58–62, 227–230.

Wilson, K. L., Zastrow, M. S., & Lee, K. K. (2001). Lamins and disease: Insights into nuclear infrastructure. *Cell*, *104*(5), 647–650.

Worman, H. J., Ostlund, C., & Wang, Y. (2010). Diseases of the nuclear envelope. *Cold Spring Harbor Perspectives in Biology*, *2*(2), a000760.

Zheng, R., Ghirlando, R., Lee, M. S., Mizuuchi, K., Krause, M., & Craigie, R. (2000). Barrier-to-autointegration factor (BAF) bridges DNA in a discrete, higher-order nucleoprotein complex. *Proceedings of the National Academy of Sciences of the United States of America*, *97*(16), 8997–9002.

CHAPTER 11

Central Nervous System Genomics

Matthew J. Gallek and Leslie Ritter

ABSTRACT

In the past 25 years, remarkable progress has been made in our understanding of genomics and its influence on central nervous system diseases. In this chapter, common diseases of the central nervous system will be reviewed along with the genomics associated with these diseases. The diseases/injuries that will be investigated include neurovascular disorders such as ischemic stroke, hemorrhagic stroke, subarachnoid hemorrhage, and traumatic brain injury. This chapter will also explore Apolipoprotein E (APOE), a 299-aminoacid protein encoded by the APOE gene, and its associations with many of the previously named diseases. APOE was first tied to the risk of Alzheimer's disease and has since then been investigated in traumatic brain injury and hemorrhagic strokes. In addition, we will discuss the future of genomic research in central nervous system diseases.

INTRODUCTION

Remarkable progress has been made in our understanding of genomics and its influence on central nervous system (CNS) disease over the last several decades. As such, this review will assist nurse practitioners and scientists synthesize information related to genomics and selected central nervous system disease. First, a brief overview of the common methodological approaches to genomics and CNS disease, including genome-wide association studies (GWAS), gene expression

http://dx.doi.org/10.1891/0739-6686.29.205

studies, and candidate gene studies, will be discussed. In the context of these approaches, selected CNS pathologies, specifically, ischemic stroke, hemorrhagic stroke, subarachnoid hemorrhage, and traumatic brain injury, will be reviewed. The authors will also specifically discuss the role of Apolipoprotein E (APOE), the 299-aminoacid protein encoded by the APOE gene that was first identified in Alzheimer's disease (Strittmatter et al., 1993), and its associations with the CNS diseases. APOE is believed to be involved with nerve growth (Pitas, Boyles, Lee, Foss, & Mahley, 1987) and nerve regeneration (Boyles et al., 1989) within the brain. There are three major isoforms of APOE (E2, E3, and E4). These isoforms arise from a single APOE genetic locus. There are three known homozygous phenotypes (E2/E2, E3/E3, and E4/E4) and three known heterozygous phenotypes (E2/E3, E2/E4, and E3/E4). All studies reviewed will be limited to human adult research, which have resulted in positive findings. Last, the authors will discuss the future of genomic research in central nervous system disease.

COMMON METHODOLOGICAL APPROACHES

A brief overview of three common methodological approaches to genetics and disease exploration will be presented. One methodological approach is the genome-wide association study, or GWAS. This approach uses two groups of subjects, a disease group and a control group (or disease-free group). GWAS investigations seek to identify single nucleotide polymorphisms (SNP) that are associated with the disease group. The benefit of this approach is that the whole genome is scanned and all genes are investigated. This approach is also hypothesis-neutral, meaning that the investigators do not choose the genes of interest and therefore avoid selection bias. Finally, GWAS is a theory-generating design. It brings to light the genes that are associated with a disease and allows investigators to focus on these genes in further studies.

Another common methodological approach is the genome-wide gene expression study. Using this approach, ribonucleic acid (RNA) is collected from subjects who have the disease and from subjects without the disease. The quantity of RNA produced between groups (disease and disease-free) is compared. It is important to note that although inferences are often made about protein levels relative to the upregulation and downregulation of genes, routine measurement of protein levels is not always conducted in gene expression studies. In the absence of protein level measurements, inferences regarding the effect of changes in protein gene expression on changes in the mature protein cannot be made. However, similar to the GWAS investigations, strengths of this approach are that the whole genome is investigated and it is hypothesis-neutral and theory-generating.

The last approach presented here is the candidate gene study. This approach also uses two groups, a disease group and a control group. Unlike the genome-wide gene expression study approach, in candidate gene studies, investigators choose what genes are to be investigated on the basis of their research interests. Thus, the candidate gene study is not hypothesis-neutral and may be subjected to selection bias. The candidate gene study is currently the most commonly used methodological approach for genetic research.

Last, commonly used indexes in genetic studies; the odds ratio (*OR*), relative risk (RR), and confidence intervals (CI), are worth mentioning. The odds ratio is the ratio of two probabilities: the probability of an event/disease occurring divided by the probability that it will not occur (Klienbaum, Kupper, Muller, & Nizam, 1998). An odds ratio of 1 means that there is not a difference in the probability of the disease occurring compared to the disease not occurring. An *OR* of 3 refers to the odds relating to disease occurrence: 3-to-1. On the other hand, an odds ratio of 0.33 clarifies an odds of 3-to-1 that a disease-free state will occur. The RR (also known as the risk ratio) represents the estimated proportion of the original risk of an adverse outcome that persists when people are exposed to the gene. Both the odds ratio and the relative risk compare the relative likelihood of an event occurring between two distinct groups. In simple terms, odds ratio will report the odds of contracting the disease, whereas the risk ratio will report the probability of the disease occurring. Most of the articles reviewed for this chapter use odds ratio. Confidence intervals are the range of values within which a population parameter is estimated to lie, at a specified probability (e.g., 95% CI; Polit & Beck, 2012). In the section that follows, GWAS, genome-wide gene association studies, and candidate gene studies in ischemic stroke, hemorrhagic stroke, subarachnoid hemorrhage, and traumatic brain injury will be reviewed.

ISCHEMIC STROKE

Stroke remains a leading cause of death and adult disability. Each year, approximately 795,000 people experience a new or recurrent stroke. Approximately 610,000 are first time strokes, and 185,000 are recurrent strokes. Ischemic stroke accounts for 87% of all reported strokes (Roger et al., 2011). Stroke subtypes include large-vessel disease, arteriosclerotic stroke, small vessel disease, cardioembolic stroke, and cryptogenic stroke (Adams et al., 1993). In the studies explored below the subtype of ischemic stroke will be identified if the information was available.

Genome-Wide Association Studies

Because of the similarities between coronary artery disease and ischemic stroke, investigations of a locus on chromosome 9p21 as a risk factor for ischemic stroke followed the many studies of the same gene in coronary artery disease (Helgadottir et al., 2007; McPherson et al., 2007; Samani et al., 2007; Wellcome Trust Case Control Consortium, 2007). Of note are two recent GWAS investigations reporting an association between 9p21 and the risk of large vessel disease (LVD) subtype of ischemic stroke ischemic stroke (Gschwendtner et al., 2009; Olsson, Jood, Blomstrand, & Jern, 2011). The Gschwendetner et al. study included 4,376 cases and 4,305 controls from six different centers across Europe and North America. The SNP associated with large vessel disease ischemic stroke in this study was rs1537378 with an *OR* of 1.20 (95%; CI = 1.07 to 1.34; $p = .0011$) when controlling for age, sex, ethnicity, CAD, and MI (Gschwendtner et al., 2009). In contrast, Olsson et al. (2011) reported that SNP rs7857345 was associated with large vessel disease ischemic stroke with an *OR* of 0.56 (95%; CI = 0.39–0.86; $p > .01$) when controlling for age, sex, hypertension, diabetes mellitus, and smoking. In this study, there were 668 controls and 111 subjects with large vessel disease, all of which were of Swedish decent. The Olson study also reported minor alleles of both SNPs rs7857345 and rs1537378 are associated with a decreased risk of death or dependency after controlling for age and sex (*OR* = 0.25, 95%; CI = 0.06–0.78; $p = .01$ and *OR* = 0.29, 95%; CI = 0.09–0.74; $p = .005$ respectively; Olsson et al., 2011). The SNP of importance in the Gschwendetner study, rs1537378, is found in a different linkage disequilibrium block than the SNP of importance in the Olson study, rs7857345, as defined in the HapMap (http://HapMap.org) CEU population. The results of these two studies indicate that the two SNPs are not linked and "travel separately" during recombination. Therefore, it may be that these findings support the notion that the SNPs represent the same signal or represent different signals that give similar risk effects.

Gretarsdottir and associates (2008) investigated locus 4q25 using the GWAS approach. They genotyped 1,661 Icelandic ischemic stroke subjects and 10,815 control subjects. A total of 310,881 SNPs were tested for association to ischemic stroke. Two SNPs, rs2200733 and rs10033464 were explored further in a European sample with 2,327 ischemic stroke subjects and 16,760 control subjects. An association was reported between rs2200733-T with ischemic stroke in general (*OR* = 1.26, 95%; CI = 1.17–1.35; $p = 2.18 \times 10^{-10}$), with the cardioembolism subtype of ischemic stroke (*OR* = 1.52, 95%; CI = 1.35–1.71; $p = 5.82 \times 10^{-12}$), and with large artery atherosclerosis subtype of ischemic stroke (*OR* = 1.23, 95%; CI = 1.09–1.39; $p = .0011$; Gretarsdottir et al., 2008). Other studies report SNP (rs2200733) was associated with atrial fibrillation (Lubitz et al., 2010). Because of the shared risk factors for atrial fibrillation and cardioembolic stroke, it is not surprising that this SNP is associated with ischemic stroke. In the Gretarsdottir et al. study, along with the aforementioned SNP, rs10033464-T has been associated with cardioembolic stroke (*OR* = 1.27, 95%; CI = 1.11–1.45; $p = .00061$; Gretarsdottir et al., 2008). In a more recent study, Lemmens and associates (2010) also found an association between the 4q25 locus and cardioembolic stroke. This study had 4,199 subjects with ischemic stroke and 3,750 control subjects. Another SNP, rs1906591-A, was associated with cardioembolic stroke (*OR* = 1.48, 95%; CI = 1.28–1.71; $p = 1.2 \times 10^{-7}$). However, unlike the Gretarsdottir et al. study, no association was found between rs10033464-T and cardioembolic stroke (Lemmens et al., 2010).

Additional genetic risk factors for atrial fibrillation and ischemic stroke were reported for a sequence variant in the zinc finger homeobox 3 gene (*ZFHX3*). This gene is located on 16q22 and is responsible for the regulation of myogenic and neuronal differentiation. In a population of European decent with 39,898 control subjects, 6,235 ischemic stroke subjects, and 1,454 cardioembolic stroke subjects, SNP rs7193343-T was associated with ischemic stroke (*OR* = 1.11, 95%; CI = 1.04–1.17; $p = .00054$) and cardioembolic stroke (*OR* = 1.22, 95%; CI = 1.10–1.35; $p = .00021$; Gudbjartsson et al., 2009).

Kubo and associates (2007) investigated the protein kinase C eta gene (PRKCH) in two independent Japanese populations. The SNP 1425G/A was found to be associated with lacunar (deep brain) infarction with an *OR* of 1.40 (95%; CI = 1.12–1.59; $p = 10^{-7}$) in the combined analysis. Additionally, this study found that PRKCH was expressed mainly in vascular endothelial cells and foamy macrophages in human atherosclerotic lesions, and its expression increased as the lesion type progressed. This gene was also associated with ischemic stroke in a Chinese population (1,209 ischemic stroke subjects and 1,174 controls subjects) with and *OR* of 1.31 (95%; CI = 1.08–1.60; $p = .0058$; Wu et al., 2009). Protein kinase C eta is a serine-threonine kinase. Protein kinase C eta is involved in the development and progression of atherosclerosis in humans. Pathological findings showed that protein kinase C eta was abundantly expressed in foamy macrophages that are essential in all phases of atherosclerosis. Macrophages contribute to the uptake of lipoproteins, release of reactive oxygen species, and immune mediators that have important roles in the development of atherosclerosis (Kubo et al., 2007).

The angiotensin receptor-like 1 (AGTRL1) gene, SNP rs9943582-G, was associated with brain infarction (*OR* = 1.30, 95%; CI = 1.14–1.47; $p = .000066$) in a study conducted by Hata et al. (2007). This study included 1,112 brain infarction subjects and age/sex matched control subjects. This SNP is located in the 5′-flanking region of the gene. Angiotensin receptor-like 1 is important in the control of blood pressure and volume in the cardiovascular system.

Gene Expression Studies

Peripheral whole blood gene expression profiling has also been commonly used in the exploration of ischemic stroke and genetics. A sample of 39 MRI-diagnosed patients with acute ischemic cerebrovascular syndrome (AICS) was compared to 25 non-stroke control subjects in a study conducted by Barr et al. (2010). Age, hypertension, and dyslipidemia were controlled for in the analysis. Nine genes were identified with a greater than 2-fold difference in expression following ischemic stroke: genes that were up-regulated included Arginase 1 (*ARG1*), carbonic anhydrase 4 (*CA4*), chondroitin sulfate proteoglycan 2 (CSPG2), IG motif-containing GT-Pase activation protein 1 (IQGAP1), lymphocyte antigen 96 (LY96), matrix metalloproteinase 9 (MMP9), orosomucoid 1 (ORM1), s100 calcium binding protein A12 (s100A12). The chemokine receptor7 (CCR7) was the only gene found to be down-regulated. Five of the 9 genes identified in this study were also found to be differently expressed in an earlier study by Tang and associates (2006). These five genes are ARG1, CA4, LY96, MMP9, and s100A12.

Candidate Gene Studies

Using candidate gene studies, numerous genes have been reported to be associated with ischemic stroke. Identification of these genes has been replicated in numerous studies. Genes associated most frequently are Factor V Leiden, angiotensinogen, interleukin-6, matrix metalloprotieinase-3, platelet glycoprotein IIb/IIIa, angiotensin-converting enzyme, apolipoprotein E, paraoxonase-1, 5, 10-Methylenetetrahydrolfolate reductase. Table 11.1 summarizes the genes that have been discovered to be associated with ischemic stroke. The authors attempted to find the earliest positive finding for each gene listed in Table 11.1.

APOE was studied relatively early (1999) in the ischemic stroke risk literature. Since then numerous studies have been released investigating the APOE gene. There are numerous studies that support the association of APOE genotypes with a higher risk of ischemic stroke (Chowdhury et al., 2001; Peng, Zhao, & Wang, 1999). Of note, there are also multiple studies that reported no association between APOE genotypes and ischemic stroke risk (Coria et al., 1995; Kessler et al., 1997; Nakata et al., 1997). These differences in study findings are most likely because of the small sample sizes that result in an underpowered study.

One of the earliest studies to investigate the influence of genetics on outcomes following ischemic stroke was completed by McCarron et al. in 1998. In this study, APOE genotyping was performed on 640 ischemic stroke subjects. Results indicated an improved survival with the presence of the APOE $\varepsilon 4$ allele present (relative hazard = 0.76 per allele; $p = .04$). Another study performed by

TABLE 11.1
Candidate Gene Studies Reporting Significant Associations With Stroke Risk

Gene	Polymorphism	RR/*OR* (95% CI)	Reference
APOE	APOE 2, 3, 4	1.68 (1.36–2.09)	(McCarron, Delong, & Alberts, 1999)
PON1	Q192R	3.6 (1.4–9.3)	(Voetsch, Benke, Damasceno, Siqueira, & Loscalzo, 2002)
Platelet glycoprotein llb/llla	PLA2	2.9 (1.6–4.9)	(Szolnoki et al., 2003)
SELP	val640leu	1.63 (1.22–2.17)	(Zee et al., 2004)
IL4	C582T	1.40 (1.13–1.73)	(Zee et al., 2004)
MMP-3	5A6A	2.4 (1.3–4.5)	(Flex et al., 2004)
IL-6	G480C	8.1 (4.0–16.4)	(Flex et al., 2004)
Factor 5 Leiden	ARG506GLN	1.33 (1.12–1.58)	(Casas, Hingorani, Bautista, & Sharma, 2004)
MTHFR	C677T	1.24 (1.08–1.42)	(Casas et al., 2004)
Prothrombin	G20210A	1.44 (1.11–1.86)	(Casas et al., 2004)
ACE	Insertion/ Deletion	1.21 (1.08–1.35)	(Casas et al., 2004)
ALOX5AP	Haplotype A	1.67 $p = .000095$[a]	(Helgadottir et al., 2004)
EPHX2	Haplotype	—[b]	(Fornage et al., 2005)
NOS3	glu298asp (rs1799983)	2.39 (1.23–4.63)	(Berger et al., 2007)
ICAM in Caucasian	G241R	2.10 (1.01–4.68)	(Volcik, Ballantyne, Hoogeveen, Folsom, & Boerwinkle, 2010)
ICAM in African Americans	G241R	7.04 (3.72–13.3)	(Volcik et al., 2010)

Note. RR = relative risk; *OR* = odds ratio; CI = confidence intervals (95%).
[a]no CI available for reference; [b]no estimates of RR/*OR* available for reference.

Carter et al. (1999) investigated platelet glycoprotein genotypes and mortality rates in 515 patients with ischemic stroke. Those with aa or ab genotype of the platelet glycoprotein IIb HPA-3 polymorphism had worse mortality when compared to those with the bb genotype. The relative risks for mortality was 2.42 (95%; CI = 1.24–4.71) for the aa genotype and 2.13 (95%; CI = 1.09–4.17) for the ab genotype.

Greisenegger et al. (2003) investigated the interleukin-6 gene in 214 subjects and outcomes following ischemic stroke. They found that subjects with severe disability at 3 months post insult were more often carriers of the 174 GG genotype. Meloperoxidase may also affect outcome after stroke. Hoy et al. (2003) studied 450 subjects and discovered that the A allele of G463A polymorphism was associated with poorer outcomes measured by the Rankin score. Because there are numerous small studies reporting associations between genes and functional outcomes after stroke, but as yet, no large prospective study to investigate the same, the evidence for any such associations remains heterogeneous, if not confusing.

INTRACEREBRAL HEMORRHAGE

Intracerebral hemorrhage (ICH) accounts for 10% of all strokes, with nearly 79,500 ICHs occurring each year (Roger et al., 2011). Like ischemic stroke, spontaneous hemorrhagic stroke is classified into a number of subtypes. ICH is classified according to the region of the brain in which it occurs. Hemorrhage in the deep structures of the brain, such as the thalamus or brain stem, are considered non-lobar hemorrhage. Lobar hemorrhages are located at the junction of the cortical gray matter and subcortical white matter.

The major risk factor for ICH is hypertension. In general, hypertension as a cause of ICH is more common in nonlobar ICH. Cerebral amyloid angiopathy (CAA) is also associated with risk of ICH (Knudsen, Rosand, Karluk, & Greenberg, 2001). CAA refers to the deposition of protein β-amyloid on the walls of the arteries in the brain and is the main cause of lobar hemorrhage (Knudsen et al., 2001).

GENOME-WIDE ASSOCIATION STUDIES

To date, there are no published genome-wide association studies (GWAS) investigations that have examined genetic influences on ICH. In addition, no genome-wide gene expression studies have been published in this area. The International Stroke Genetics Consortium continues to assemble large-scale multi-center collaborative efforts to assemble large sample sizes for both GWAS and gene expression studies (http://www.strokegenetics.com). As these finding become available, novel genetic risk factors for ICH will be elucidated.

Candidate Gene Studies

Most of genetic information about ICH has been investigated using candidate gene studies. These studies have focused on genes that influence vascular reactivity, inflammation, lipid binding molecules, and clotting factors. In Table 11.2 is a list of genes associated with different subtypes of ICH. The genes described in Table 11.2 were studied in numerous populations and have not necessarily been replicated since their publication. These studies need to be repeated with large cohorts to confirm that they are truly associated with ICH (see Table 11.2).

Once again APOE has been studied for over 10 years in the genetic research of ICH. Not only is APOE implicated with the risk of numerous types of ICH, it is also implicated in outcomes following ICH. A meta-analysis suggests that those with the APOE ε4 genotype are at an increased risk of poor outcome (death or dependency, or death alone) several months after ICH (Martínez-González & Sudlow, 2006).

Outcome studies in this area focus mainly on hematoma size and amount of cerebral edema. However, Gomez-Sanchez et al. (2011) investigated the Tp53 Arg72Pro polymorphism in 176 subjects. They found that subjects with the Arg/Arg genotype were more likely to have poor outcomes as measured by the modified Rankin scale. As with ischemic stroke, large prospective studies are needed to find genes that truly influence outcome following ICH.

SUBARACHNOID HEMORRHAGE

Subarachnoid hemorrhage accounts for 3% of all strokes or approximately 24,000 cases each year (Roger et al., 2011). This type of stroke mostly occurs in women at an average age of 55 years old (Sobey & Faraci, 1998). Ninety-five percent of cases of spontaneous subarachnoid hemorrhage result from ruptured cerebral aneurysms (Sobey & Faraci, 1998). A major complication of aneurymal subarachnoid hemorrhage (aSAH) is cerebral vasospasm. Cerebral vasospasm is the constriction of cerebral vessels that results in diminished blood flow and subsequent cerebral ischemia. Cerebral vasospasm is the most common cause of mortality and morbidity after aSAH (Sobey & Faraci, 1998). There have been numerous candidate gene studies investigating the genetics implicated in the development and rupture of intracranial aneurysms. These findings have been reviewed and summarized by Ruigrok and Rinkel (2008) and Nahed et al. (2007). Most of these studies are related to structural proteins of the extracellular matrix. There appears to be a deficiency in this matrix that may be the cause of many aneurysms. The focus here is cerebral vasospasm and delayed cerebral ischemia following aneurysm rupture. To date, no genome-wide association studies or genome-wide gene association studies have been published investigating cerebral vasospasm following aSAH.

TABLE 11.2
Candidate Gene Studies Reporting Of Significant Associations With Hemorrhagic Stroke

Gene	Variant	RR/*OR* (95% CI)	Reference
ICH			
ACE	DD (In/Del intron 16)	2.13 (1.10–4.14)	(Slowik et al., 2004)
Apo H	(341) G/A (Ser88ASN)	—[b]	(Xia et al., 2004)
APOE	ε2	1.32 (1.01–1.74)	(Sudlow et al., 2006)
	ε4	1.16 (0.93–1.44)	
β-1-tubulin (Tubb1)	Q43P	2.78 (1.16–6.63)	(Navarro-Núñez et al., 2007)
CD-14	C/T	1.62 (1.17–2.29)	(Yamada, 2006)
DDAH2	c449G (rs805305	0.51 (0.38–0.68)	(Bai et al., 2009)
Endoglin	6-bp intronic insertion between exons 7 and 8	4.76 (1.28–21.6)	(Alberts et al., 1997)
ESR-1	(c.454–397) T/T	2.31 (1.16–4.60)	(Strand et al., 2007)
Factor VII	(323) 10-bp In/Del	1.54 (1.03–2.72)	(Corral, Iniesta, González-Conejero, Villalón, & Vicente, 2001)
Factor XII	(G/T) Val 34 Leu	3.79[a]	(Catto et al., 1998)
FBN1	T/C	1.47 (1.09–1.99)	(Yamada, 2006)
IL-6	(572) G/C	1.57 (1.21–2.07)	(Yamada et al., 2006)
LIPC	G/A	1.43 (1.04–2.01)	(Yamada, 2006)
PECAM1	C/G	1.49 (1.08–2.09)	(Yamada, 2006)
PRKCH	G1425A	1.94 (1.21–3.10)	(Wu et al., 2009)
TNF-α in men	1031C	1.9 (1.1–3.4)	(Chen et al., 2010)
	308A	2.6 (1.3–5.3)	(Chen et al., 2010)
TNF-α in women	863A	0.5 (0.2–0.9)	(Chen et al., 2010)
VKORC1	(2255) T/C	1.68 (1.29–2.20)	(Wang et al., 2006)

(Continued)

TABLE 11.2
Candidate Gene Studies Reporting Of Significant Associations With Hemorrhagic Stroke (Continued)

Gene	Variant	RR/*OR* (95% CI)	Reference
Lobar ICH			
APOE	ε2	1.8 (0.8–3.7)	(Woo, et al., 2002)
	ε4	1.7 (0.9–3.2)	
Recurrent Lobar ICH			
APOE	ε2	4.7 (1.4–15.9)	(O'Donnell et al., 2000)
	ε4	3.7 (1.1–11.7)	
Warfarin-related ICH			
APOE	ε2	3.8 (1.0–14.6)	(Rosand, Hylek, O'Donnell, & Greenberg, 2000)
Sporadic CAA			
APOE	ε2	(p = .003)[b]	(Nicoll et al., 1997)
	ε4	2.9 (1.1–7.4)	(Greenberg, Rebeck, Vonsattel, Gomez-Isla, & Hyman, 1995)
NEP	GT repeat	(p = .005)[b]	(Yamada et al., 2003)
PS-1	1/2 polymorphism (intron 8)	(p = .013)[b]	(Yamada et al., 1997)
TGF-β1	T/C (exon 1)	(p = .003)[b]	(Hamaguchi et al., 2005)
Hemorrhagic stroke			
α-1-antichymotrypsin	A/T	2.8 (1.19–6.58)	(Obach, Revilla, Vila, Cervera, & Chamorro, 2001)
APO (a)	TTTTA repeat (PNTR)	1.24 (0.91–1.68)	(Sun et al., 2003)

Note. RR = relative risk; *OR* = odds ratio; CI = confidence intervals (95%)
[a]no CI available for reference; [b]no estimates of RR/*OR* available for reference.

Candidate Gene Studies

Once again, most of the work in the area of on cerebral vasospasm and delayed cerebral ischemia following aneurysm rupture has used the candidate gene study approach. Genes that are suspected of being involved with cerebral vasospasm and delayed cerebral ischemia fall into categories of pathophysiologic processes such as inflammation, vascular reactivity, inflammatory genes, coagulation (fibrinolysis), and neuronal repair genes. There are very few studies that have reported an association between genes and cerebral vasospasm, delayed cerebral ischemia, and delayed ischemic neurologic deficit (Table 11.3). The lack of studies with larger sample sizes likely reflects the fact that aSAH is a relatively rare occurrence (see Table 11.3).

Compared to the ICH literature, there is a larger number studies examining genes and outcomes following subarachnoid hemorrhage. Once again, there is a preponderance of studies examining the association of APOE and aSAH. In 2002, Leung et al. (2002) investigated 72 subjects with aSAH. APOE ε4 was associated with unfavorable outcomes at 6 months (*OR* = 6.0, 95%; CI = 1.7–21.3). These findings were replicated in later studies (Gallek et al., 2009; Tang et al.,

TABLE 11.3

Genes Associated With Cerebral Vasospasm, Delayed Cerebral Ischemia, or Delayed Ischemic Neurologic Deficit

Gene	Variant	RR/*OR* (95% CI)	Reference
Cerebral Vasospasm			
Endothelial nitric oxide synthase	T786C	—[a]	(Khurana et al., 2004)
Haptoglobin	α2	—[a]	(Borsody, Burke, Coplin, Miller-Lotan, & Levy, 2006)
Delayed Cerebral Ischemia			
Plasminogen activator inhibitor-1 (PAI-1)	4G allele in the 4G/5G promoter	3.3 (1.1–10.0)	(Vergouwen et al., 2004)
Delayed Ischemic Neurologic Deficit			
APOE	ε4	9.35 (2.05–42.66)	(Lanterna et al., 2005)

Note. RR = relative risk; *OR* = odds ratio; CI = confidence intervals (95%).
[a]no estimates of RR/*OR* available for reference.

2003). Similar to the findings in other CNS diseases discussed, there is a body of work in aSAH that demonstrates no association between APOE and outcomes following aSAH (Fontanella et al., 2007; Juvela, Siironen, & Lappalainen, 2009). Other genes that are associated with outcomes following aSAH are interleukin-1 (Fontanella et al., 2010), endothelial nitric oxide synthase (Alexander et al., 2009), and brain derived neurotrophic factor (Siironen et al., 2007).

TRAUMATIC BRAIN INJURY

It is estimated that there is 1.7 million cases of traumatic brain injury (TBI) each year (Faul, Xu, Wald, & Coronado, 2010). Of this 1.7 million, 52,000 die, 275,000 are hospitalized, and 1.365 million are treated and released from the emergency department. TBI accounts for one third of all injury-related deaths (Faul et al., 2010). Seventy-five percent of TBIs are concussions or other forms of mild TBI ("Centers for Disease Control and Prevention, National Center for Injury Prevention and Control. Report to Congress on mild traumatic brain injury in the United States: Steps to prevent a serious public health problem," 2003). TBI is more common in males than females. People are at the highest risk of TBI between the ages of 0 to 4, 15 to 19, and greater than 65. Adults aged 75 years and older have the highest rate of TBI related hospitalizations and death. This chapter will focus on moderate to severe TBI.

Gene Expression

Michael et al. (2005) used microarray techniques that had the capacity to measure gene expression of 5,000 genes in brain tissue from 6 subjects, 4 TBI subjects, 1 subject with vasculitis, and 1 subject with normal brain tissue that were removed during craniotomy for meningioma. Of these 5,000 genes, approximately 1,200 gene segments showed evidence of significant expression. Differences in expression were observed in 104 transcripts on the microarray. Gene expression changes were most prominently observed for factors relating to physiological, cellular, and molecular regulation (such as calcium channels). Although this study identified over 25 genes that were differentially regulated in more than one of the TBI subjects, there were numerous genes that were only differentially regulated in one individual with TBI. The differences between subjects in this study remind us that individual variation in people with the same diagnosis is a variable that requires consideration.

Candidate Gene Studies

The study of TBI is unique among the CNS diseases previously discussed, in that candidate gene studies involving TBI focus primarily on outcomes following the injury. These studies are summarized in Table 11.4. When reviewing the data in

TABLE 11.4
Genes Associated With Outcomes Following TBI

Gene	Variant	Phenotype	RR/*OR* (95% CI)	Reference
APOE	ε4	Poor GOS score at 6 months	3.01 (1.02–8.88)	(Chiang, Chang, & Hu, 2003)
p53	Arg72Pro	Poor GOS score at discharge	2.9 (1.05–8.31)	(Martínez-Lucas et al., 2005)
ACE	insertion/ deletion	Neuropsych tests	—[a]	(Ariza et al., 2006)
IL-1β	C3953T	Poor GOS score at 6 months	0.25 (0.12–0.55)	(Ariza et al., 2006)
DRD2	rs1800497	Neuropsych tests	—[a]	(McAllister et al., 2005)
COMT	Val158Met	Neuropsych tests	—[a]	(Lipsky et al., 2005)
BCL2	4 tagging SNPs	Functional outcomes Mortality	—[a]	(Hoh et al., 2010)
NGB	rs3783988 (TT)	GOS score at 12 months	2.65 (1.11–6.30)	(Chuang et al., 2010)

Note. RR = relative risk; *OR* = odds ratio; CI = confidence intervals (95%).
[a]no estimates of RR/*OR* available for reference.

this table, keep in mind that the authors attempted to report the earliest positive association study. The authors did not report any negative findings or findings that support the original positive association. As observed in other CNS diseases, the study of APOE in the TBI population predominates. Comparable to the ICH and aSAH literature, the APOE ε4 allele is associated with worse outcomes in the TBI population. Also of note is that there are numerous studies that do not support the association of APOE polymorphisms and outcomes following TBI. In the TBI literature, there have been more than 40 articles published between 1995 and 2008 on APOE and TBI, a little over half found positive associations with outcomes and APOE genotype (see Table 11.4).

SUMMARY

This chapter reviewed studies in genetics and genomics of chosen central nervous system diseases. Specifically, we focused on ischemic stroke, ICH, aSAH, and TBI. In general, the number of studies in any one type of stroke mirrored the occurrence of that type of stroke. Thus, most genetic research is in ischemic stroke, whereas ICH and aSAH have significantly less representation. In conducting this review, it became readily apparent that APOE represented a major percentage of the genetic studies in all of the central nervous system diseases reviewed. The reasons for this emphasis will be explored in the following paragraphs.

Apolipoprotein E

APOE has not only been investigated in each of the four diseases that were reviewed, but it is one of the most commonly studied genes in neurological research. APOE is located on chromosome 19q13.2. APOE is a 299 amino acid long protein (Rall, Weisgraber, & Mahley, 1982). Its main function is to transport lipoproteins, fat-soluble vitamins, and cholesterol facilitating movement into and out of cells. In the central nervous system, APOE is the major apolipoprotein. It is found both in plasma and in cerebrospinal fluid (CSF). When there is neuronal cell loss in the CNS large amounts of lipids are released from the damaged cells. Astrocytes then increase the production and release of the APOE protein that scavenge the cholesterol and phospholipids from cellular debris. This cholesterol is then used to reinnervate the CNS. The different isoforms of APOE have different affinities for their receptors. Therefore, a person with the APOE ε4 allele will have less circulating cholesterol and phospholipids compared to other people with the APOE ε2 or ε3 alleles. These differences have an influence on both risk and outcomes following central nervous system diseases.

THE FUTURE OF GENETICS AND THE CENTRAL NERVOUS SYSTEM

After the review of the literature for these central nervous system diseases, what have we learned? First, most research to date used a candidate gene approach, which is inherently biased. As we move forward, more GWAS and gene expression studies are required, especially in the ICH, aSAH, and TBI populations. Candidate gene studies are important, but these studies need to be performed with larger sample sizes than prior studies to identify associations that we can use in practice.

Second, most studies that were reviewed in this chapter have odds ratios of 1.2–3.6. Although some studies reported much higher odds ratios, these findings were accompanied by large confidence intervals. Therefore, the relatively small odds ratios and large confidence intervals must be interpreted with caution, because of the large variability that is associated with these measures.

We must also keep in mind that diseases are complex and the influence of one gene on disease risk or outcome may be relatively small compared to the influence of other physiologic/pathophysiologic factors or clusters of factors. Although not reviewed here, there are studies that have examined the notion that gene groups or clusters may exert greater influence within a disease process than do single genes (Improgo, Scofield, Tapper, & Gardner, 2010). Of course, the larger number of genes examined requires that a larger sample size be used. In addition, proteomics need to be taken into account. Future studies need to be designed that look at not only genetics and proteomics, but the interaction between proteomics and genetics.

Progress has been made in examining the role of environment-gene interactions (Offit, 2011). Hypothetically, when 80% of the disease is influenced by the environment, we are not going to see large effect sizes from a single gene. Examination of gene–environment interactions will no doubt be more prominent in future studies.

Finally, a word of caution is offered with respect to the balance of positive and negative studies in central nervous system genomics. The relative balance of positive and negative reports that support associations of a certain genes with a disease process may be mainly because of the small sample size of the studies. As larger samples are used in studies, earlier hypotheses may be rejected. An example of this has already occurred. Phosphodiesterase 4D was originally thought to be a risk factor for stroke (Gretarsdottir et al., 2003). After a meta-analysis was performed, there was no association found and this protein has now been abandoned as a risk factor for ischemic stroke (Meschia, 2011).

In summary, this review reveals that although indeed remarkable progress has been made in our understanding of genomics and its influence on central nervous system (CNS) disease, much remains to be explored and understood. As nurses, we are in the position to translate this genetic research into clinical practice. As scientific discovery in this fascinating area of science continues, we will elucidate the genes that are involved with these disease processes. As nurses, we are in the position to deliver individualized care based on the genetic differences in patients. This individualized care will evolve as discoveries are made in the genetics/genomics of the central nervous system.

REFERENCES

Adams, H. P., Jr., Bendixen, B. H., Kappelle, L. J., Biller, J., Love, B. B., Gordon, D. L., & Marsh, E. E., III. (1993). Classification of subtype of acute ischemic stroke. Definitions for use in a multicenter clinical trial. TOAST. Trial of Org 10172 in Acute Stroke Treatment. *Stroke*, *24*(1), 35–41.

Alberts, M. J., Davis, J. P., Graffagnino, C., McClenny, C., Delong, D., Granger, C., . . . Roses, A. D. (1997). Endoglin gene polymorphism as a risk factor for sporadic intracerebral hemorrhage. *Annals of Neurology*, *41*(5), 683–686.

Alexander, S., Poloyac, S., Hoffman, L., Gallek, M., Dianxu, R., Balzer, J., . . . Conley, Y. (2009). Endothelial nitric oxide synthase tagging single nucleotide polymorphisms and recovery from aneurysmal subarachnoid hemorrhage. *Biological Research for Nursing*, *11*(1), 42–52.

Ariza, M., Matarin, M. D., Junqué, C., Mataró, M., Clemente, I., Moral, P., . . . Sahuquillo, J. (2006). Influence of Angiotensin-converting enzyme polymorphism on neuropsychological subacute performance in moderate and severe traumatic brain injury. *Journal of Neuropsychiatry and Clinical Neurosciences*, *18*(1), 39–44.

Bai, Y., Chen, J., Sun, K., Xin, Y., Liu, J., & Hui, R. (2009). Common genetic variation in DDAH2 is associated with intracerebral haemorrhage in a Chinese population: A multi-centre case-control study in China. *Clinical Science*, *117*(7), 273–279.

Barr, T. L., Conley, Y., Ding, J., Dillman, A., Warach, S., Singleton, A., & Matarin, M. (2010). Genomic biomarkers and cellular pathways of ischemic stroke by RNA gene expression profiling. *Neurology*, *75*(11), 1009–1014.

Berger, K., Stögbauer, F., Stoll, M., Wellmann, J., Huge, A., Cheng, S., . . . Funke, H. (2007). The glu298asp polymorphism in the nitric oxide synthase 3 gene is associated with the risk of ischemic stroke in two large independent case-control studies. *Human Genetics*, *121*(2), 169–178.

Borsody, M., Burke, A., Coplin, W., Miller-Lotan, R., & Levy, A. (2006). Haptoglobin and the development of cerebral artery vasospasm after subarachnoid hemorrhage. *Neurology*, *66*(5), 634–640.

Boyles, J. K., Zoellner, C. D., Anderson, L. J., Kosik, L. M., Pitas, R. E., Weisgraber, K. H., . . . Ignatius, M. J. (1989). A role for apolipoprotein E, apolipoprotein A-I, and low density lipoprotein receptors in cholesterol transport during regeneration and remyelination of the rat sciatic nerve. *Journal of Clinical Investigation*, *83*(3), 1015–1031.

Carter, A. M., Catto, A. J., Bamford, J. M., & Grant, P. J. (1999). Association of the platelet glycoprotein IIb HPA-3 polymorphism with survival after acute ischemic stroke. *Stroke*, *30*(12), 2606–2611.

Casas, J. P., Hingorani, A. D., Bautista, L. E., & Sharma, P. (2004). Meta-analysis of genetic studies in ischemic stroke: Thirty-two genes involving approximately 18,000 cases and 58,000 controls. *Archives of Neurology*, *61*(11), 1652–1661.

Catto, A. J., Kohler, H. P., Bannan, S., Stickland, M., Carter, A., & Grant, P. J. (1998). Factor XIII Val 34 Leu: A novel association with primary intracerebral hemorrhage. *Stroke*, *29*(4), 813–816.

Centers for Disease Control and Prevention, National Center for Injury Prevention and Control. (2003). *Report to congress on mild traumatic brain injury in the United States: Steps to prevent a serious public health problem*. Atlanta, GA: Centers for Disease Control and Prevention.

Chen, Y. C., Hu, F. J., Chen, P., Wu, Y. R., Wu, H. C., Chen, S. T., . . . Chen, C. M. (2010). Association of TNF-alpha gene with spontaneous deep intracerebral hemorrhage in the Taiwan population: A case control study. *BMC Neurology*, *10*, 41.

Chiang, M. F., Chang, J. G., & Hu, C. J. (2003). Association between apolipoprotein E genotype and outcome of traumatic brain injury. *Acta Neurochirurgica*, *145*(8), 649–653, discussion 653–644.

Chowdhury, A. H., Yokoyama, T., Kokubo, Y., Zaman, M. M., Haque, A., & Tanaka, H. (2001). Apolipoprotein E genetic polymorphism and stroke subtypes in a Bangladeshi hospital-based study. *Journal of Epidemiology*, *11*(3), 131–138.

Chuang, P. Y., Conley, Y. P., Poloyac, S. M., Okonkwo, D. O., Ren, D., Sherwood, P. R., . . . Alexander, S. A. (2010). Neuroglobin genetic polymorphisms and their relationship to functional outcomes after traumatic brain injury. *Journal of Neurotrauma*, *27*(6), 999–1006.

Coria, F., Rubio, I., Nuñez, E., Sempere, A. P., SantaEngarcia, N., Bayón, C., & Cuadrado, N. (1995). Apolipoprotein E variants in ischemic stroke. *Stroke*, *26*(12), 2375–2376.

Corral, J., Iniesta, J. A., González-Conejero, R., Villalón, M., & Vicente, V. (2001). Polymorphisms of clotting factors modify the risk for primary intracranial hemorrhage. *Blood*, *97*(10), 2979–2982.

Faul, M., Xu, L., Wald, M. M., & Coronado, V. G. (2010). *Traumatic brain injury in the United States: Emergency department visits, hospitalizations, and deaths 2002–2006*. Atlanta, GA: Centers for Disease Control and Prevention, National Center for Injury Prevention and Control.

Flex, A., Gaetani, E., Papaleo, P., Straface, G., Proia, A. S., Pecorini, G., . . . Pola, R. (2004). Proinflammatory genetic profiles in subjects with history of ischemic stroke. *Stroke, 35*(10), 2270–2275.

Fontanella, M., Rainero, I., Gallone, S., Rubino, E., Fornaro, R., Fenoglio, P., . . . Pinessi, L. (2010). Interleukin-1 cluster gene polymorphisms and aneurysmal subarachnoid hemorrhage. *Neurosurgery, 66*(6), 1058–1062, discussion 1062–1053.

Fontanella, M., Rainero, I., Gallone, S., Rubino, E., Rivoiro, C., Valfrè, W., . . . Pinessi, L. (2007). Lack of association between the apolipoprotein E gene and aneurysmal subarachnoid hemorrhage in an Italian population. *Journal of Neurosurgery, 106*(2), 245–249.

Fornage, M., Lee, C. R., Doris, P. A., Bray, M. S., Heiss, G., Zeldin, D. C., & Boerwinkle, E. (2005). The soluble epoxide hydrolase gene harbors sequence variation associated with susceptibility to and protection from incident ischemic stroke. *Human Molecular Genetics, 14*(19), 2829–2837.

Gallek, M. J., Conley, Y. P., Sherwood, P. R., Horowitz, M. B., Kassam, A., & Alexander, S. A. (2009). APOE genotype and functional outcome following aneurysmal subarachnoid hemorrhage. *Biological Research for Nursing, 10*(3), 205–212.

Gomez-Sanchez, J. C., Delgado-Esteban, M., Rodriguez-Hernandez, I., Sobrino, T., Perez de la Ossa, N., Reverte, S., . . . Almeida, A. (2011). The human Tp53 Arg72Pro polymorphism explains different functional prognosis in stroke. *Journal of Experimental Medicine, 208*(3), 429–437.

Greenberg, S. M., Rebeck, G. W., Vonsattel, J. P., Gomez-Isla, T., & Hyman, B. T. (1995). Apolipoprotein E epsilon 4 and cerebral hemorrhage associated with amyloid angiopathy. *Annals of Neurology, 38*(2), 254–259.

Greisenegger, S., Endler, G., Haering, D., Schillinger, M., Lang, W., Lalouschek, W., & Mannhalter, C. (2003). The (-174) G/C polymorphism in the interleukin-6 gene is associated with the severity of acute cerebrovascular events. *Thrombosis Research, 110*(4), 181–186.

Gretarsdottir, S., Thorleifsson, G., Manolescu, A., Styrkarsdottir, U., Helgadottir, A., Gschwendtner, A., . . . Stefansson, K. (2008). Risk variants for atrial fibrillation on chromosome 4q25 associate with ischemic stroke. *Annals of Neurology, 64*(4), 402–409.

Gretarsdottir, S., Thorleifsson, G., Reynisdottir, S. T., Manolescu, A., Jonsdottir, S., Jonsdottir, T., . . . Gulcher, J. R. (2003). The gene encoding phosphodiesterase 4D confers risk of ischemic stroke. *Nature Genetics, 35*(2), 131–138.

Gschwendtner, A., Bevan, S., Cole, J. W., Plourde, A., Matarin, M., Ross-Adams, H., . . . Dichgans, M. (2009). Sequence variants on chromosome 9p21.3 confer risk for atherosclerotic stroke. *Annals of Neurology, 65*(5), 531–539.

Gudbjartsson, D. F., Holm, H., Gretarsdottir, S., Thorleifsson, G., Walters, G. B., Thorgeirsson, G., . . . Stefansson, K. (2009). A sequence variant in ZFHX3 on 16q22 associates with atrial fibrillation and ischemic stroke. *Nature Genetics, 41*(8), 876–878.

Hamaguchi, T., Okino, S., Sodeyama, N., Itoh, Y., Takahashi, A., Otomo, E., . . . Yamada, M. (2005). Association of a polymorphism of the transforming growth factor-beta1 gene with cerebral amyloid angiopathy. *Journal of Neurology, Neurosurgery, and Psychiatry, 76*(5), 696–699.

Hata, J., Matsuda, K., Ninomiya, T., Yonemoto, K., Matsushita, T., Ohnishi, Y., . . . Kubo, M. (2007). Functional SNP in an Sp1-binding site of AGTRL1 gene is associated with susceptibility to brain infarction. *Human Molecular Genetics, 16*(6), 630–639.

Helgadottir, A., Manolescu, A., Thorleifsson, G., Gretarsdottir, S., Jonsdottir, H., Thorsteinsdottir, U., . . . Stefansson, K. (2004). The gene encoding 5-lipoxygenase activating protein confers risk of myocardial infarction and stroke. *Nature Genetics, 36*(3), 233–239.

Helgadottir, A., Thorleifsson, G., Manolescu, A., Gretarsdottir, S., Blondal, T., Jonasdottir, A., . . . Stefansson, K. (2007). A common variant on chromosome 9p21 affects the risk of myocardial infarction. *Science*, *316*(5830), 1491–1493.

Hoh, N. Z., Wagner, A. K., Alexander, S. A., Clark, R. B., Beers, S. R., Okonkwo, D. O., . . . Conley, Y. P. (2010). BCL2 genotypes: Functional and neurobehavioral outcomes after severe traumatic brain injury. *Journal of Neurotrauma*, *27*(8), 1413–1427.

Hoy, A., Leininger-Muller, B., Poirier, O., Siest, G., Gautier, M., Elbaz, A., . . . Visvikis, S. (2003). Myeloperoxidase polymorphisms in brain infarction. Association with infarct size and functional outcome. *Atherosclerosis*, *167*(2), 223–230.

Improgo, M. R., Scofield, M. D., Tapper, A. R., & Gardner, P. D. (2010). The nicotinic acetylcholine receptor CHRNA5/A3/B4 gene cluster: Dual role in nicotine addiction and lung cancer. *Progress in Neurobiology*, *92*(2), 212–226.

Juvela, S., Siironen, J., & Lappalainen, J. (2009). Apolipoprotein E genotype and outcome after aneurysmal subarachnoid hemorrhage. *Journal of Neurosurgery*, *110*(5), 989–995.

Kessler, C., Spitzer, C., Stauske, D., Mende, S., Stadlmüller, J., Walther, R., & Rettig, R. (1997). The apolipoprotein E and beta-fibrinogen G/A-455 gene polymorphisms are associated with ischemic stroke involving large-vessel disease. *Arteriosclerosis, Thrombosis, and Vascular Biology*, *17*(11), 2880–2884.

Khurana, V. G., Sohni, Y. R., Mangrum, W. I., McClelland, R. L., O'Kane, D. J., Meyer, F. B., & Meissner, I. (2004). Endothelial nitric oxide synthase gene polymorphisms predict susceptibility to aneurysmal subarachnoid hemorrhage and cerebral vasospasm. *Journal of Cerebral Blood Flow and Metabolism*, *24*(3), 291–297.

Klienbaum, D. G., Kupper, L. L., Muller, K. E., & Nizam, A. (1998). *Applied regression analysis and other multivariavle methods* (3rd ed.). Pacific Grove, CA: Duxbury Press.

Knudsen, K. A., Rosand, J., Karluk, D., & Greenberg, S. M. (2001). Clinical diagnosis of cerebral amyloid angiopathy: Validation of the Boston criteria. *Neurology*, *56*(4), 537–539.

Kubo, M., Hata, J., Ninomiya, T., Matsuda, K., Yonemoto, K., Nakano, T., . . . Kiyohara, Y. (2007). A nonsynonymous SNP in PRKCH (protein kinase C eta) increases the risk of cerebral infarction. *Nature Genetics*, *39*(2), 212–217.

Lanterna, L. A., Rigoldi, M., Tredici, G., Biroli, F., Cesana, C., Gaini, S. M., & Dalprà, L. (2005). APOE influences vasospasm and cognition of noncomatose patients with subarachnoid hemorrhage. *Neurology*, *64*(7), 1238–1244.

Lemmens, R., Buysschaert, I., Geelen, V., Fernandez-Cadenas, I., Montaner, J., Schmidt, H., . . . Thijs, V. (2010). The association of the 4q25 susceptibility variant for atrial fibrillation with stroke is limited to stroke of cardioembolic etiology. *Stroke*, *41*(9), 1850–1857.

Leung, C. H., Poon, W. S., Yu, L. M., Wong, G. K., & Ng, H. K. (2002). Apolipoprotein e genotype and outcome in aneurysmal subarachnoid hemorrhage. *Stroke*, *33*(2), 548–552.

Lipsky, R. H., Sparling, M. B., Ryan, L. M., Xu, K., Salazar, A. M., Goldman, D., & Warden, D. L. (2005). Association of COMT Val158Met genotype with executive functioning following traumatic brain injury. *Journal of Neuropsychiatry and Clinical Neurosciences*, *17*(4), 465–471.

Lubitz, S. A., Sinner, M. F., Lunetta, K. L., Makino, S., Pfeufer, A., Rahman, R., . . . Ellinor, P. T. (2010). Independent susceptibility markers for atrial fibrillation on chromosome 4q25. *Circulation*, *122*(10), 976–984.

Martínez-González, N. A., & Sudlow, C. L. M. (2006). Effects of apolipoprotein E genotype on outcome after ischaemic stroke, intracerebral haemorrhage and subarachnoid haemorrhage. *Journal of Neurology, Neurosurgery, and Psychiatry*, *77*(12), 1329–1335.

Martínez-Lucas, P., Moreno-Cuesta, J., García-Olmo, D. C., Sánchez-Sánchez, F., Escribano-Martínez, J., del Pozo, A. C., . . . García-Olmo, D. (2005). Relationship between the Arg72Pro polymorphism of p53 and outcome for patients with traumatic brain injury. *Intensive Care Medicine*, *31*(9), 1168–1173.

McAllister, T. W., Rhodes, C. H., Flashman, L. A., McDonald, B. C., Belloni, D., & Saykin, A. J. (2005). Effect of the dopamine D2 receptor T allele on response latency after mild traumatic brain injury. *American Journal of Psychiatry*, *162*(9), 1749–1751.

McCarron, M. O., Delong, D., & Alberts, M. J. (1999). APOE genotype as a risk factor for ischemic cerebrovascular disease: A meta-analysis. *Neurology*, *53*(6), 1308–1311.

McCarron, M. O., Muir, K. W., Weir, C. J., Dyker, A. G., Bone, I., Nicoll, J. A., & Lees, K. R. (1998). The apolipoprotein E epsilon4 allele and outcome in cerebrovascular disease. *Stroke*, *29*(9), 1882–1887.

McPherson, R., Pertsemlidis, A., Kavaslar, N., Stewart, A., Roberts, R., Cox, D. R., . . . Cohen, J. C. (2007). A common allele on chromosome 9 associated with coronary heart disease. *Science*, *316*(5830), 1488–1491.

Meschia, J. F. (2011). New information on the genetics of stroke. *Current Neurology and Neuroscience Reports*, *11*(1), 35–41.

Michael, D. B., Byers, D. M., & Irwin, L. N. (2005). Gene expression following traumatic brain injury in humans: Analysis by microarray. *Journal of Clinical Neuroscience*, *12*(3), 284–290.

Nahed, B. V., Bydon, M., Ozturk, A. K., Bilguvar, K., Bayrakli, F., & Gunel, M. (2007). Genetics of intracranial aneurysms. *Neurosurgery*, *60*(2), 213–225, discussion 225–216.

Nakata, Y., Katsuya, T., Rakugi, H., Takami, S., Sato, N., Kamide, K., . . . Ogihara, T. (1997). Polymorphism of angiotensin converting enzyme, angiotensinogen, and apolipoprotein E genes in a Japanese population with cerebrovascular disease. *American Journal of Hypertension*, *10*(12, Pt. 1), 1391–1395.

Navarro-Núñez, L., Lozano, M. L., Rivera, J., Corral, J., Roldán, V., González-Conejero, R., . . . Martínez, C. (2007). The association of the beta1-tubulin Q43P polymorphism with intracerebral hemorrhage in men. *Haematologica*, *92*(4), 513–518.

Nicoll, J. A., Burnett, C., Love, S., Graham, D. I., Dewar, D., Ironside, J. W., . . . Vinters, H. V. (1997). High frequency of apolipoprotein E epsilon 2 allele in hemorrhage due to cerebral amyloid angiopathy. *Annals of Neurology*, *41*(6), 716–721.

O'Donnell, H. C., Rosand, J., Knudsen, K. A., Furie, K. L., Segal, A. Z., Chiu, R. I., . . . Greenberg, S. M. (2000). Apolipoprotein E genotype and the risk of recurrent lobar intracerebral hemorrhage. *New England Journal of Medicine*, *342*(4), 240–245.

Obach, V., Revilla, M., Vila, N., Cervera, A. A., & Chamorro, A. A. (2001). alpha(1)-antichymotrypsin polymorphism: A risk factor for hemorrhagic stroke in normotensive subjects. *Stroke*, *32*(11), 2588–2591.

Offit, K. (2011). Personalized medicine: New genomics, old lessons. *Human Genetics*, *130*(1), 3–14.

Olsson, S., Jood, K., Blomstrand, C., & Jern, C. (2011). Genetic variation on chromosome 9p21 shows association with the ischaemic stroke subtype large-vessel disease in a Swedish sample aged ≤ 70. *European Journal of Neurology*, *18*(2), 365–367.

Peng, D. Q., Zhao, S. P., & Wang, J. L. (1999). Lipoprotein (a) and apolipoprotein E epsilon 4 as independent risk factors for ischemic stroke. *Journal of Cardiovascular Risk*, *6*(1), 1–6.

Pitas, R. E., Boyles, J. K., Lee, S. H., Foss, D., & Mahley, R. W. (1987). Astrocytes synthesize apolipoprotein E and metabolize apolipoprotein E-containing lipoproteins. *Biochimica et Biophysica Acta*, *917*(1), 148–161.

Polit, D. F., & Beck, C. T. (2012). *Nursing research: Generating and assessing evidence for nursing practice* (9th ed.). Philadelphia, PA: Wolters Kluwer/Lippincott Williams & Wilkins.

Rall, S. C., Jr., Weisgraber, K. H., & Mahley, R. W. (1982). Human apolipoprotein E. The complete amino acid sequence. *Journal of Biological Chemistry*, *257*(8), 4171–4178.

Roger, V. L., Go, A. S., Lloyd-Jones, D. M., Adams, R. J., Berry, J. D., Brown, T. M., . . . Wylie-Rosett, J. (2011). Heart disease and stroke statistics—2011 update: A report from the American Heart Association. *Circulation*, *123*(4), e18–e209.

Rosand, J., Hylek, E. M., O'Donnell, H. C., & Greenberg, S. M. (2000). Warfarin-associated hemorrhage and cerebral amyloid angiopathy: A genetic and pathologic study. *Neurology*, *55*(7), 947–951.

Ruigrok, Y. M., & Rinkel, G. J. (2008). Genetics of intracranial aneurysms. *Stroke*, *39*(3), 1049–1055.

Samani, N. J., Erdmann, J., Hall, A. S., Hengstenberg, C., Mangino, M., Mayer, B., . . . Schunkert, H. (2007). Genomewide association analysis of coronary artery disease. *New England Journal of Medicine*, *357*(5), 443–453.

Siironen, J., Juvela, S., Kanarek, K., Vilkki, J., Hernesniemi, J., & Lappalainen, J. (2007). The Met allele of the BDNF Val66Met polymorphism predicts poor outcome among survivors of aneurysmal subarachnoid hemorrhage. *Stroke*, *38*(10), 2858–2860.

Slowik, A., Turaj, W., Dziedzic, T., Haefele, A., Pera, J., Malecki, M. T., . . . Szczudlik, A. (2004). DD genotype of ACE gene is a risk factor for intracerebral hemorrhage. *Neurology*, *63*(2), 359–361.

Sobey, C. G., & Faraci, F. M. (1998). Subarachnoid haemorrhage: What happens to the cerebral arteries? *Clinical and Experimental Pharmacology & Physiology*, *25*(11), 867–876.

Strand, M., Söderström, I., Wiklund, P. G., Hallmans, G., Weinehall, L., Söderberg, S., & Olsson, T. (2007). Estrogen receptor alpha gene polymorphisms and first-ever intracerebral hemorrhage. *Cerebrovascular Diseases*, *24*(6), 500–508.

Strittmatter, W. J., Saunders, A. M., Schmechel, D., Pericak-Vance, M., Enghild, J., Salvesen, G. S., & Roses, A. D. (1993). Apolipoprotein E: High-avidity binding to beta-amyloid and increased frequency of type 4 allele in late-onset familial Alzheimer disease. *Proceedings of the National Academy of Sciences of the United States of America*, *90*(5), 1977–1981.

Sudlow, C., Martínez González, N. A., Kim, J., & Clark, C. (2006). Does apolipoprotein E genotype influence the risk of ischemic stroke, intracerebral hemorrhage, or subarachnoid hemorrhage? Systematic review and meta-analyses of 31 studies among 5961 cases and 17,965 controls. *Stroke*, *37*(2), 364–370.

Sun, L., Li, Z., Zhang, H., Ma, A., Liao, Y., Wang, D., . . . Hui, R. (2003). Pentanucleotide TTTTA repeat polymorphism of apolipoprotein(a) gene and plasma lipoprotein(a) are associated with ischemic and hemorrhagic stroke in Chinese: A multicenter case-control study in China. *Stroke*, *34*(7), 1617–1622.

Szolnoki, Z., Somogyvári, F., Kondacs, A., Szabó, M., Bene, J., Havasi, V., . . . Melegh, B. (2003). Increased prevalence of platelet glycoprotein IIb/IIIa PLA2 allele in ischaemic stroke associated with large vessel pathology. *Thrombosis Research*, *109*(5–6), 265–269.

Tang, J., Zhao, J., Zhao, Y., Wang, S., Chen, B., & Zeng, W. (2003). Apolipoprotein E epsilon4 and the risk of unfavorable outcome after aneurysmal subarachnoid hemorrhage. *Surgical Neurology*, *60*(5), 391–396, discussion 396–397.

Tang, Y., Xu, H., Du, X., Lit, L., Walker, W., Lu, A., . . . Sharp, F. R. (2006). Gene expression in blood changes rapidly in neutrophils and monocytes after ischemic stroke in humans: A microarray study. *Journal of Cerebral Blood Flow and Metabolism*, *26*(8), 1089–1102.

Vergouwen, M. D., Frijns, C. J., Roos, Y. B., Rinkel, G. J., Baas, F., & Vermeulen, M. (2004). Plasminogen activator inhibitor-1 4G allele in the 4G/5G promoter polymorphism increases the occurrence of cerebral ischemia after aneurysmal subarachnoid hemorrhage. *Stroke*, *35*(6), 1280–1283.

Voetsch, B., Benke, K. S., Damasceno, B. P., Siqueira, L. H., & Loscalzo, J. (2002). Paraoxonase 192 Gln→Arg polymorphism: An independent risk factor for nonfatal arterial ischemic stroke among young adults. *Stroke*, *33*(6), 1459–1464.

Volcik, K. A., Ballantyne, C. M., Hoogeveen, R., Folsom, A. R., & Boerwinkle, E. (2010). Intercellular adhesion molecule-1 G241R polymorphism predicts risk of incident ischemic stroke: Atherosclerosis Risk in Communities study. *Stroke*, *41*(5), 1038–1040.

Wang, Y., Zhang, W., Zhang, Y., Yang, Y., Sun, L., Hu, S., . . . Hui, R. (2006). VKORC1 haplotypes are associated with arterial vascular diseases (stroke, coronary heart disease, and aortic dissection). *Circulation*, *113*(12), 1615–1621.

Wellcome Trust Case Control Consortium. (2007). Genome-wide association study of 14,000 cases of seven common diseases and 3,000 shared controls. *Nature*, *447*(7145), 661–678.

Woo, D., Sauerbeck, L. R., Kissela, B. M., Khoury, J. C., Szaflarski, J. P., Gebel, J., . . . Broderick, J. P. (2002). Genetic and environmental risk factors for intracerebral hemorrhage: Preliminary results of a population-based study. *Stroke*, *33*(5), 1190–1195.

Wu, L., Shen, Y., Liu, X., Ma, X., Xi, B., Mi, J., . . . Wang, X. (2009). The 1425G/A SNP in PRKCH is associated with ischemic stroke and cerebral hemorrhage in a Chinese population. *Stroke*, *40*(9), 2973–2976.

Xia, J., Yang, Q. D., Yang, Q. M., Xu, H. W., Liu, Y. H., Zhang, L., . . . Cao, G. F. (2004). Apolipoprotein H gene polymorphisms and risk of primary cerebral hemorrhage in a Chinese population. *Cerebrovascular Diseases*, *17*(2–3), 197–203.

Yamada, M., Sodeyama, N., Itoh, Y., Suematsu, N., Otomo, E., Matsushita, M., & Mizusawa, H. (1997). Association of presenilin-1 polymorphism with cerebral amyloid angiopathy in the elderly. *Stroke*, *28*(11), 2219–2221.

Yamada, M., Sodeyama, N., Itoh, Y., Takahashi, A., Otomo, E., Matsushita, M., & Mizusawa, H. (2003). Association of neprilysin polymorphism with cerebral amyloid angiopathy. *Journal of Neurology, Neurosurgery, and Psychiatry*, *74*(6), 749–751.

Yamada, Y. (2006). Identification of genetic factors and development of genetic risk diagnosis systems for cardiovascular diseases and stroke. *Circulation Journal*, *70*(10), 1240–1248.

Yamada, Y., Metoki, N., Yoshida, H., Satoh, K., Ichihara, S., Kato, K., . . . Nozawa, Y. (2006). Genetic risk for ischemic and hemorrhagic stroke. *Arteriosclerosis, Thrombosis, and Vascular Biology*, *26*(8), 1920–1925.

Zee, R. Y., Cook, N. R., Cheng, S., Reynolds, R., Erlich, H. A., Lindpaintner, K., & Ridker, P. M. (2004). Polymorphism in the P-selectin and interleukin-4 genes as determinants of stroke: a population-based, prospective genetic analysis. *Human Molecular Genetics*, *13*(4), 389–396.

CHAPTER 12

Shared Genomics of Type 2 and Gestational Diabetes Mellitus

Shu-Fen Wung and Pei-Chao Lin

ABSTRACT

Gestational diabetes mellitus (GDM) is one of the most common complications of pregnancy and the prevalence of GDM is increasing worldwide. Short- and long-term complications of GDM on mothers and fetuses are well-recognized. These include more than seven-fold higher risk for type 2 diabetes mellitus (T2DM) later in life in women with GDM than those without. Evidence supports that GDM shares several risk factors with T2DM, including genetic risks. This chapter reviewed studies on candidate genes shared by T2DM and GDM published from 1990 to 2011. At least 20 susceptible genes of T2DM have been studied in women with GDM in various races. Results from current association studies on T2DM susceptible genes in GDM have shown significant heterogeneity. There may be primary evidence that polymorphisms of susceptible genes of T2DM such as transcription factor 7-like 2 (TCF7L2) gene, potassium channel voltage-gate KQT-like subfamily member 1 (KCNQ1) gene, and cyclin-dependent kinase 5 regulatory subunit-associated protein 1-like 1 (CDKAL1) gene, may increase risk of GDM. Associations between GDM and many genetic variants have led to different findings across populations. Many genetic polymorphisms related to GDM were investigated in a single study or a single population. Replication studies to verify contributions of both common and rare genetic variants for GDM and T2DM in specific racial/ethnic groups are needed.

http://dx.doi.org/10.1891/0739-6686.29.227

INTRODUCTION

Gestational diabetes mellitus (GDM) is defined as glucose intolerance with onset or first recognition during pregnancy (American Diabetes Association, 2009). GDM is one of the most common complications of pregnancy, affecting 1%–14% of all pregnancies (Mulla, Henry, & Homko, 2010). In the United States, GDM affects approximately 135,000 cases annually (American Diabetes Association, 2009). Prevalence of GDM is increasing worldwide; data from past 2 decades show that GDM has increased by 16%–127% in several race/ethnicity groups (Ferrara, 2007; Hunt & Schuller, 2007).

There are racial/ethnic differences in the prevalence of GDM. Based on population-based studies worldwide, the prevalence of GDM is 2%–6% in general public and 10%–22% in high risk population such as Asian Indians (Galtier, 2010). A recent review of 3,108,877 births in the United States, Asian and Pacific Islander women had a substantially age-adjusted higher prevalence of GDM (6.3%) than Whites (3.8%), Blacks (3.5%), or Hispanics (3.6%; Chu et al., 2009). Prevalence of age-adjusted GDM also varied significantly among Asian and Pacific Islander women, from 3.7% among women of Japanese descent to 8.6% among women of Asian Indian descent. Asian and Pacific Islanders born outside of United States had significant higher GDM rates than those born in the United States, except among women of Japanese and Korean ancestry (Chu et al., 2009).

Complications of GDM on mothers and fetus are well recognized (Yogev & Visser, 2009). Maternal complications include preterm labor, preeclampsia, nephropathy, birth trauma, cesarean section, and postoperative wound complications. Evidence suggests that even mild maternal hyperglycemia has a substantial impact on maternal health, such as risk of cesarean section, gestational hypertension, and preeclampsia (Landon et al., 2009; Vambergue et al., 2000; Yang, Hsu-Hage, Zhang, Zhang, & Zhang, 2002). Perinatal complications include miscarriage, congenital anomalies, macrosomia (fetal overgrowth), shoulder dystocia, stillbirth, growth restriction, hypoglycemia, hyperbilirubinemia, polycythemia (erythrocytosis), and birth trauma such as bone fractures and nerve palsies (Crowther et al., 2005; Yogev & Visser, 2009). Neonatal adiposity is also positively associated with maternal glucose level (HAPO Study Cooperative Research Group, 2009).

It is becoming evident that in utero exposure to GDM has long-lasting consequence in children born to mothers with GDM. Literature on long-term adverse health outcomes among mothers diagnosed with GDM and their offspring have been summarized (Dabelea, 2007; Metzger, 2007; Wroblewska-Seniuk, Wender-Ozegowska, & Szczapa, 2009). Studies have shown that offspring of mothers with GDM have increased cardiometabolic risks, such as obesity, impaired glucose tolerance/type 2 diabetes mellitus (T2DM), and hypertension

(Clausen et al., 2009; Dabelea et al., 2008; Tam et al., 2008). GDM may also adversely affect attention span and motor functions, but not cognitive ability, of the offspring and these effects are negatively correlated with the degree of maternal glycemic control (Ornoy, 2005).

For women, GDM is a strong risk factor for T2DM. Although postpartum glucose metabolism and insulin resistance return to normal for most women with GDM, the risk for T2DM later in life is more than seven-fold higher in women with GDM than those without (Bellamy, Casas, Hingorani, & Williams, 2009). Half of women with GDM are expected to develop T2DM within 5 years of the index pregnancy (Kjos et al., 1995; Metzger, Cho, Roston, & Radvany, 1993). A high proportion of women (> 50%) have recurrent GDM in subsequent pregnancy (Metzger, 2007). Additional pregnancy after GDM may accelerate the risk of progression to T2DM (Peters, Kjos, Xiang, & Buchanan, 1996). In addition, women with a history of GDM may be at an increased risk for cardiovascular disease (King, Gerich, Guzick, King, & McDermott, 2009; Metzger, 2007).

T2DM is a multi-factorial disease, including many genetic and environmental factors. Recently, Baptiste-Roberts and associates (2009) conducted a systematic review on risk factors for the development of T2DM among women with history of GDM. These authors concluded that there is substantial evidence that anthropometric measures of obesity, gestational age at GDM diagnosis, and method of glucose control were three major risk factors for the subsequent development of T2DM among women with previous GDM. For example, high prepreganancy body mass index (BMI) > 27 kg/m^2 is associated with an eight-fold increased risk of T2DM (Pallardo et al., 1999). High postpartum anthropometric measures, such as BMI, body weight, waist circumference, waist–hip ratio, skin fold thickness at subscapular, suprailiac, or tricep, are all positively associated with risk of T2DM (Cho, Jang, Park, & Cho, 2006). Among these anthropometric measures, waist circumference is one of the key risk factors for the onset of T2DM in Korean women with history of GDM (Cho et al., 2006). Postpartum BMI ≥ 27 kg/m^2 is also associated with a four-fold increased risk of T2DM (Dacus et al., 1994). For each week of increase in gestational age at GDM diagnosis, there was a 0.99 decrease in the odds of developing T2DM (Jang et al., 2003). Women who used insulin during pregnancy had three- to five-fold higher risk of developing T2DM than those did not use insulin (Cheung & Helmink, 2006; Löbner et al., 2006).

One important physiologic change during normal pregnancy is increased insulin resistance allowing continuous glucose transfer to the fetus. Insulin sensitivity usually declines to one-third of the nonpregnant state by late pregnancy (Buchanan & Xiang, 2005). Although the exact pathophysiology is still unclear, similar to T2DM, GDM is thought to occur when increased insulin resistance is coupled with pancreatic beta cell insufficiency (Buchanan, 2001).

Evidence supports that GDM shares several risk factors with T2DM, such as high BMI. Studies have consistently shown that a family history of diabetes mellitus contributes to an increased risk of GDM (Solomon et al., 1997; Williams, Qiu, Dempsey, & Luthy, 2003). In addition, researchers have demonstrated that GDM shares some genetic risk factors with T2DM (Robitaille & Grant, 2008; Shaat & Groop, 2007).

Because GDM shares several risk factors and similar pathophysiology with T2DM, research to explore candidate genes of type 1 diabetes mellitus (T1DM) and T2DM in GDM (Robitaille & Grant, 2008; Watanabe et al., 2007) is increasing. For example, using the whole genome microarray, Zhao and associates (2011) examined genetic expression profiles in blood and placenta from seven patients with GDM and six healthy pregnant Chinese women. These authors found immune related pathways and inflammatory functional categories were associated with GDM in this small group of Chinese subjects. In addition, there were 8 overlapping pathways (335 genes) between GDM and T1DM and 11 overlapping pathways (48 genes) between GDM and T2DM. This study illustrates that GDM shares important common characteristics with T1DM and T2DM, including autoimmune destruction and insulin resistance.

Genetic exploration of GDM is in its initial stage. The genetics of GDM, focusing on human association studies with candidate genes common to both T2DM and GDM is elegantly summarized by Robitaille and Grant (2008). The purpose of this chapter is to provide a comprehensive overview to include recent literature on susceptible gene variants that may contribute to both GDM and T2DM.

SEARCH STRATEGIES

A systematic literature search using PubMed was performed to identify studies on genetic variants related to GDM, published from 1990 to 2011. Search terms used include "gestational diabetes," "type 2 diabetes," in combination with "genetics," "gene," and "polymorphism." Studies examining association between GDM and T2DM susceptible genes were abstracted. Research reports and doctoral dissertations in English and Chinese languages were reviewed. Studies excluded from review were those focused on immune-related genes (e.g., human leukocyte antigen [HLA]), which are more likely associated with T1DM. Of the articles identified, 24 were included in this review.

RESULTS

Since 2007, genome-wide association studies have identified many novel susceptible genes for T2DM. These genes have also been studied in women with GDM of various races, including Americans, Europeans, Mediterraneans, Asians, and Arabians. The number of women with GDM included in these studies ranged greatly from 25 (Low, Mohd Tohit, Chong, & Idris, 2010) to 930 (Shin et al., 2010). Sample size powered at least 80% was reported in only 6 of 24 studies (King, 2007; Lauenborg et al., 2009; Pappa et al., 2011; Shaat et al., 2005; Shaat et al., 2007; Zhou et al., 2009). Most studies compared genetic variants between pregnant women with GDM (cases) and pregnant women with normoglycemia (controls). Three studies recruited women with previous GDM as cases (Lauenborg et al., 2009; Rissanen et al., 2000; Watanabe et al., 2007) and controls were women who have gone through pregnancy without GDM (Lauenborg et al., 2009; Watanabe et al., 2007) or normoglycemic men and women (Rissanen et al., 2000). Two studies recruited older nondiabetic women and men (mean age > 60) as controls (Cho et al., 2009; Kwak et al., 2010). The various diagnostic criteria for GDM used in these studies are summarized in Table 12.1.

Genes reviewed were categorized into three groups: genes affecting insulin secretion, genes affecting insulin resistance, and genes affecting mitochondria function. Findings from these studies are summarized in Tables 12.2–12.4. Polymorphisms of genes, such as plasminogen activator inhibitor type 1 (PAI-1) gene and forkhead box C2 (FOXC2) gene, studied in women with GDM (Leipold, Knoefler, Gruber, Klein, et al., 2006; Pappa et al., 2011; Shaat et al., 2007) but not associated with T2DM (Carlsson, Groop, & Ridderstråle, 2005; Osawa et al., 2003) were not included in this review.

Genes Affecting Insulin Secretion Function

ABCC8 and KCNJ11

The pancreatic beta cell adenosine triphosphate (ATP) sensitive potassium (K_{ATP}) channel is a key component in insulin secretion. Elevated blood glucose level leads to increased glucose metabolism and results in closure of the K_{ATP} channel. Closure of the K_{ATP} channel leads to beta cell membrane depolarization and activates voltage dependent calcium channels, subsequently triggering insulin granule exocytosis. Pharmacological agents, such as sulfonylureas, bind to and close the K_{ATP} channel leading to insulin secretion. The beta cell K_{ATP} channel consists of two essential subunits, Kir6.2 (inwardly rectifying potassium channel, subfamily J, member 11 [KCNJ11]) and sulfonylurea receptor 1 (SUR1; ATP-binding cassette, subfamily C, member 8 [ABCC8]; Inagaki et al., 1995). Mutations in the genes encoding for the two essential subunits of the K_{ATP} channel can result in hypoglycemia and hyperglycemia (Flanagan et al., 2009). Gene mapping data showed that these two potassium channel subunit genes, KCNJ11 and ABCC8, are clustered on human chromosome 11 at position 11p15.1, located only 4.5-kilobases apart (Inagaki et al., 1995).

TABLE 12.1
Genetics Studies of GDM and Sample Characteristics

	Cases				Controls		
Author	*n*	Age (yrs)	BMI (kg/m²)	Dx Criteria	*n*	Age (yrs)	BMI (kg/m²)
Pappa et al., 2011	148	32.5 ± 4.5	26 ± 5	other	107	26.7 ± 3.9	24.3 ± 2.1
Kwak et al., 2010	869	32.0 ± 3.9	23.1 ± 3.6	other	632[N]	64.7 ± 3.6	23.3 ± 3.0
Low et al., 2010	26	-	-	other	53	-	-
Shin et al., 2010	930	33.17	23.32 ± 4.0	other	1260[N]	-	-
Cho et al., 2009	869	32.0 ± 3.9	23.1 ± 3.6	other	278[M]	64.9 ± 3.8	22.9 ± 2.7
					345[N]	64.4 ± 3.3	23.9 ± 3.3
Lauenborg et al., 2009	283[N]	43.1	-	WHO	2446[N]	45.2	-
Zhou et al., 2009	520	32.5 ± 3.9	-	other	275	-	-
					641	30.7 ± 3.9	-
Litou et al., 2007	161	32.9 ± 5.1	27.4 ± 5.8	ADA	111	30.8 ± 5.1	24.9 ± 4.2
King, 2007	157	29.8 ± 5.3	27.4 ± 6.3	other	157	29.8 ± 5.2	27.3 ± 6.3
Shaat et al., 2007	649	32.3 ± 0.2	-	other	1232	30.5 ± 0.1	-
Watanabe et al., 2007	94[N]	35.0 ± 8.6	30.9 ± 8.9	other	58[N]	33.4 ± 7.6	27.2 ± 6.6
Fallucca et al., 2006	309	34.1 ± 4.6	25.4 ± 5.0	other	277	32.7 ± 4.3	23.8 ± 4.0
Leipold, Knoefler, Gruber, Huber, et al., 2006	100	32.2 ± 5.5	28.0 ± 7.1	GSD	100	29.7 ± 6.1	25.0 ± 5.7
Tok et al., 2006b	62	-	-	NDDG	100	-	-

Cauza et al., 2005	98	-	-	other	102	-	-
Niu et al., 2005	35	29 ± 2	-	other	35	27 ± 2	-
Shaat et al., 2005	588	32.2 ± 0.2	-	other	1189	30.5 ± 0.1	-
Leipold et al., 2004	40	33.6 ± 4.8	27.6 ± 6.1	GSD	40	31.0 ± 6.2	24.9 ± 5.7
Shaat et al., 2004	400	32.4 ± 0.4	28.9 ± 0.5	other	428	-	-
	100	31.9 ± 0.6	30.9 ± 0.6	other	122	-	-
Tsai et al., 2004	34	32.9 ± 4.3	22.9 ± 3.4	other	189	30.8 ± 4.2	21.4 ± 2.9
	7	31.0 ± 2.0	23.7 ± 2.6		69	30.7 ± 4.1	21.1 ± 2.9
Alevizaki et al., 2000	180	17–48	17.5–47.8	ADA	131	18–45	16.5–50.8
Chen et al., 2000	137	33 ± 4.2	-	ADA	292	32 ± 4.3	
Rissanen et al., 2000	42^{N}	38 ± 1	29.1 ± 1.2	other	82^{M}	54 ± 1	23.6 ± 1.4
					150^{M}; 145^{N}	44 ± 1	25.6 ± 0.2
Festa et al., 1999	70	-	-	other	109	-	-

Note. M = men; N = nonpregnant women; Dx Criteria = diagnostic criteria of GDM; GSD = German Society for Diabetes; NDDG = National Diabetes Data Group; WHO = World Health Organization; ADA = American Diabetes Association.

TABLE 12.2

Summary of Study Findings Related to Insulin Secretion Genes

Gene	Variant	Population	Frequency (%)		At-risk		
			GDM	Control	Genotype	Statistic *p* or *OR*	Reference
ABCC8	exon 16, −3C>T; rs1799854	Finn	55.0	43.0	tag GCC	$p = .024$*	Rissanen et al., 2000
	exon 31, R1273R;	Finn	87.0	74.0	G	$p = .009$*	Rissanen et al., 2000
	rs1799859	Chinese	41.4	24.3	A	$p < .05$*	Niu et al., 2005
	intron 24, −3T>C	Chinese	70.0	52.9	C	$p < .05$*	Niu et al., 2005
KCNJ11	rs5219 (E23K; Glu23Lys)	Scandinavian	42.2	38.3	K	1.17 (1.02–1.35)*	Shaat et al., 2005
		Greek	21.0	19.2	K	1.12 (0.72–1.74)	Pappa et al., 2011
		Korean	40.8	37.9	A	$p = .13$	Cho et al., 2009
		Danish White	40.0	27.7	T	1.17 (0.97–1.41)	Lauenborg et al., 2009 Cho et al., 2009
TCF7L2	rs7903146 (IVS3 C>T)	Scandinavian	31.9	23.8	T	1.49 (1.28–1.75)*	Shaat et al., 2007
		Korean	3.9	2.5	T	$p = .038$*	Cho et al., 2009
		Danish White	34.6	26.8	T	1.45 (1.19–1.75)*	Lauenborg et al., 2009
		Greek	39.5	24.3	T	2.04 (1.38–3.00)*	Pappa et al., 2011
		Mexican-American	-	-	-	$p = .601$	Watanabe et al., 2007
	rs12255372 (IVS4 G>T)	Mexican-American	39.4	20.7	T	2.49 (1.17–5.31)*	Watanabe et al., 2007
		Korean	0.4	0.2	T	2.56 (0.53–12.34)	Cho et al., 2009
	rs7100927	Mexican-American	-	-	G	$p = .627$	Watanabe et al., 2007
KCNQ1	rs2237892	Korean	66.0	61.0	C	1.25 (1.08–1.45)*	Shin et al., 2010
		Korean	65.9	60.8	C	1.24 (1.07–1.45)*	Kwak et al., 2010
		Chinese	70.5	67.7	C	1.14 (0.97–1.35)	Zhou et al., 2009
	rs2074196 (in LD with	Korean	61.8	58.1	G	1.17 (1.01–1.36)*	Kwak et al., 2010
	rs2237892)	Korean	60.0	58.0	G	1.10 (0.95–1.27)	Shin et al., 2010
	rs2237895	Korean	35.0	30.0	C	1.24 (1.07 – 1.45)*	Shin et al., 2010
		Chinese	34.3	30.3	C	1.20 (1.02 – 1.41)*	Zhou et al., 2009
		Korean	32.9	30.1	C	1.15 (0.98 – 1.36)	Kwak et al., 2010
	rs2237896	Chinese	67.8	63.2	G	1.23 (1.05 – 1.44)*	Zhou et al., 2009

UCP2	rs659366 (−866 G>A)	Scandinavian	61.8	60.2	G	1.07 (0.92 – 1.23)	Shaat et al., 2005
CDKAL1	rs7756992 (G>A)	Korean	60.9	52.6	G	$p = 9.14 \times 10^{-9}$*	Cho et al., 2009
		Danish White	31.8	27.6	G	1.22 (1.01 – 1.49)*	Lauenborg et al., 2009
	rs7754840 (C>G)	Korean	57.7	46.4	C	$p = 4.17 \times 10^{-9}$*	Cho et al., 2009
CDKN2A/2B	rs564398	Korean	13.2	12.9	C	$p = .78$	Cho et al., 2009
	rs1333040	Korean	33.9	31.6	C	$p = .20$	Cho et al., 2009
	rs10757278	Korean	46.6	45.3	G	$p = .50$	Cho et al., 2009
	rs10811661	Korean	61.2	51.2	T	$p = 1.05 \times 10^{-7}$*	Cho et al., 2009
		Danish White	84.9	83.4	T	1.12 (0.87 – 1.45)	Lauenborg et al., 2009
HHEX/IDE	rs5015480	Korean	21.7	18.6	C	$p = .035$*	Cho et al., 2009
	rs7923837	Korean	25.0	21.1	G	$p = .011$*	Cho et al., 2009
	rs1111875	Korean	35.6	30.5	C	$p = .003$*	Cho et al., 2009
		Danish White	62.4	58.9	C	1.18 (0.98 – 1.43)	Lauenborg et al., 2009
IGF2BP2	rs4402960	Korean	33.3	29.6	T	$p = .034$*	Cho et al., 2009
		Danish White	33.9	30.4	T	1.18 (0.97 – 1.42)	Lauenborg et al., 2009
SLC30A8	rs13266634	Korean	63.8	58.5	C	$p = .005$*	Cho et al., 2009
		Danish White	70.8	67.4	C	1.17 (0.97 – 1.43)	Lauenborg et al., 2009
WFS1	rs10010131 (in LD with rs6446482)	Danish White	54.2	57.6	G	0.87 (0.73 – 1.05)	Lauenborg et al., 2009
HFE	C282Y (Cys282Tyr; G845A)	Northern & Central European	7.7	2.9	Y	$p = .04$*	Cauza et al., 2005
	H63D (His63Asp; C187G)	Mediterranean	1.0	0.8	Y	NS	Cauza et al., 2005
		Northern & Central European	11.7	13.7	D	NS	Cauza et al., 2005
		Mediterranean	6.8	3.0	D	NS	Cauza et al., 2005

NS = nonsignificant; *OR* = odds ratio; * = statistical significant; - = missing information.

TABLE 12.3

Summary of Study Findings Related to Insulin Resistance Genes

Gene	Variant	Population	Frequency (%)		At-risk		
			GDM	Control	Genotype	Statistic$p$ or OR	Reference
ADIPOQ	rs1501299 (+276 G>T)	Scandinavian	30.8	27.6	T	1.17	Shaat et al., 2007
	rs2241766 (+45 T>G)	Malaysian	32.7	17.9	G	(1.01–1.36)* p = .038*	Low et al., 2010
PPARG	rs1801282 (Pro12Ala, C > G)	Scandinavian	14.6	13.7	Pro	NS	Shaat et al., 2004
		Arabian	4.5	7.0	Pro	NS	Shaat et al., 2004
		Scandinavian	14.1	13.4	Pro	1.06 (0.87–1.29)	Shaat et al., 2007
		Americans	4.6	5.6	Pro	NS	King, 2007
		Greek	1.7	3.3	Pro	0.51 (0.16–1.62)	Pappa et al., 2011
		Danish White	87.2	86.5	C	1.06 (0.81–1.41)	Lauenborg et al., 2009
		Korean	95.8	94.7	C	p = .17	Cho et al., 2009
		Turkish	81.6	84	Pro	p = .30	Tok et al., 2006b
	rs3856806 (1431C > T)	Korean	83.7	82.5	C	p = .37	Cho et al., 2009
PPARGC1A	rs8192678 (Gly482Ser; G > A)	Austrian White	35.0	36.0	A	NS	Leipold, Knoefler, Gruber, Huber, et al., 2006
		Scandinavian					
		Austrian White	33.1	34.1	Ser	0.96 (0.83–1.10)	Shaat et al., 2007
	rs2970847 (Thr394Thr; G > A)		18.0	14.5	A	NS	Leipold, Knoefler, Gruber, Huber, et al., 2006
CAPN10	UCSNP-19 (rs3842570; allele 1 = 2 repeats and allele 2 = 3 repeats of 32-bp sequence)	Americans	50.0	75.0	1	p < .001*	King, 2007
		Austrian White	43.8	38.8	1	NS	Leipold et al., 2004

	UCSNP-43 (rs3792267; G/A; allele 1 = G; allele 2 = A)	Austrian White	21.3	26.3	G	NS	Leipold et al., 2004
		Scandinavian	71.9	72.6	G	0.96 (0.82–1.13)	Shaat et al., 2005
		Americans	86.0	82.0	G	NS	King, 2007
	UCSNP-44 (rs2975760; T/C; allele 1 = T; allele 2 = C)	Scandinavian	18.0	18.5	C	0.97 (0.81–1.16)	Shaat et al., 2005
	UCSNP-63 (rs5030952; C/T; allele 1 = C; allele 2 = T)	Austrian White	26.7	28.8	T	NS	Leipold et al., 2004
		Americans	14.5	13.3	T	NS	King, 2007
ADRB3	rs4994 (Trp64Arg)	Austrian White	25.7	11.0	Arg	$p = .01$*	Festa et al., 1999
		Greek	3.3	3.4	Arg	NS	Alevizaki et al., 2000
		Taiwanese	9.8	14.5	Arg	$p = .086$	Tsai et al., 2004
		Scandinavian	8.6	7.2	Arg	1.22 (0.95–1.56)	Shaat et al., 2007
		Italian White	6.3	5.2	Arg	NS	Fallucca et al., 2006
		Americans	6.7	9.0	Arg	NS	King, 2007
FTO	rs8050136	Korean	13.6	12.2	A	$p = .30$	Cho et al., 2009
	rs9939609	Danish White	46.2	40.6	A	1.15 (0.95–1.38)	Lauenborg et al., 2009
INS	VNTR	Greek	23.0	12.6	III	$p = .003$*	Litou et al., 2007
		Scandinavian	28.4	28.4	III	NS	Shaat et al., 2004
		Arabian	22.0	18.9	III	NS	Shaat et al., 2004
IRS1	rs1801278 (Gly972Arg, G972R)	Italian White	6.8	3.9	Arg	$p = .039$*	Fallucca et al., 2006
		Greek	36.1	25.2	Arg	1.67	Pappa et al., 2011
		Scandinavian	4.8	4.7	Arg	(1.14–2.47)*	Shaat et al., 2005
		Turkish	14.5	11	Arg	1.04 (0.75–1.44) $p = .51$	Tok et al., 2006a

NS = nonsignificant; *OR* = odds ratio.

TABLE 12.4

Summary of Study Findings Related to Mitochondria Function

Gene	Variant	Population	Frequency (%)		At-risk	Statistic	Reference
			GDM	Control	Genotype	p or *OR*	
tRNA-Leu (UUR)	A3243G	Arabian	1.0	0	-	NS	Shaat et al., 2004
		Scandinavian	0.3	0	-	NS	Shaat et al., 2004
		Singaporean	0	0	-	-	Chen et al., 2000
	C3254A	Singaporean	0.7	0	-	$p = .31$	Chen et al., 2000
ND1	G3316A (Ala > Thr)	Singaporean	3.6	1.0	A	$p = .117$	Chen et al., 2000
	T3394C (Tyr > His)	Singaporean	2.9	0.7	C	$p = .085$	Chen et al., 2000
	T3398C (Met > Thr)	Singaporean	2.9	0	T	$p = .01$*	Chen et al., 2000
	A3399T (Met > Ile)	Singaporean	0.7	0	T	$p = .31$	Chen et al., 2000

NS = nonsignificant; *OR* = odds ratio.

Rissanen and colleagues (2000) compared eight variants of the SUR1 (ABCC8) gene in 42 subjects with GDM, 40 subjects with T2DM, and 377 normoglycemic subjects. They found that a G allele of the R1273R polymorphism (rs1799859) in exon 31 and a tagGCC allele of exon 16 splice acceptor site were associated with both GDM and T2DM among Finns. Despite the reported associations, these variations were not associated with altered insulin secretion. These researchers further found that the R1273R and cagGCC > tagGCC were in linkage disequilibrium in patients with GDM, but were not in normoglycemic subjects, suggesting a functional variant contributing to the risk of GDM and T2DM may locate close to the SUR1 gene. Niu and colleagues (2005) compared two variants of the SUR1 (ABCC8) gene in 35 women with GDM, 35 women with T2DM, and 35 healthy women. They reported that the A allele of the R1273R in exon 31 and the C allele of −3 T/C polymorphism in intron 24 were significantly implicated in the susceptibility of GDM among Chinese pregnant women.

The difference in results regarding polymorphism of the R1273R in exon 31 and GDM between Finns and Chinese is unclear. Goksel and colleagues (1998) have shown that A allele of the R1273R in exon 31 is associated with hyperinsulinemia, a risk factor for T2DM (Haffner, Stern, Mitchell, Hazuda, & Patterson, 1990) and obesity in nondiabetic Mexican Americans. Studies in larger populations with more detailed phenotypes are required to investigate properly the interactions of ABCC8 variants with these traits in women with GDM.

Hani and colleagues (1998) conducted a meta-analysis in combined Whites showed an association between T2DM and the E23K variant (Glu23Lys; rs5219) of the KCNJ11, resulting in substitution of a lysine for a glutamic acid. Shaat and colleagues (2005) found an increased frequency of the K genotype among 588 Scandinavian women with GDM as compared to 1,189 nondiabetic pregnant women. However, this association is not supported by a subsequent study comparing 867 Korean women with GDM to 632 older women and men without T2DM (Cho et al., 2009). The reported frequency of the at-risk K genotype was 40.0% and 27.6% in 283 Danish women with GDM and 2,446 glucose-tolerant women, respectively (Lauenborg et al., 2009). Because of limited studies available, the contribution of the E23K variant of KCNJ11 gene to GDM is difficult to evaluate.

TCF7L2

Transcription factor 7-like 2 (TCF7L2, formerly TCF-4) gene, locating at chromosome 10q25.3, is implicated in blood glucose homeostasis via repression of preglucagon gene expression and the synthesis of the incretin hormone, glucagon-like peptide-1 (GLP-1), in the enteroendocrine cells (Yi, Brubaker, & Jin, 2005). Glucagon, a counter-regulatory hormone to insulin, is produced by pancreatic alpha cells. GLP-1 stimulates insulin secretion, inhibits glucagon release and gastric emptying, and enhances peripheral insulin sensitivity, inducing satiety (Kieffer & Habener, 1999). In mouse islet cells, TCF7L2 is required for maintaining glucose-stimulated insulin secretion and beta-cell survival (Shu et al., 2008). Changes in the level of active TCF7L2 in pancreatic beta cells from carriers of at-risk allele may be the cause for defective insulin secretion and progression of T2DM.

Grant and colleagues (2006) demonstrated a strong genetic association between variants of the TCF7L2 gene, rs7903146 and rs12255372, and T2DM in Icelandic individuals, a Danish cohort, and a European White cohort in the United States. In a meta-analysis involving 35,843 cases of T2DM and 39,123 controls, TCF7L2 genetic variants have been associated with T2DM in various ethnicities (Tong et al., 2009).

Recently, rs7903146 (IVS3 C > T) of the TCF7L2 gene was associated with GDM in Koreans (Cho et al., 2009), Danish Whites (Lauenborg et al., 2009), Scandinavians (Shaat et al., 2007), and Greeks (Pappa et al., 2011). However, this association is not supported in a small sample of Mexican Americans (tagged by rs7901695; Watanabe et al., 2007). Compared with wild-type (CC-genotype), heterozygous (CT-genotype), and homozygous (TT-genotype) carriers had a 1.6-fold and 2.1-fold increased risk of GDM, respectively (Shaat et al., 2007).

In a small group of Mexican Americans, the minor T allele of rs12255372 (IVS4 G > T) was significantly more prevalent in women with history of GDM (39.4%, $n = 94$) compared to those without GDM (20.7%, $n = 58$; *OR* 2.49, 95%; CI = 1.17 − 5.31; $p = .018$; Watanabe et al., 2007). In Koreans, even though the minor T allele of rs12255372 is more prevalent in women with GDM (0.8%, $n = 867$) than controls (0.3%, $n = 630$), this difference did not reach statistical significance (*OR* = 2.56, 95%; CI = 0.53 − 12.34; $p = .24$; Cho et al., 2009). In addition, there is no evidence for association between rs7100927 and GDM in Mexican-Americans (Watanabe et al., 2007).

KCNQ1

Potassium channel, voltage-gate, KQT-like subfamily, member 1 (KCNQ1) gene is located at chromosome 11p15.5. Loss-of-function mutations in the KCNQ1 gene are known to cause hereditary long QT syndrome. KCNQ1 is also found to regulate cell volume (Grunnet et al., 2003), which may be a critical element in the regulation of metabolism and signaling of insulin (Schliess, Reissmann, Reinehr, vom Dahl, & Häussinger, 2004). In pancreatic beta cells, acidic proteases cleave proinsulin to yield insulin, a function likely compromised by cell swelling and fostered by cell shrinkage (Lang et al., 1998).

Polymorphisms such as rs2074196, rs2237892, rs2237895, and rs2237896 of the KCNQ1 gene are strongly associated with T2DM and impaired insulin secretion (Unoki et al., 2008; Yasuda et al., 2008). More recently, susceptibility variants, rs2237895 and rs2237892, in KCNQ1 were associated with T2DM risk in Han Chinese (Tsai et al., 2010) and Japanese populations (Takeuchi et al., 2009), respectively.

Three studies have investigated association between KCNQ1 gene and GDM in Koreans (Kwak et al., 2010; Shin et al., 2010) and in Chinese (Zhou et al., 2009). The C allele of the rs2237892 was associated with GDM in two Korean studies (Kwak et al., 2010; Shin et al., 2010) but not in the Chinese study (Zhou et al., 2009). Instead, rs2237896 and rs2237895 were associated with GDM in a Chinese population (Zhou et al., 2009). Both rs2237892 and rs2237895 were associated with decreased insulin secretion but not with insulin resistance, indicating KCNQ1 conferring a risk for GDM by altering pancreatic beta cell function (Kwak et al., 2010). Haplotype analysis of rs2237892 and rs2237895 showed that the protective T-A haplotype was significantly associated with decreased risk for GDM (Kwak et al., 2010).

UCP2

Uncoupling protein 2 (UCP2) gene, located at 11q13, encodes mitochondrial uncoupling protein 2. In pancreatic islet beta cells, UCP2 has the potential to play a role in the pathogenesis of diet-related T2DM as UCP2 inhibits insulin secretion (Chan, Saleh, Koshkin, & Wheeler, 2004). The A allele of the rs659366 polymorphism of UCP2 gene had been associated with a lower risk of T2DM (Wang et al., 2004). However, there is no significant difference in allelic frequency between the at-risk G allele in Scandinavian pregnant women with and without GDM (Shaat et al., 2005).

CDKAL1

Cyclin-dependent kinase 5 (CDK5) regulatory subunit-associated protein 1-like 1 (CDKAL1) gene, located at chromosome 6p22.3, encodes a 65-kDa protein that has property of inhibiting activation of CDK5. CDK5 is present in pancreatic beta cells and acts as a regulator of insulin exocytosis (Lilja et al., 2001). Using an in vitro model, activation of CDK5 was shown to enhance transcriptional activation of the insulin gene (Ubeda, Rukstalis, & Habener, 2006).

CDKAL1 variants are related to impaired insulin secretion in multiple racial/ethnic groups, including individuals of European ancestry, Europeans (French, Danish, Finnish, German, Italian, Swedish, and Dutch), Hispanic Americans, African Americans, and Han Chinese (Dehwah, Wang, & Huang, 2010). Similarly, CDKAL1 gene was shown to confer susceptibility to T2DM in multiple ethnic groups. In a recent meta-analysis, there was a significant association of rs7756992 and rs7754840 in CDKAL1 gene with T2DM ($OR = 1.15$, 95%; $CI = 1.07 - 1.23$; $p < .0001$; $OR = 1.14$, 95%; $CI = 1.06 - 1.24$; $p = .001$, respectively). However, these associations vary in different ethnic populations (Dehwah et al., 2010).

Both rs7756992 and rs7754840 in CDKAL1 gene were associated with GDM as well as reduced insulin secretory capacity in Koreans (Cho et al., 2009). The at-risk C allele of rs7754840 was also associated with insulin resistance (Cho et al., 2009). The at-risk G allele of rs7756992 was associated with GDM in Danish Whites (Lauenborg et al., 2009).

CDKN2A/2B

Cyclin-dependent kinase inhibitor 2A and 2B (CDKN2A/2B) genes, located at 9p21, may play a role in beta-cell function and pancreatic islet regenerative capacity through inhibition of cyclin-dependent kinase 4 and 6 (Cdk4 and Cdk6, respectively). In murine model, loss of Cdk4 reduces the number of beta-islet cells and activation of Cdk4 results in beta-cell hyperplasia (Rane et al., 1999). The rs10811661 polymorphism, located 125 kb upstream of the CDKN2A/2B genes, has been associated with T2DM in three of the genome-wide association studies (GWAS; Saxena et al., 2007; Scott et al., 2007; Zeggini et al., 2007). Using three nondiabetic White samples of European ancestry, Hribal and colleagues (2011) found that the rs10811661 polymorphism was significantly associated with impaired glucose-stimulated insulin release. However, the rs10811661 polymorphism was not associated with either T2DM or insulin release in Pima Indians (Rong et al., 2009), suggesting ethnic differences.

Interestingly, the rs10811661 polymorphism was significantly associated with GDM in Koreans (Cho et al., 2009) but not in Danish Whites (Lauenborg et al., 2009). Other polymorphisms of CDKN2A/2B, rs564398, rs1333040, and rs10757278, were not associated with GDM in Koreans (Cho et al., 2009).

HHEX/IDE

A 270-kb linkage disequilibrium block on chromosome 10 contains two genes of biological significance, hematopoietically expressed homebox (HHEX) and insulin degrading enzyme (IDE). These genes have roles in the pancreatic development (Bort, Martinez-Barbera, Beddington, & Zaret, 2004) and function (Farris et al., 2003). Among several polymorphisms identified in HHEX/IDS region, the rs1111875 polymorphism, located near the outside of a HHEX exon, has been studied for its association with T2DM. A recent meta-analysis demonstrated that the C allele of the rs1111875 polymorphism is a risk factor for developing T2DM, particularly in East Asian population (Wang, Qiao, Zhao, & Tao, 2011). single nucleotide polymorphism (SNP) rs1111875 does not change amino acid sequence; therefore, further study is needed to determine the biological function of this polymorphism.

The rs1111875 polymorphism in HHEX was significantly associated with GDM in Koreans (Cho et al., 2009) but not in Danish Whites (Lauenborg et al.,

2009). Polymorphisms of the HHEX gene, rs5015480 and rs7923837, were also associated with GDM in Koreans (Cho et al., 2009). In addition, risk alleles of rs1111875, rs5015480, and rs7923837, were all strongly associated with a reduced insulin level in Korean women with GDM (Cho et al., 2009). In this sample of Koreans, the protective allele of rs1111875 in HHEX was modestly associated with increased insulin resistance (adjusted $p = .026$) but strongly associated with increased insulin secretory capacity (adjusted $p = .0000002$). This suggests rs1111875 may be involved in beta cell compensatory insulin secretion due to pregnancy induced insulin resistance.

IGF2BP2

Insulin-like growth factor 2 mRNA-binding protein 2 (IGF2BP2) gene, located at 3q28, belongs to a family of mRNA-binding proteins (IMP1, IMP2, and IMP3; Christiansen, Kolte, Hansen, & Nielsen, 2009). Based on a series of GWAS, subsequent replicated studies, and meta-analyses, IMP2 has been implicated in T2DM, even though there is not yet a compelling molecular mechanism (Christiansen et al., 2009). In a recent retrospective analysis of the Age, Gene/Environment Susceptibility (AGES)-Reykjavik Study ($n \sim 2{,}500$), Rodriguez and colleagues (2010) reported significant associations between polymorphisms, rs1470579 and rs4402960 (in strong linkage disequilibrium [D' = 0.996]), and lower fasting insulin levels as well as impaired beta-cell function, independent of obesity phenotype. The rs4402960 polymorphism was significantly associated with GDM in Koreans (Cho et al., 2009) but not in Danish Whites (Lauenborg et al., 2009). In a sample of Korean women with GDM, rs4402960 polymorphism was not associated with increased insulin resistance or secretory function (Cho et al., 2009).

SLC30A8

Zinc plays an important role in all processes of insulin trafficking, including synthesis, storage, and secretion, and is important in islet of Langerhans cell communication (Chimienti, Devergnas, Favier, & Seve, 2004). Solute carrier family 30, member 8 (SLC30A8) gene, located at 8q24.11, encodes a zinc transporter protein (ZnT-8), which is related to insulin maturation and/or storage processes in insulin-secreting pancreatic beta-cells (Chimienti et al., 2004). SLC30A is strictly expressed in the pancreas islet cells (Chimienti et al., 2004).

In a GWAS of the French population, polymorphism rs13266634 of SLC30A8 gene has been associated with T2DM (Sladek et al., 2007). In a large meta-analysis including 42,609 cases and 69,564 controls from various ethnic groups from Europe, Asia, and Africa, polymorphism rs13266634 was also associated with T2DM in both Europeans and Asians (Jing, Sun, Bi, Shen, & Zhu, 2011).

The rs13266634 polymorphism was significantly associated with GDM in Koreans (Cho et al., 2009) but not in Danish Whites (Lauenborg et al., 2009). In the Korean study, the rs13266634 polymorphism was not associated with increased insulin resistant or secretory function (Cho et al., 2009).

WFS1

Wolfram syndrome (WFS), encompassing diabetes insipidus, diabetes mellitus, optic atrophy, and deafness, is caused by WFS1 gene located at 4p16.1 (Inoue et al., 1998). WFS1, encodes wolframin, is expressed in neurons and pancreatic beta cells and regulates calcium fluxes in the endoplasmic reticulum (Inoue et al., 1998). In a murine model, deletion of WFS1 resulted in a reduction in beta cell mass because of endoplasmic reticulum stress-mediated apoptosis, suggesting WFS1 may be involved in beta cell survival (Riggs et al., 2005).

WFS1 polymorphisms, rs10010131, rs6446482, rs734312, and rs752854, were strongly associated with T2DM, with the minor allele conferring protection against T2DM (Sandhu et al., 2007). However, polymorphism rs10010131 was not statistically significant different among Danish White women with and without GDM (Lauenborg et al., 2009).

HFE

Hemochromatosis (HFE) gene, located at 6p21.3, is responsible for hereditary hemochromatosis (Feder et al., 1996), one of the most common genetic disorder among individuals of European descent. Mutations in HFE gene, C282Y and H63D, have been associated with iron overload (Neghina & Anghel, 2011). Approximately 10%–50% of patients with hemochromatosis develop T2DM (Fargion et al., 1992; Niederau et al., 1996), possibly because of accumulation of iron in the pancreas affecting insulin synthesis and secretion (Hatunic et al., 2010). Mutations in the HFE gene have been associated with an increased risk for T2DM (Kwan, Leber, Ahuja, Carter, & Gerstein, 1998; Moczulski, Grzeszczak, & Gawlik, 2001), however, this association has not been supported in a recent meta-analysis (Ellervik, Birgens, Tybjaerg-Hansen, & Nordestgaard, 2007).

Cauza and associates (2005) studied HFE gene mutations, C282Y and H63D, in 208 pregnant women with GDM and 170 pregnant women without GDM and found that the heterozygous C282Y allele frequency was significantly higher in women with GDM (7.7%) than in healthy pregnant women (2.9%, $p = .04$) of Northern and Central European origin but not those of Southern European or non-European origin. H63D allelic frequency was not different between pregnant women with and without GDM (11.7% and 13.7%, respectively, $p = .45$).

Genes Affecting Insulin Resistance Function

ADIPOQ

Adipocyte-, C1q-, and collagen domain-containing (ADIPOQ) gene, located at 3q27, encodes adiponectin, which is a protein hormone that modulates several metabolic processes, including glucose regulation and fatty acid catabolism (Yamauchi et al., 2002). Although research findings are inconclusive, many investigators have reported that at-risk SNPs in the adiponectin gene are associated with T2DM in different populations (Gibson & Froguel, 2004; Gu et al., 2004; Hara et al., 2002; Li et al., 2011). A recent meta-analysis showed that the rs2241766 (+45G, or SNP45) but not rs1501299 (+276T, or SNP276) might be a susceptibility allele for T2DM in Han Chinese (Li et al., 2011).

T allele of rs1501299 of ADIPOQ was associated with a slightly increased risk of GDM in Scandinavian women (Shaat et al., 2007). G allele of rs2241766 is significantly more frequent in women with GDM (32.7%) than those with normal glucose (17.9%, $p = .038$) in Malaysians (Low et al., 2010). Similar to prior work (Ranheim et al., 2004), Low and colleagues (2010) also reported a significantly lower plasma adiponectin level during early pregnancy in women with GDM than those without ($p < .05$). The plasma adponectin level is related to the genotype of rs2241766. Women with GDM carrying the G allele of rs2241766 had lowest level of plasma adiponectin as compared to other groups indicating a role of rs2241766 in circulating plasma adiponectin levels and its subsequent risk of GDM.

PPARG

Peroxisome proliferator-activated receptor gamma (PPARG) gene, located at 3p25, encodes peroxisome proliferator-activated receptor gamma protein, which is important in the control of insulin sensitivity, glucose homeostasis, and blood pressure (Barroso et al., 1999). Similar to previous meta-analyses, data from a recent meta-analysis involving 32,849 cases and 47,456 controls in 60 studies showed that PPARG polymorphism rs1801282 (Pro12Ala) was associated with a reduction in T2DM risk ($OR = 0.86$, 95%; $CI = 0.81 - 0.90$; Gouda et al., 2010). Most recently, other variant genotypes, including rs3856806 (1431C > T), have been associated with T2DM in a Chinese Han population (Lu et al., 2011).

Although PPARG gene have been explored in different populations, variants rs1801282 and rs3856806 were not different between women with GDM and controls (Cho et al., 2009; King, 2007; Lauenborg et al., 2009; Pappa et al., 2011; Shaat et al., 2004; Shaat et al., 2007; Tok et al., 2006b). In Korean (Cho et al., 2009) and Greek (Pappa et al., 2011) women with GDM, insulin resistance and secretory measures were not associated with PPARG variants, rs1801282 and rs3856806. However, Shaat and associates (2004) reported a significant higher homeostatic model assessment insulin resistant (HOMA-IR) in women with homozygous Pro/Pro (2.5 ± 0.2) than Ala carrier (1.9 ± 0.1; one-tailed $p < .05$) of rs1801282 among Scandinavian and Arabian women.

PPARGC1A

Peroxisome proliferator activated receptor gamma coactivator 1-alpha (PPARGC1A) gene, located at 4p15.1, codes peroxisome proliferator activated receptor gamma coactivator 1 protein. Its expression might influence insulin sensitivity as well as energy expenditure, thereby contributing to the development of obesity, a risk factor for diabetes mellitus (Esterbauer, Oberkofler, Krempler, & Patsch, 1999). The most recent meta-analysis showed that rs8192678 (Gly482Ser) and rs2970847 (Thr394Thr) polymorphisms of PPARGC1A were significantly associated with the risk of T2DM, especially in the Asian Indian population (Yang, Mo, Chen, Lu, & Gu, 2011). Studies on PPARGC1A genetic polymorphisms and GDM are limited. Polymorphisms, rs8192678 and rs2970847, have not been associated with GDM in European Whites living in Vienna (Leipold, Knoefler, Gruber, Huber, et al., 2006) or Scandinavian women (Shaat et al., 2007).

CAPN10

Calpain 10 (CAPN10) gene, located at 2q37.3, encodes a nonlysosomal cysteine protease that is expressed in tissues, including pancreatic islets, liver, skeletal muscle, and adipose that play an important role in the regulation of glucose homeostasis (Horikawa et al., 2000).

In Mexican Americans, UCSNP-43 polymorphism in the CAPN10 gene accounted for 14% of the population attributable risk to T2DM, with the common G allele significantly increased in affected subjects (Horikawa et al., 2000). The rare allele C in the UCSNP-44 locus has also been associated with risk of T2DM (Song, Niu, Manson, Kwiatkowski, & Liu, 2004). In addition, there is evidence for the association between risk haplotypes from UCSNP-43, -19, and -63 and T2DM (Song et al., 2004).

The association between polymorphisms (UCSNP-19, -43, -44 and -63) of CAPN10 and GDM was explored (King, 2007; Leipold et al., 2004; Shaat et al., 2005). There was a significant association between GDM and UCSNP-19 but not UCSNP-43 or -64 in a group of women living in the United States (78% European Whites and 13% African American; King, 2007). In an Austrian study, there was a significant association between GDM and UCSNP-63 ($p = .02$) as well as a risk haplotype from UCSNPs-43, -19, and -63 (121/221) with GDM (Leipold et al., 2004). However, no association was found between UCSNPs-43 or -44, in strong linkage disequilibrium with -63, with GDM in Scandinavian women (Shaat et al., 2005).

ADRB3

Beta 3-adrenergic receptor (ADRB3) gene, located at 8p11.23, is expressed in various tissues, including pancreatic beta cells (Perfetti et al., 2001). ADRB3 plays a key role in energy metabolism. Replacement of tryptophan by arginine at position 64 (Trp64Arg) in the ADRB3 gene has been associated with increased BMI (Kurokawa et al., 2008), earlier onset of T2DM (Fujisawa et al., 1996), and clinical features of the insulin resistance syndrome (Zhan & Ho, 2005) in particular ethnic groups. The frequency of ADRB3Arg64 varies greatly across different populations, from 4% in Greek and Swedish, 8%–15% in other European populations, and 31%–37% in the Pima Indian and Japanese (Alevizaki et al., 2000).

The Arg64 of ADRB3 was associated with mild GDM and increased weight gain in pregnancy among Austrians (Festa et al., 1999) but not in other populations (Alevizaki et al., 2000; Fallucca et al., 2006; King, 2007; Shaat et al., 2007; Tsai et al., 2004). In Greek women with GDM, fasting insulin resistance index was significantly higher in Arg64 carriers than those without GDM; however, this difference was no longer observed when obesity was considered (Alevizaki et al., 2000). Taiwanese women with GDM carrying the Arg64 variant had higher fasting and postload (oral glucose tolerance test) insulin levels but not plasma glucose, BMI or body weight gain than those carrying the homozygous Try64 allele, suggesting a possible role of ADRB3 polymorphism in insulin resistance in GDM (Tsai et al., 2004).

FTO

Fat mass- and obesity-associated (FTO) gene, located at 16q12.2, has been associated with obesity in human (Wu, Saunders, Szkudlarek-Mikho, Serna Ide, & Chin, 2010). Two variants, rs9939609 and rs8050136, in the FTO gene are associated with T2DM in specific populations (Hertel et al., 2011; Liu et al., 2010). However, rs9939609 and rs8050136 were not associated with GDM in Koreans (Cho et al., 2009) and Danish Whites (Lauenborg et al., 2009), respectively. In Korean women with GDM, polymorphism rs9939609 was not associated with measures of insulin resistance and insulin secretory capacity (Cho et al., 2009).

INS

Variable number of tandem repeats (VNTR) polymorphism of the insulin gene (INS), located at 11p15.5, affects transcription rate of the INS gene (Litou et al., 2007). Class I (28–44 repeats) is the most common VNTR type and the longer class III (138–159 repeats) is the less common allele. In a meta-analysis, the III/III genotype was associated with a 40% increased relative risk for T2DM (Ong et al., 1999), although this was not supported by a subsequent large study in Dutch Whites (Hansen et al., 2004).

The III/III genotype of INS-VNTR was significantly more frequent in Greek women with GDM than those without, 8.7% and 2.7%, respectively (Litou et al., 2007). In Greek women with GDM, carriers of the class III allele tend to have lower basal insulin levels than other genotypes (16.01 ± 1.09, 14.2 ± 0.97, and 11.4 ± 1.28 mU/mL for genotypes I/I, I/III and III/III, respectively); however, this difference did not reach statistical significance (p = .09; Litou et al., 2007). The INS-VNTR genotype was not associated with GDM in Arabians and Scandinavians (Shaat et al., 2004).

IRS1

Insulin receptor substrate 1 (IRS1) gene, located at 2q36, encodes IRS-1 protein that plays a pivotal role in insulin signaling. The function of IRS-1 protein is impaired in individuals with insulin resistant (Schmitz-Peiffer & Whitehead, 2003). In a meta-analysis of 27 studies, including multi-ethnic populations, carriers of the 972Arg variant of the IRS-1 gene had a 25% increased risk of having T2DM compared with noncarriers (Jellema, Zeegers, Feskens, Dagnelie, & Mensink, 2003).

Polymorphic allele 972Arg of IRS-1 was significantly associated with GDM in Greeks (Pappa et al., 2011) and Italians (Fallucca et al., 2006) but not in Turkish (Tok et al., 2006a) or Scandinavians (Shaat et al., 2005). In addition, homozygous 972Arg polymorphism was found exclusively in women with GDM by two groups of investigators (Fallucca et al., 2006; Shaat et al., 2005). In addition, Tok et al. (2006a) reported that Turkish women with 972Arg were more obese at the beginning of pregnancy, had higher serum fasting insulin and glucose levels; however, the association between 972Arg with fasting glucose and insulin levels was not observed in Greek women with GDM (Pappa et al., 2011).

Genes Affecting Mitochondria Function

tRNA-Leu (UUR)

Substantial evidence from clinical and animal studies has provided a strong relationship between mitochondrial dysfunction and insulin insensitivity or T2DM (Wang, Wang, & Wei, 2010). In 1992, van den Ouweland and associates first described a large pedigree with maternally transmitted T2DM and deafness caused by an A to G point mutation at nucleotide position 3243 in the mitochondrial transfer ribosome nucleic acid (tRNA) for Leucine 1 (MTTL1, tRNA-Leu, or UUR) gene. Beta cell dysfunction and decreased insulin sensitivity were found to be associated with mitochondrial DNA (mtDNA) mutations (de Andrade et al., 2006; Lindroos et al., 2009). Among studies in GDM, the A3243G mutation was very rare, found in only one Arabian (1%) and one Scandinavian (0.3%; Shaat et al., 2004) women with GDM. A3243 mutation was not found in any Singaporean

(Chen, Liao, Roy, Loganath, & Ng, 2000), Spanish (Albareda et al., 2000), or American (72% European White and 28% Black; Allan et al., 1997) women with GDM. A3243 mutation was not found in any controls in the aforementioned studies. The C3254A mutation was also very rare, found in only one Singaporean woman with GDM (0.7%) but not in controls (Chen et al., 2000).

ND1

Nicotinamide adenine dinucleotide (NADH) dehydrogenase is the first enzyme (Complex I) in the electron transport chain of mitochondrial oxidation phosphorylation (OXPHOS) system. Pathogenic mutations in the mitochondrially encoded NADH dehydrogenase subunit 1 (Mt-ND1 or ND1) gene could impair the OXPHOS activity, thus insulin secretory capacity of pancreatic beta cells (Wollheim, 2000). Mutations in mtDNA ND1 region, such as G3316A and T3394C, were associated with T2DM in Japanese (Hirai et al., 1996; Nakagawa et al., 1995) and Chinese (Yu, Yu, Liu, Wang, & Tang, 2004).

Chen and associates (2000) screened four-point mutations of the ND1 gene and found that the T3398C mutation was present at a significant higher frequency in Singaporean GDM patients (2.9%, 4/137) than in controls (0/292; $p = .01$). In addition, having diabetic mothers occur more frequently among GDM patients carrying the T3398C mutation (75%; 3/4) than those without (34.3%, 47/137), suggesting a strong maternal influence on the development of GDM in those carrying the mutation (Chen et al., 2000).

DISCUSSION

This review is focused on shared genetic variants associated with both GDM and T2DM. This review shows that numbers of studies investigating genetic contribution to GDM are increasing. However, existing studies are usually conducted in women of European origin or in specific Asian groups, such as Koreans, Japanese, and Chinese. Studies on high-risk populations, such as Asian Indians, Iranians, and Sardinians, are still lacking.

All genetic variants associated with GDM included in this review have also been involved in the development of T2D. Among genes affecting insulin secretion function, several susceptible T2DM genes, such as TCF7L2 and KCNQ1, seem to be associated with GDM in different populations. For example, the rs7903146 of TCF7L2 was shown to be associated with GDM in various populations, including Scandinavians, Koreans, Danish Whites, and Greeks (Cho et al., 2009; Lauenborg et al., 2009; Pappa et al., 2011; Shaat et al., 2007). This result is similar to a recent large meta-analysis suggesting TCF7L2 as the most common susceptible gene for T2DM among various ethnic groups in the world (Tong et al., 2009). The mechanism that TCF7L2 affects the susceptibility to T2DM and GDM remain to be elucidated. It is hypothesized that TCF7L2 activates many genes downstream of the Wnt signaling cascade, including proglucagon, which encodes the insulinotropic hormone glucagon-like peptide 1 that plays a critical role in blood glucose homeostasis (Yi et al., 2005). Variants of KCNQ1 gene (rs2237892, rs2237895, rs2237896) increased the risk of GDM in Koreans and Chinese (Kwak et al., 2010; Shin et al., 2010; Zhou et al., 2009), although data on other ethnic groups are not available. Polymorphism rs1799859 of ABCC8 was related to GDM in Chinese (Niu et al., 2005) and Finns (Rissanen et al., 2000); however, the sample size was relatively small in these studies. Polymorphism rs7756992 of CDKAL1 was related to GDM in Koreans (Cho et al., 2009) and in Danish Whites (Lauenborg et al., 2009). PPARG genetic variants were not associated with GDM in any of populations studied, including Scandinavians, Arabians, Americans, Greeks, Danish Whites, Turkish, and Koreans. Associations between GDM and KCNJ11, CDKN2A/2B, HHEX/IDE, IGF2BP2, SLC30A8, HFE, ADRB3, INS and IRS-1 genetic variants have led to different findings across populations. Many genetic polymorphisms related to GDM were investigated in a single study or a single population, thus too premature to reject or support a relationship with GDM in a specific ethic group. This accentuates the need for replication and additional larger studies in the same and different populations.

It is likely that multiple susceptibility variants predispose manifestation of complex genetic conditions, like GDM. In the studies reviewed, gene–gene interaction is rarely reported. Fallucca and coworkers (2006) reported that women with normal glucose tolerance who are carriers of at-risk polymorphisms of IRS-1 and ADRB3 genes had a higher prepregnancy BMI than carriers of the IRS-1 at-risk variant alone ($p = .0034$), the ADRB3 at-risk variant alone ($p = .039$), or neither ($p = .048$), suggesting a possible synergistic effect of the two gene polymorphisms. In addition, Lauenborg and colleagues (2009) tested the additive effect of multiple alleles on risk of GDM by combined analysis of 11 variants of T2DM susceptible genes in a group of Danish women with and without GDM. Disregard GDM status, all women carry between 5 and 19 risk alleles. Women carrying 15 or more T2DM risk alleles had 3.30-fold increase risk of having GDM than those with nine or fewer risk alleles.

Most genetic association studies on GDM have been based on the common-disease common-variant hypothesis, in which susceptibility is because of variants occurring at relatively high frequency but low effect size. Interestingly, some investigators have identified rare T2DM genotypes exclusively occurring in women with GDM. For example, homozygous 972Arg polymorphism of IRS-1 was found exclusively in women with GDM by two groups of investigators (Fallucca et al., 2006; Shaat et al., 2005). The A3243G mutation of tRNA-Leu gene was only found in women with GDM but not in controls (Shaat et al., 2004). The T3398C mutation of the ND1 gene was only found in Singaporean women (2.9%) but not in controls (Chen et al., 2000). It is likely that these rare variants with high effect size contribute to common disease like GDM, congruent with the common disease rare-variant hypothesis. Thus, it is imperative to investigate these and additional rare genetic variants along side of common variants.

CONCLUSION

Identification of shared genetic variants linked to GDM and T2DM will contribute to our understanding of the physiopathology of these conditions. Results from current association studies on T2DM susceptible genes in GDM have shown significant heterogeneity. There may be preliminary evidence that polymorphisms of susceptible genes of T2DM, such as TCF7L2, KCNQ1, and CDKAL1, may increase risk of GDM. Associations between GDM and many genetic variants have led to different findings across populations. Many genetic polymorphisms related to GDM were investigated in a single study or a single population. Replication studies to verify contributions of both common and rare genetic variants for GDM and T2DM in specific racial/ethnic groups are needed. Analyses on gene–gene and gene–environment interactions should also be carried out to fully understand predisposing factors for GDM as well as subsequent development of T2DM.

REFERENCES

Albareda, M., Gallart, L., Mato, M. E., Ortiz, A., Puig-Domingo, M., de Leiva, A., & Corcoy, R. (2000). Mitochondrial gene transfer ribonucleic acid (tRNA)Leu(UUR) 3243 is not a common cause of gestational diabetes mellitus in Spanish women. *Endocrine Journal*, *47*(6), 805–806.

Alevizaki, M., Thalassinou, L., Grigorakis, S. I., Philippou, G., Lili, K., Souvatzoglou, A., & Anastasiou, E. (2000). Study of the Trp64Arg polymorphism of the beta3-adrenergic receptor in Greek women with gestational diabetes. *Diabetes Care*, *23*(8), 1079–1083.

Allan, C. J., Argyropoulos, G., Bowker, M., Zhu, J., Lin, P. M., Stiver, K., . . . Garvey, W. T. (1997). Gestational diabetes mellitus and gene mutations which affect insulin secretion. *Diabetes Research and Clinical Practice*, *36*(3), 135–141.

American Diabetes Association. (2009). Diagnosis and classification of diabetes mellitus. *Diabetes Care*, *32*(Suppl. 1), S62–S67.

Baptiste-Roberts, K., Barone, B. B., Gary, T. L., Golden, S. H., Wilson, L. M., Bass, E. B., & Nicholson, W. K. (2009). Risk factors for type 2 diabetes among women with gestational diabetes: A systematic review. *American Journal of Medicine*, *122*(3), 207–214.e4.

Barroso, I., Gurnell, M., Crowley, V. E., Agostini, M., Schwabe, J. W., Soos, M. A., . . . O'Rahilly, S. (1999). Dominant negative mutations in human PPARgamma associated with severe insulin resistance, diabetes mellitus and hypertension. *Nature*, *402*(6764), 880–883.

Bellamy, L., Casas, J. P., Hingorani, A. D., & Williams, D. (2009). Type 2 diabetes mellitus after gestational diabetes: A systematic review and meta-analysis. *Lancet*, *373*(9677), 1773–1779.

Bort, R., Martinez-Barbera, J. P., Beddington, R. S., & Zaret, K. S. (2004). Hex homeobox gene-dependent tissue positioning is required for organogenesis of the ventral pancreas. *Development, 131*(4), 797–806.

Buchanan, T. A. (2001). Pancreatic B-cell defects in gestational diabetes: Implications for the pathogenesis and prevention of type 2 diabetes. *Journal of Clinical Endocrinology and Metabolism, 86*(3), 989–993.

Buchanan, T. A., & Xiang, A. H. (2005). Gestational diabetes mellitus. *Journal of Clinical Investigation, 115*(3), 485–491.

Carlsson, E., Groop, L., & Ridderstråle, M. (2005). Role of the FOXC2 −512C>T polymorphism in type 2 diabetes: Possible association with the dysmetabolic syndrome. *International Journal of Obesity, 29*(3), 268–274.

Cauza, E., Hanusch-Enserer, U., Bischof, M., Spak, M., Kostner, K., Tammaa, A., . . . Ferenci, P. (2005). Increased C282Y heterozygosity in gestational diabetes. *Fetal Diagnosis and Therapy, 20*(5), 349–354.

Chan, C. B., Saleh, M. C., Koshkin, V., & Wheeler, M. B. (2004). Uncoupling protein 2 and islet function. *Diabetes, 53*(Suppl. 1), S136–S142.

Chen, Y., Liao, W. X., Roy, A. C., Loganath, A., & Ng, S. C. (2000). Mitochondrial gene mutations in gestational diabetes mellitus. *Diabetes Research and Clinical Practice, 48*(1), 29–35.

Cheung, N. W., & Helmink, D. (2006). Gestational diabetes: The significance of persistent fasting hyperglycemia for the subsequent development of diabetes mellitus. *Journal of Diabetes and Its Complications, 20*(1), 21–25.

Chimienti, F., Devergnas, S., Favier, A., & Seve, M. (2004). Identification and cloning of a beta-cell-specific zinc transporter, ZnT-8, localized into insulin secretory granules. *Diabetes, 53*(9), 2330–2337.

Cho, N. H., Jang, H. C., Park, H. K., & Cho, Y. W. (2006). Waist circumference is the key risk factor for diabetes in Korean women with history of gestational diabetes. *Diabetes Research and Clinical Practice, 71*(2), 177–183.

Cho, Y. M., Kim, T. H., Lim, S., Choi, S. H., Shin, H. D., Lee, H. K., . . . Jang, H. C. (2009). Type 2 diabetes-associated genetic variants discovered in the recent genome-wide association studies are related to gestational diabetes mellitus in the Korean population. *Diabetologia, 52*(2), 253–261.

Christiansen, J., Kolte, A. M., Hansen, T. O., & Nielsen, F. C. (2009). IGF2 mRNA-binding protein 2: Biological function and putative role in type 2 diabetes. *Journal of Molecular Endocrinology, 43*(5), 187–195.

Chu, S. Y., Abe, K., Hall, L. R., Kim, S. Y., Njoroge, T., & Qin, C. (2009). Gestational diabetes mellitus: All Asians are not alike. *Preventive Medicine, 49*(2–3), 265–268.

Clausen, T. D., Mathiesen, E. R., Hansen, T., Pedersen, O., Jensen, D. M., Lauenborg, J., . . . Damm, P. (2009). Overweight and the metabolic syndrome in adult offspring of women with diet-treated gestational diabetes mellitus or type 1 diabetes. *Journal of Clinical Endocrinology and Metabolism, 94*(7), 2464–2470.

Crowther, C. A., Hiller, J. E., Moss, J. R., McPhee, A. J., Jeffries, W. S., & Robinson, J. S. (2005). Effect of treatment of gestational diabetes mellitus on pregnancy outcomes. *New England Journal of Medicine, 352*(24), 2477–2486.

Dabelea, D. (2007). The predisposition to obesity and diabetes in offspring of diabetic mothers. *Diabetes Care, 30*(Suppl. 2), S169–S174.

Dabelea, D., Mayer-Davis, E. J., Lamichhane, A. P., D'Agostino, R. B., Jr., Liese, A. D., Vehik, K. S., . . . Hamman, R. F. (2008). Association of intrauterine exposure to maternal diabetes and obesity with type 2 diabetes in youth: The SEARCH Case-Control Study. *Diabetes Care, 31*(7), 1422–1426.

Dacus, J. V., Meyer, N. L., Muram, D., Stilson, R., Phipps, P., & Sibai, B. M. (1994). Gestational diabetes: Postpartum glucose tolerance testing. *American Journal of Obstetrics and Gynecology*, *171*(4), 927–931.

de Andrade, P. B., Rubi, B., Frigerio, F., van den Ouweland, J. M., Maassen, J. A., & Maechler, P. (2006). Diabetes-associated mitochondrial DNA mutation A3243G impairs cellular metabolic pathways necessary for beta cell function. *Diabetologia*, *49*(8), 1816–1826.

Dehwah, M. A., Wang, M., & Huang, Q. Y. (2010). CDKAL1 and type 2 diabetes: A global meta-analysis. *Genetics and Molecular Research*, *9*(2), 1109–1120.

Ellervik, C., Birgens, H., Tybjaerg-Hansen, A., & Nordestgaard, B. G. (2007). Hemochromatosis genotypes and risk of 31 disease endpoints: Meta-analyses including 66,000 cases and 226,000 controls. *Hepatology*, *46*(4), 1071–1080.

Esterbauer, H., Oberkofler, H., Krempler, F., & Patsch, W. (1999). Human peroxisome proliferator activated receptor gamma coactivator 1 (PPARGC1) gene: cDNA sequence, genomic organization, chromosomal localization, and tissue expression. *Genomics*, *62*(1), 98–102.

Fallucca, F., Dalfrà, M. G., Sciullo, E., Masin, M., Buongiorno, A. M., Napoli, A., . . . Lapolla, A. (2006). Polymorphisms of insulin receptor substrate 1 and beta3-adrenergic receptor genes in gestational diabetes and normal pregnancy. *Metabolism: Clinical and Experimental*, *55*(11), 1451–1456.

Fargion, S., Mandelli, C., Piperno, A., Cesana, B., Fracanzani, A. L., Fraquelli, M., . . . Conte, D. (1992). Survival and prognostic factors in 212 Italian patients with genetic hemochromatosis. *Hepatology*, *15*(4), 655–659.

Farris, W., Mansourian, S., Chang, Y., Lindsley, L., Eckman, E. A., Frosch, M. P., . . . Guenette, S. (2003). Insulin-degrading enzyme regulates the levels of insulin, amyloid beta-protein, and the beta-amyloid precursor protein intracellular domain in vivo. *Proceedings of the National Academy of Sciences of the United States of America*, *100*(7), 4162–4167.

Feder, J. N., Gnirke, A., Thomas, W., Tsuchihashi, Z., Ruddy, D. A., Basava, A., . . . Wolff, R. K. (1996). A novel MHC class I-like gene is mutated in patients with hereditary haemochromatosis. *Nature Genetics*, *13*(4), 399–408.

Ferrara, A. (2007). Increasing prevalence of gestational diabetes mellitus: A public health perspective. *Diabetes Care*, *30*(Suppl. 2), S141–S146.

Festa, A., Krugluger, W., Shnawa, N., Hopmeier, P., Haffner, S. M., & Schernthaner, G. (1999). Trp64Arg polymorphism of the beta3-adrenergic receptor gene in pregnancy: Association with mild gestational diabetes mellitus. *Journal of Clinical Endocrinology and Metabolism*, *84*(5), 1695–1699.

Flanagan, S. E., Clauin, S., Bellanné-Chantelot, C., de Lonlay, P., Harries, L. W., Gloyn, A. L., & Ellard, S. (2009). Update of mutations in the genes encoding the pancreatic beta-cell K(ATP) channel subunits Kir6.2 (KCNJ11) and sulfonylurea receptor 1 (ABCC8) in diabetes mellitus and hyperinsulinism. *Human Mutation*, *30*(2), 170–180.

Fujisawa, T., Ikegami, H., Yamato, E., Takekawa, K., Nakagawa, Y., Hamada, Y., . . . Ogihara, T. (1996). Association of Trp64Arg mutation of the beta3-adrenergic-receptor with NIDDM and body weight gain. *Diabetologia*, *39*(3), 349–352.

Galtier, F. (2010). Definition, epidemiology, risk factors. *Diabetes & Metabolism*, *36*(6, Pt. 2), 628–651.

Gibson, F., & Froguel, P. (2004). Genetics of the APM1 locus and its contribution to type 2 diabetes susceptibility in French Caucasians. *Diabetes*, *53*(11), 2977–2983.

Goksel, D. L., Fischbach, K., Duggirala, R., Mitchell, B. D., Aguilar-Bryan, L., Blangero, J., . . . O'Connell, P. (1998). Variant in sulfonylurea receptor-1 gene is associated with high insulin concentrations in non-diabetic Mexican Americans: SUR-1 gene variant and hyperinsulinemia. *Human Genetics*, *103*(3), 280–285.

Gouda, H. N., Sagoo, G. S., Harding, A. H., Yates, J., Sandhu, M. S., & Higgins, J. P. (2010). The association between the peroxisome proliferator-activated receptor-gamma2 (PPARG2) Pro12Ala gene variant and type 2 diabetes mellitus: A HuGE review and meta-analysis. *American Journal of Epidemiology*, *171*(6), 645–655.

Grant, S. F., Thorleifsson, G., Reynisdottir, I., Benediktsson, R., Manolescu, A., Sainz, J., . . . Stefansson, K. (2006). Variant of transcription factor 7-like 2 (TCF7L2) gene confers risk of type 2 diabetes. *Nature Genetics*, *38*(3), 320–323.

Grunnet, M., Jespersen, T., MacAulay, N., Jørgensen, N. K., Schmitt, N., Pongs, O., . . . Klaerke, D. A. (2003). KCNQ1 channels sense small changes in cell volume. *Journal of Physiology*, *549* (Pt. 2), 419–427.

Gu, H. F., Abulaiti, A., Ostenson, C. G., Humphreys, K., Wahlestedt, C., Brookes, A. J., & Efendic, S. (2004). Single nucleotide polymorphisms in the proximal promoter region of the adiponectin (APM1) gene are associated with type 2 diabetes in Swedish caucasians. *Diabetes*, *53*(Suppl. 1), S31–S35.

Haffner, S. M., Stern, M. P., Mitchell, B. D., Hazuda, H. P., & Patterson, J. K. (1990). Incidence of type II diabetes in Mexican Americans predicted by fasting insulin and glucose levels, obesity, and body-fat distribution. *Diabetes*, *39*(3), 283–288.

Hani, E. H., Boutin, P., Durand, E., Inoue, H., Permutt, M. A., Velho, G., & Froguel, P. (1998). Missense mutations in the pancreatic islet beta cell inwardly rectifying K+ channel gene (KIR6.2/BIR): A meta-analysis suggests a role in the polygenic basis of Type II diabetes mellitus in Caucasians. *Diabetologia*, *41*(12), 1511–1515.

Hansen, S. K., Gjesing, A. P., Rasmussen, S. K., Glümer, C., Urhammer, S. A., Andersen, G., . . . Pedersen, O. (2004). Large-scale studies of the HphI insulin gene variable-number-of-tandem-repeats polymorphism in relation to Type 2 diabetes mellitus and insulin release. *Diabetologia*, *47*(6), 1079–1087.

HAPO Study Cooperative Research Group. (2009). Hyperglycemia and Adverse Pregnancy Outcome (HAPO) Study: Associations with neonatal anthropometrics. *Diabetes*, *58*(2), 453–459.

Hara, K., Boutin, P., Mori, Y., Tobe, K., Dina, C., Yasuda, K., . . . Kadowaki, T. (2002). Genetic variation in the gene encoding adiponectin is associated with an increased risk of type 2 diabetes in the Japanese population. *Diabetes*, *51*(2), 536–540.

Hatunic, M., Finucane, F. M., Brennan, A. M., Norris, S., Pacini, G., & Nolan, J. J. (2010). Effect of iron overload on glucose metabolism in patients with hereditary hemochromatosis. *Metabolism: Clinical and Experimental*, *59*(3), 380–384.

Hertel, J. K., Johansson, S., Sonestedt, E., Jonsson, A., Lie, R. T., Platou, C. G., . . . Njølstad, P. R. (2011). FTO, type 2 diabetes, and weight gain throughout adult life: A meta-analysis of 41,504 subjects from the Scandinavian HUNT, MDC, and MPP studies. *Diabetes*, *60*(5), 1637–1644.

Hirai, M., Suzuki, S., Onoda, M., Hinokio, Y., Ai, L., Hirai, A., . . . Toyota, T. (1996). Mitochondrial DNA 3394 mutation in the NADH dehydrogenase subunit 1 associated with non-insulin-dependent diabetes mellitus. *Biochemical and Biophysical Research Communications*, *219*(3), 951–955.

Horikawa, Y., Oda, N., Cox, N. J., Li, X., Orho-Melander, M., Hara, M., . . . Bell, G. I. (2000). Genetic variation in the gene encoding calpain-10 is associated with type 2 diabetes mellitus. *Nature Genetics*, *26*(2), 163–175.

Hribal, M. L., Presta, I., Procopio, T., Marini, M. A., Stancakova, A., Kuusisto, J., . . . Sesti, G. (2011). Glucose tolerance, insulin sensitivity and insulin release in European non-diabetic carriers of a polymorphism upstream of CDKN2A and CDKN2B. *Diabetologia*, *54*(4), 795–802.

Hunt, K. J., & Schuller, K. L. (2007). The increasing prevalence of diabetes in pregnancy. *Obstetrics and Gynecology Clinics of North America*, *34*(2), 173–199, vii.

Inagaki, N., Gonoi, T., Clement, J. P., IV, Namba, N., Inazawa, J., Gonzalez, G., . . . Bryan, J. (1995). Reconstitution of IKATP: An inward rectifier subunit plus the sulfonylurea receptor. *Science, 270*(5239), 1166–1170.

Inoue, H., Tanizawa, Y., Wasson, J., Behn, P., Kalidas, K., Bernal-Mizrachi, E., . . . Permutt, M. A. (1998). A gene encoding a transmembrane protein is mutated in patients with diabetes mellitus and optic atrophy (Wolfram syndrome). *Nature Genetics, 20*(2), 143–148.

Jang, H. C., Yim, C. H., Han, K. O., Yoon, H. K., Han, I. K., Kim, M. Y., . . . Cho, N. H. (2003). Gestational diabetes mellitus in Korea: Prevalence and prediction of glucose intolerance at early postpartum. *Diabetes Research and Clinical Practice, 61*(2), 117–124.

Jellema, A., Zeegers, M. P., Feskens, E. J., Dagnelie, P. C., & Mensink, R. P. (2003). Gly972Arg variant in the insulin receptor substrate-1 gene and association with Type 2 diabetes: A meta-analysis of 27 studies. *Diabetologia, 46*(7), 990–995.

Jing, Y. L., Sun, Q. M., Bi, Y., Shen, S. M., & Zhu, D. L. (2011). SLC30A8 polymorphism and type 2 diabetes risk: Evidence from 27 study groups. *Nutrition, Metabolism, and Cardiovascular Diseases, 21*(6), 398–405.

Kieffer, T. J., & Habener, J. F. (1999). The glucagon-like peptides. *Endocrine Reviews, 20*(6), 876–913.

King, K. B., Gerich, J. E., Guzick, D. S., King, K. U., & McDermott, M. P. (2009). Is a history of gestational diabetes related to risk factors for coronary heart disease? *Research in Nursing and Health, 32*(3), 298–306.

King, K. U. (2007). *Association testing of type 2 diabetes mellitus implicated allelic variants in women with gestational diabetes mellitus and matched controls* (Doctoral dissertation). University of Rochester.

Kjos, S. L., Peters, R. K., Xiang, A., Henry, O. A., Montoro, M., & Buchanan, T. A. (1995). Predicting future diabetes in Latino women with gestational diabetes. Utility of early postpartum glucose tolerance testing. *Diabetes, 44*(5), 586–591.

Kurokawa, N., Young, E. H., Oka, Y., Satoh, H., Wareham, N. J., Sandhu, M. S., & Loos, R. J. (2008). The ADRB3 Trp64Arg variant and BMI: A meta-analysis of 44 833 individuals. *International Journal of Obesity, 32*(8), 1240–1249.

Kwak, S. H., Kim, T. H., Cho, Y. M., Choi, S. H., Jang, H. C., & Park, K. S. (2010). Polymorphisms in KCNQ1 are associated with gestational diabetes in a Korean population. *Hormone Research in Paediatrics, 74*(5), 333–338.

Kwan, T., Leber, B., Ahuja, S., Carter, R., & Gerstein, H. C. (1998). Patients with type 2 diabetes have a high frequency of the C282Y mutation of the hemochromatosis gene. *Clinical and Investigative Medicine. Medecine Clinique et Experimentale, 21*(6), 251–257.

Landon, M. B., Spong, C. Y., Thom, E., Carpenter, M. W., Ramin, S. M., Casey, B., . . . Anderson, G. B. (2009). A multicenter, randomized trial of treatment for mild gestational diabetes. *New England Journal of Medicine, 361*(14), 1339–1348.

Lang, F., Busch, G. L., Ritter, M., Völkl, H., Waldegger, S., Gulbins, E., & Häussinger, D. (1998). Functional significance of cell volume regulatory mechanisms. *Physiological Reviews, 78*(1), 247–306.

Lauenborg, J., Grarup, N., Damm, P., Borch-Johnsen, K., Jørgensen, T., Pedersen, O., & Hansen, T. (2009). Common type 2 diabetes risk gene variants associate with gestational diabetes. *Journal of Clinical Endocrinology and Metabolism, 94*(1), 145–150.

Leipold, H., Knoefler, M., Gruber, C., Huber, A., Haslinger, P., & Worda, C. (2006). Peroxisome proliferator-activated receptor gamma coactivator-1alpha gene variations are not associated with gestational diabetes mellitus. *Journal of the Society for Gynecologic Investigation, 13*(2), 104–107.

Leipold, H., Knoefler, M., Gruber, C., Klein, K., Haslinger, P., & Worda, C. (2006). Plasminogen activator inhibitor 1 gene polymorphism and gestational diabetes mellitus. *Obstetrics and Gynecology, 107*(3), 651–656.

Leipold, H., Knöfler, M., Gruber, C., Haslinger, P., Bancher-Todesca, D., & Worda, C. (2004). Calpain-10 haplotype combination and association with gestational diabetes mellitus. *Obstetrics and Gynecology*, *103*(6), 1235–1240.

Li, Y., Li, X., Shi, L., Yang, M., Yang, Y., Tao, W., . . . Yao, Y. (2011). Association of adiponectin SNP+45 and SNP+276 with type 2 diabetes in Han Chinese populations: A meta-analysis of 26 case-control studies. *PLoS One*, *6*(5), e19686.

Lilja, L., Yang, S. N., Webb, D. L., Juntti-Berggren, L., Berggren, P. O., & Bark, C. (2001). Cyclin-dependent kinase 5 promotes insulin exocytosis. *Journal of Biological Chemistry*, *276*(36), 34199–34205.

Lindroos, M. M., Majamaa, K., Tura, A., Mari, A., Kalliokoski, K. K., Taittonen, M. T., . . . Nuutila, P. (2009). m.3243A>G mutation in mitochondrial DNA leads to decreased insulin sensitivity in skeletal muscle and to progressive beta-cell dysfunction. *Diabetes*, *58*(3), 543–549.

Litou, H., Anastasiou, E., Thalassinou, L., Sarika, H. L., Philippou, G., & Alevizaki, M. (2007). Increased prevalence of VNTR III of the insulin gene in women with gestational diabetes mellitus (GDM). *Diabetes Research and Clinical Practice*, *76*(2), 223–228.

Liu, Y., Liu, Z., Song, Y., Zhou, D., Zhang, D., Zhao, T., . . . Xu, H. (2010). Meta-analysis added power to identify variants in FTO associated with type 2 diabetes and obesity in the Asian population. *Obesity (Silver Spring)*, *18*(8), 1619–1624.

Löbner, K., Knopff, A., Baumgarten, A., Mollenhauer, U., Marienfeld, S., Garrido-Franco, M., . . . Ziegler, A. G. (2006). Predictors of postpartum diabetes in women with gestational diabetes mellitus. *Diabetes*, *55*(3), 792–797.

Low, C. F., Mohd Tohit, E. R., Chong, P. P., & Idris, F. (2010). Adiponectin SNP45TG is associated with gestational diabetes mellitus. *Archives of Gynecology and Obstetrics*, *283*(6), 1255–1260.

Lu, Y., Ye, X., Cao, Y., Li, Q., Yu, X., Cheng, J., . . . Zhou, L.(2011). Genetic variants in peroxisome proliferator-activated receptor-γ and retinoid X receptor-α gene and type 2 diabetes risk: A case-control study of a Chinese Han population. *Diabetes Technology & Therapeutics*, *13*(2), 157–164.

Metzger, B. E. (2007). Long-term outcomes in mothers diagnosed with gestational diabetes mellitus and their offspring. *Clinical Obstetrics and Gynecology*, *50*(4), 972–979.

Metzger, B. E., Cho, N. H., Roston, S. M., & Radvany, R. (1993). Prepregnancy weight and antepartum insulin secretion predict glucose tolerance five years after gestational diabetes mellitus. *Diabetes Care*, *16*(12), 1598–1605.

Moczulski, D. K., Grzeszczak, W., & Gawlik, B. (2001). Role of hemochromatosis C282Y and H63D mutations in HFE gene in development of type 2 diabetes and diabetic nephropathy. *Diabetes Care*, *24*(7), 1187–1191.

Mulla, W. R., Henry, T. Q., & Homko, C. J. (2010). Gestational diabetes screening after HAPO: Has anything changed? *Current Diabetes Reports*, *10*(3), 224–228.

Nakagawa, Y., Ikegami, H., Yamato, E., Takekawa, K., Fujisawa, T., Hamada, Y., . . . Kumahara, Y. (1995). A new mitochondrial DNA mutation associated with non-insulin-dependent diabetes mellitus. *Biochemical and Biophysical Research Communications*, *209*(2), 664–668.

Neghina, A. M., & Anghel, A. (2011). Hemochromatosis genotypes and risk of iron overload—A meta-analysis. *Annals of Epidemiology*, *21*(1), 1–14.

Niederau, C., Fischer, R., Purschel, A., Stremmel, W., Haussinger, D., & Strohmeyer, G. (1996). Long-term survival in patients with hereditary hemochromatosis. *Gastroenterology*, *110*(4), 1107–1119.

Niu, X. M., Yang, H., Zhang, H. Y., Li, N. J., Qi, X. M., Chang, Y., . . . Zhang, Y. (2005). [Study on association between gestational diabetes mellitus and sulfonylurea receptor-1 gene polymorphism]. *Zhonghua Fu Chan Ke Za Zhi*, *40*(3), 159–163.

Ong, K. K., Phillips, D. I., Fall, C., Poulton, J., Bennett, S. T., Golding, J., . . . Dunger, D. B. (1999). The insulin gene VNTR, type 2 diabetes and birth weight. *Nature Genetics*, *21*(3), 262–263.

Ornoy, A. (2005). Growth and neurodevelopmental outcome of children born to mothers with pregestational and gestational diabetes. *Pediatric Endocrinology Reviews*, *3*(2), 104–113.

Osawa, H., Onuma, H., Murakami, A., Ochi, M., Nishimiya, T., Kato, K., . . . Makino, H. (2003). Systematic search for single nucleotide polymorphisms in the FOXC2 gene: The absence of evidence for the association of three frequent single nucleotide polymorphisms and four common haplotypes with Japanese type 2 diabetes. *Diabetes*, *52*(2), 562–567.

Pallardo, F., Herranz, L., Garcia-Ingelmo, T., Grande, C., Martin-Vaquero, P., Jañez, M., & Gonzales, A. (1999). Early postpartum metabolic assessment in women with prior gestational diabetes. *Diabetes Care*, *22*(7), 1053–1058.

Pappa, K. I., Gazouli, M., Economou, K., Daskalakis, G., Anastasiou, E., Anagnou, N. P., & Antsaklis, A. (2011). Gestational diabetes mellitus shares polymorphisms of genes associated with insulin resistance and type 2 diabetes in the Greek population. *Gynecological Endocrinology*, *27*(4), 267–272.

Perfetti, R., Hui, H., Chamie, K., Binder, S., Seibert, M., McLenithan, J., . . . Walston, J. D. (2001). Pancreatic beta-cells expressing the Arg64 variant of the beta(3)-adrenergic receptor exhibit abnormal insulin secretory activity. *Journal of Molecular Endocrinology*, *27*(2), 133–144.

Peters, R. K., Kjos, S. L., Xiang, A., & Buchanan, T. A. (1996). Long-term diabetogenic effect of single pregnancy in women with previous gestational diabetes mellitus. *Lancet*, *347*(8996), 227–230.

Rane, S. G., Dubus, P., Mettus, R. V., Galbreath, E. J., Boden, G., Reddy, E. P., & Barbacid, M. (1999). Loss of Cdk4 expression causes insulin-deficient diabetes and Cdk4 activation results in beta-islet cell hyperplasia. *Nature Genetics*, *22*(1), 44–52.

Ranheim, T., Haugen, F., Staff, A. C., Braekke, K., Harsem, N. K., & Drevon, C. A. (2004). Adiponectin is reduced in gestational diabetes mellitus in normal weight women. *Acta Obstetricia et Gynecologica Scandinavica*, *83*(4), 341–347.

Riggs, A. C., Bernal-Mizrachi, E., Ohsugi, M., Wasson, J., Fatrai, S., Welling, C., . . . Permutt, M. A. (2005). Mice conditionally lacking the Wolfram gene in pancreatic islet beta cells exhibit diabetes as a result of enhanced endoplasmic reticulum stress and apoptosis. *Diabetologia*, *48*(11), 2313–2321.

Rissanen, J., Markkanen, A., Kärkkäinen, P., Pihlajamäki, J., Kekäläinen, P., Mykkänen, L., . . . Laakso, M. (2000). Sulfonylurea receptor 1 gene variants are associated with gestational diabetes and type 2 diabetes but not with altered secretion of insulin. *Diabetes Care*, *23*(1), 70–73.

Robitaille, J., & Grant, A. M. (2008). The genetics of gestational diabetes mellitus: Evidence for relationship with type 2 diabetes mellitus. *Genetics in Medicine*, *10*(4), 240–250.

Rodriguez, S., Eiriksdottir, G., Gaunt, T. R., Harris, T. B., Launer, L. J., Gudnason, V., & Day, I. N. (2010). IGF2BP1, IGF2BP2 and IGF2BP3 genotype, haplotype and genetic model studies in metabolic syndrome traits and diabetes. *Growth Hormone and IGF Research*, *20*(4), 310–318.

Rong, R., Hanson, R. L., Ortiz, D., Wiedrich, C., Kobes, S., Knowler, W. C., . . . Baier, L. J. (2009). Association analysis of variation in/near FTO, CDKAL1, SLC30A8, HHEX, EXT2, IGF2BP2, LOC387761, and CDKN2B with type 2 diabetes and related quantitative traits in Pima Indians. *Diabetes*, *58*(2), 478–488.

Sandhu, M. S., Weedon, M. N., Fawcett, K. A., Wasson, J., Debenham, S. L., Daly, A., . . . Barroso, I. (2007). Common variants in WFS1 confer risk of type 2 diabetes. *Nature Genetics*, *39*(8), 951–953.

Saxena, R., Voight, B. F., Lyssenko, V., Burtt, N. P., de Bakker, P. I., Chen, H., . . . Purcell, S. (2007). Genome-wide association analysis identifies loci for type 2 diabetes and triglyceride levels. *Science*, *316*(5829), 1331–1336.

Schliess, F., Reissmann, R., Reinehr, R., vom Dahl, S., & Häussinger, D. (2004). Involvement of integrins and Src in insulin signaling toward autophagic proteolysis in rat liver. *Journal of Biological Chemistry*, *279*(20), 21294–21301.

Schmitz-Peiffer, C., & Whitehead, J. P. (2003). IRS-1 regulation in health and disease. *IUBMB Life*, *55*(7), 367–374.

Scott, L. J., Mohlke, K. L., Bonnycastle, L. L., Willer, C. J., Li, Y., Duren, W. L., . . . Boehnke, M. (2007). A genome-wide association study of type 2 diabetes in Finns detects multiple susceptibility variants. *Science*, *316*(5829), 1341–1345.

Shaat, N., Ekelund, M., Lernmark, A., Ivarsson, S., Almgren, P., Berntorp, K., & Groop, L. (2005). Association of the E23K polymorphism in the KCNJ11 gene with gestational diabetes mellitus. *Diabetologia*, *48*(12), 2544–2551.

Shaat, N., Ekelund, M., Lernmark, A., Ivarsson, S., Nilsson, A., Perfekt, R., . . . Groop, L. (2004). Genotypic and phenotypic differences between Arabian and Scandinavian women with gestational diabetes mellitus. *Diabetologia*, *47*(5), 878–884.

Shaat, N., & Groop, L. (2007). Genetics of gestational diabetes mellitus. *Current Medicinal Chemistry*, *14*(5), 569–583.

Shaat, N., Lernmark, A., Karlsson, E., Ivarsson, S., Parikh, H., Berntorp, K., & Groop, L. (2007). A variant in the transcription factor 7-like 2 (TCF7L2) gene is associated with an increased risk of gestational diabetes mellitus. *Diabetologia*, *50*(5), 972–979.

Shin, H. D., Park, B. L., Shin, H. J., Kim, J. Y., Park, S., Kim, B., & Kim, S. H. (2010). Association of KCNQ1 polymorphisms with the gestational diabetes mellitus in Korean women. *Journal of Clinical Endocrinology and Metabolism*, *95*(1), 445–449.

Shu, L., Sauter, N. S., Schulthess, F. T., Matveyenko, A. V., Oberholzer, J., & Maedler, K. (2008). Transcription factor 7-like 2 regulates beta-cell survival and function in human pancreatic islets. *Diabetes*, *57*(3), 645–653.

Sladek, R., Rocheleau, G., Rung, J., Dina, C., Shen, L., Serre, D., . . . Froguel, P. (2007). A genome-wide association study identifies novel risk loci for type 2 diabetes. *Nature*, *445*(7130), 881–885.

Solomon, C. G., Willett, W. C., Carey, V. J., Rich-Edwards, J., Hunter, D. J., Colditz, G. A., . . . Manson, J. E. (1997). A prospective study of pregravid determinants of gestational diabetes mellitus. *Journal of Allied Medical Association*, *278*(13), 1078–1083.

Song, Y., Niu, T., Manson, J. E., Kwiatkowski, D. J., & Liu, S. (2004). Are variants in the CAPN10 gene related to risk of type 2 diabetes? A quantitative assessment of population and family-based association studies. *American Journal of Human Genetics*, *74*(2), 208–222.

Takeuchi, F., Serizawa, M., Yamamoto, K., Fujisawa, T., Nakashima, E., Ohnaka, K., . . . Kato, N. (2009). Confirmation of multiple risk Loci and genetic impacts by a genome-wide association study of type 2 diabetes in the Japanese population. *Diabetes*, *58*(7), 1690–1699.

Tam, W. H., Ma, R. C., Yang, X., Ko, G. T., Tong, P. C., Cockram, C. S., . . . Chan, J. C. (2008). Glucose intolerance and cardiometabolic risk in children exposed to maternal gestational diabetes mellitus in utero. *Pediatrics*, *122*(6), 1229–1234.

Tok, E. C., Ertunc, D., Bilgin, O., Erdal, E. M., Kaplanoglu, M., & Dilek, S. (2006a). Association of insulin receptor substrate-1 G972R variant with baseline characteristics of the patients with gestational diabetes mellitus. *American Journal of Obstetrics and Gynecology*, *194*(3), 868–872.

Tok, E. C., Ertunc, D., Bilgin, O., Erdal, E. M., Kaplanoglu, M., & Dilek, S. (2006b). PPAR-gamma2 Pro12Ala polymorphism is associated with weight gain in women with gestational diabetes mellitus. *European Journal of Obstetrics, Gynecology, and Reproductive Biology*, *129*(1), 25–30.

Tong, Y., Lin, Y., Zhang, Y., Yang, J., Liu, H., & Zhang, B. (2009). Association between TCF7L2 gene polymorphisms and susceptibility to type 2 diabetes mellitus: A large Human Genome Epidemiology (HuGE) review and meta-analysis. *BMC Medical Genetics*, *10*, 15.

Tsai, F. J., Yang, C. F., Chen, C. C., Chuang, L. M., Lu, C. H., Chang, C. T., . . . Wu, J. Y. (2010). A genome-wide association study identifies susceptibility variants for type 2 diabetes in Han Chinese. *PLoS Genetics*, *6*(2), e1000847.

Tsai, P. J., Ho, S. C., Tsai, L. P., Lee, Y. H., Hsu, S. P., Yang, S. P., . . . Yu, C. H. (2004). Lack of relationship between beta3-adrenergic receptor gene polymorphism and gestational diabetes mellitus in a Taiwanese population. *Metabolism: Clinical and Experimental*, *53*(9), 1136–1139.

Ubeda, M., Rukstalis, J. M., & Habener, J. F. (2006). Inhibition of cyclin-dependent kinase 5 activity protects pancreatic beta cells from glucotoxicity. *Journal of Biological Chemistry*, *281*(39), 28858–28864.

Unoki, H., Takahashi, A., Kawaguchi, T., Hara, K., Horikoshi, M., Andersen, G., . . . Maeda, S. (2008). SNPs in KCNQ1 are associated with susceptibility to type 2 diabetes in East Asian and European populations. *Nature Genetics*, *40*(9), 1098–1102.

Vambergue, A., Nuttens, M. C., Verier-Mine, O., Dognin, C., Cappoen, J. P., & Fontaine, P. (2000). Is mild gestational hyperglycaemia associated with maternal and neonatal complications? The Diagest Study. *Diabetic Medicine*, *17*(3), 203–208.

Wang, C. H., Wang, C. C., & Wei, Y. H. (2010). Mitochondrial dysfunction in insulin insensitivity: Implication of mitochondrial role in type 2 diabetes. *Annals of the New York Academy of Sciences*, *1201*, 157–165.

Wang, H., Chu, W. S., Lu, T., Hasstedt, S. J., Kern, P. A., & Elbein, S. C. (2004). Uncoupling protein-2 polymorphisms in type 2 diabetes, obesity, and insulin secretion. *American Journal of Physiology. Endocrinology and Metabolism*, *286*(1), E1–E7.

Wang, Y., Qiao, W., Zhao, X., & Tao, M. (2011). Quantitative assessment of the influence of hematopoietically expressed homeobox variant (rs1111875) on type 2 diabetes risk. *Molecular Genetics and Metabolism*, *102*(2), 194–199.

Watanabe, R. M., Black, M. H., Xiang, A. H., Allayee, H., Lawrence, J. M., & Buchanan, T. A. (2007). Genetics of gestational diabetes mellitus and type 2 diabetes. *Diabetes Care*, *30*(Suppl. 2), S134–S140.

Williams, M. A., Qiu, C., Dempsey, J. C., & Luthy, D. A. (2003). Familial aggregation of type 2 diabetes and chronic hypertension in women with gestational diabetes mellitus. *Journal of Reproductive Medicine*, *48*(12), 955–962.

Wollheim, C. B. (2000). Beta-cell mitochondria in the regulation of insulin secretion: A new culprit in type II diabetes. *Diabetologia*, *43*(3), 265–277.

Wroblewska-Seniuk, K., Wender-Ozegowska, E., & Szczapa, J. (2009). Long-term effects of diabetes during pregnancy on the offspring. *Pediatric Diabetes*, *10*(7), 432–440.

Wu, Q., Saunders, R. A., Szkudlarek-Mikho, M., Serna Ide, L., & Chin, K. V. (2010). The obesity-associated Fto gene is a transcriptional coactivator. *Biochemical and Biophysical Research Communications*, *401*(3), 390–395.

Yamauchi, T., Kamon, J., Minokoshi, Y., Ito, Y., Waki, H., Uchida, S., . . . Kadowaki, T. (2002). Adiponectin stimulates glucose utilization and fatty-acid oxidation by activating AMP-activated protein kinase. *Nature Medicine*, *8*(11), 1288–1295.

Yang, X., Hsu-Hage, B., Zhang, H., Zhang, C., & Zhang, Y. (2002). Women with impaired glucose tolerance during pregnancy have significantly poor pregnancy outcomes. *Diabetes Care*, *25*(9), 1619–1624.

Yang, Y., Mo, X., Chen, S., Lu, X., & Gu, D. (2011). Association of peroxisome proliferator-activated receptor gamma coactivator 1 alpha (PPARGC1A) gene polymorphisms and type 2 diabetes mellitus: A meta-analysis. *Diabetes/Metabolism Research and Reviews*, *27*(2), 177–184.

Yasuda, K., Miyake, K., Horikawa, Y., Hara, K., Osawa, H., Furuta, H., . . . Kasuga, M. (2008). Variants in KCNQ1 are associated with susceptibility to type 2 diabetes mellitus. *Nature Genetics*, *40*(9), 1092–1097.

Yi, F., Brubaker, P. L., & Jin, T. (2005). TCF-4 mediates cell type-specific regulation of proglucagon gene expression by beta-catenin and glycogen synthase kinase-3beta. *Journal of Biological Chemistry*, *280*(2), 1457–1464.

Yogev, Y., & Visser, G. H. (2009). Obesity, gestational diabetes and pregnancy outcome. *Seminars in Fetal & Neonatal Medicine*, *14*(2), 77–84.

Yu, P., Yu, D. M., Liu, D. M., Wang, K., & Tang, X. Z. (2004). Relationship between mutations of mitochondrial DNA ND1 gene and type 2 diabetes. *Chinese Medical Journal*, *117*(7), 985–989.

Zeggini, E., Weedon, M. N., Lindgren, C. M., Frayling, T. M., Elliott, K. S., Lango, H., . . . Hattersley, A. T. (2007). Replication of genome-wide association signals in UK samples reveals risk loci for type 2 diabetes. *Science*, *316*(5829), 1336–1341.

Zhan, S., & Ho, S. C. (2005). Meta-analysis of the association of the Trp64Arg polymorphism in the beta3 adrenergic receptor with insulin resistance. *Obesity Research*, *13*(10), 1709–1719.

Zhao, Y. H., Wang, D. P., Zhang, L. L., Zhang, F., Wang, D. M., & Zhang, W. Y. (2011). Genomic expression profiles of blood and placenta reveal significant immune-related pathways and categories in Chinese women with gestational diabetes mellitus. *Diabetic Medicine*, *28*(2), 237–246.

Zhou, Q., Zhang, K., Li, W., Liu, J. T., Hong, J., Qin, S. W., . . . Nie, M. (2009). Association of KCNQ1 gene polymorphism with gestational diabetes mellitus in a Chinese population. *Diabetologia*, *52*(11), 2466–2468.

CHAPTER 13

Genetics and Gastrointestinal Symptoms

Margaret M. Heitkemper, Ruth Kohen, Sang-Eun Jun, and
Monica E. Jarrett

ABSTRACT

Gastrointestinal (GI) symptoms including nausea, vomiting, diarrhea, constipation, abdominal discomfort/pain, and heartburn are ubiquitous and as such are often the focus of nursing interventions. The etiologies of these symptoms include GI pathology (e.g., cancer, inflammation), dietary factors (e.g., lactose intolerance), infection, stress, autonomic nervous system dysregulation, medications, as well as a host of diseases outside the GI tract. This review focuses on a common condition (irritable bowel syndrome [IBS]) that is linked with both bowel pattern and abdominal discomfort/pain symptoms. Family and twin studies give evidence for a role of genetic factors in IBS. Whether genes are directly associated with IBS or influence disease risk indirectly by modulating the response to environmental factors remains unknown at this time. Given the multifactorial nature of IBS, it is unlikely that a single genetic factor is responsible for IBS. In addition, gene–gene (epistatic) interactions are also likely to play a role. Four genes coding for proteins involved in neurotransmission (i.e., the serotonin reuptake transporter [SERT], tryptophan hydroxylase [TPH], alpha$_2$-adrenergic receptor [α_2-ADR], catechol-o-methyl transferase [COMT]) and their potential relevance to GI symptoms and IBS will be reviewed. Further research using genome-wide association approaches with samples well characterized by ethnicity and standardized symptom subgrouping is needed.

http://dx.doi.org/10.1891/0739-6686.29.261

THE CHALLENGE OF IRRITABLE BOWEL SYNDROME

Gastrointestinal (GI) symptoms including nausea, vomiting, diarrhea, constipation, abdominal pain, and heartburn are ubiquitous and as such are often the focus of nursing interventions. Both gastrointestinal and non-gastrointestinal etiologies contribute to these symptoms. Common GI etiologies include GI pathology (e.g., cancer, inflammation, enteric infections), dietary factors (e.g., lactose intolerance, food allergies, gluten intolerance), stress, autonomic nervous system dysregulation (e.g., low vagal tone), medications (e.g., anticholinergics), as well as a host of diseases and disorders outside the GI tract. This review focuses on genetic research related to one of the most common GI conditions diagnosed by the presence of both an abnormal bowel pattern and abdominal pain symptoms.

Irritable bowel syndrome (IBS) is a common chronic GI disorder characterized by symptoms of abdominal pain or discomfort associated with disturbed defecation, for example, diarrhea, constipation, or mixed (Drossman, 2006; Drossman, Camilleri, Mayer, & Whitehead, 2002). IBS is a functional GI disorder. At this time there is no single biomarker that distinguishes all IBS patients either from healthy individuals or into different subgroups characterized by their predominant bowel pattern. Its diagnosis is based on the absence of detectable organic causes and the use of symptom-based diagnostic criteria, currently the Rome III criteria (Mayer, 2008): abdominal pain relieved by a bowel movement or associated with changes in stool frequency or consistency. In the United States, as well as other industrialized nations, IBS is more frequently diagnosed in women than men. Whether this gender difference is related to hormones, stress reactivity, psychological distress, health care seeking behavior, or cultural bias requires further explication (Hungin, Chang, Lock, Dennis, & Barghout, 2005).

The prevalence of IBS based on large population surveys conducted in the United States ranges from 12% to 14% (Hungin et al., 2005). In a nationwide survey, U.S. gastroenterologists identified IBS as the most common diagnosis in their clinical practice, seen in 19% of patients (Russo, Gaynes, & Drossman, 1999). In a recent web-based survey of IBS patients in southeastern United States, 55% of those who responded (n = 1,966) indicated that their symptoms were so severe that it interfered with their lifestyle (Drossman et al., 2009). Compared to healthy controls, it has been shown that IBS sufferers have more sick days, are less productive while at work, and feel that their IBS symptoms had a significant impact on their well-being and daily activities (Heitkemper, Carter, Ameen, Olden, & Cheng, 2002; Hungin et al., 2005; Russo et al., 1999). In addition, there is a higher prevalence of psychological distress and psychiatric disorders, in particular depression and anxiety (e.g., post-traumatic stress disorder [PTSD]; Folks, 2004) in patients with IBS. The treatment of IBS is limited by the lack of understanding of what causes it.

SYMPTOMS AND IRRITABLE BOWEL SYNDROME

As noted previously, a range of GI bowel pattern symptoms as well as discomfort/pain severity are included in the current diagnostic criteria for IBS. The first challenge in studying the role of genetics in IBS is to determine the relevance or importance of sub-grouping of IBS patients based on symptoms and/or their severity and duration. The use of categories based on predominant bowel pattern symptoms (diarrhea-predominant, constipation-predominant, alternating or mixed; IBS-D, IBS-C, IBS-M, respectively; Rome III criteria) has received acceptance by most clinicians and investigators (Drossman, 2006). Indeed, this categorization commonly dictates drug testing and management. However, the influence of pain severity and other symptoms (intestinal and extra-intestinal) may also differentiate the response of patients to treatments. Beyond stool pattern, limited attention has been paid to further differentiate IBS patients by other symptoms (e.g., pain, urgency, bloating, intestinal gas) or their severity. Pain and discomfort severity may be relevant to the subtyping IBS patients as well as understanding mechanisms responsible for symptom reports. When symptoms were tracked through the use of a daily diary, women with medically diagnosed IBS reported more abdominal pain/discomfort, bloating, and intestinal gas than bowel pattern symptoms of constipation and diarrhea (Cain et al., 2009). In addition, abdominal pain/discomfort symptoms correlated more strongly with reductions in quality of life, increased psychological distress, and interference with work and school than bowel symptoms. Others concur that categorizing subgroups of patients with functional GI disorders on bowel patterns alone may not sufficiently identify clinically distinct entities (Talley, Boyce, & Jones, 1998; Wong et al., 2010).

The report of abdominal pain or discomfort is a fundamental symptom in the diagnosis of IBS using the ROME III criteria (Drossman, 2006). In a study of premenopausal women with IBS, Cain and colleagues (2009) noted that study participants reported moderate to severe abdominal pain on an average of 37% of days in a 28-day diary. Patients who reported more frequent or severe abdominal pain or discomfort also endorsed greater psychological distress, more negative cognitions about IBS, reduced quality of life, and more interruptions in daily activities (Cain et al., 2009).

The study of abdominal pain and discomfort within distinct bowel pattern (IBS-C, IBS-D, IBS-M) subgroups is challenged by the various methodological approaches that are typically taken. Often, in clinical practice or research protocols, retrospective questionnaires are used. But this information may not capture the typical bowel pattern. Using respondents from urogynecology, gynecology, and colorectal clinics, Digesu et al., (2010) found that women had difficulty recalling bowel symptoms that had occurred even within a 6-month period.

With retrospective measures recall bias has the potential to influence responses to bowel questionnaires (Bharucha, Seide, Zinsmeister, & Melton, 2008; Halder et al., 2007; Pamuk, Pamuk, & Celik, 2003; Varma et al., 2008). The finding that IBS bowel symptoms wax and wane over time, taken together with patient's recall of their predominant bowel pattern, make it challenging to determine the role of genetic risk factors as well as treatment outcomes using retrospective evaluations of bowel patterns only. The study of genetics in IBS patients should therefore not only be informed by determination of bowel pattern subgroup by the patient or a clinician, but also through the use of measures that track daily symptoms, and include the assessment of parameters such as discomfort or pain along with stool pattern.

GENETICS AND IRRITABLE BOWEL SYNDROME

Numerous studies have shown that IBS aggregates in families (Bellentani et al., 1990; Kanazawa et al., 2004; Levy, Whitehead, Von Korff, & Feld, 2000; Locke, Zinsmeister, Talley, Fett, & Melton, 2000; Morris-Yates, Talley, Boyce, Nandurkar, & Andrews, 1998; Saito et al., 2008; Saito et al., 2010; Whorwell, McCallum, Creed, & Roberts, 1986). However, this aggregation could result from shared environmental factors and learned illness behavior as well as genetic factors (Saito, Mitra, & Mayer, 2010). Definite proof of a genetic basis of IBS has to rely on twin studies, with a greater concordance rate among monozygotic (MZ) than dizygotic (DZ) twins being evidence for heritability. A study in 186 MZ and 157 DZ Australian twin pairs showed 33% vs. 13% concordance, resulting in an estimated genetic liability of 20% (Morris-Yates et al., 1998). A study on 6,060 twins from the Virginia Twin Registry reported 17% concordance for MZ and 8% concordance for DZ twins (Levy et al., 2001). In another U.S. study using 986 twin pairs from the Minnesota Twin Registry, concordance rates for IBS of 17% among MZ and 1% among DZ twins were found (Lembo, Zaman, Jones, & Talley, 2007). Similar results were observed in a study of 3,199 Norwegian twin pairs with 22% concordance for MZ and 9% concordance for DZ twins (Bengtson, Ronning, Vatn, & Harris, 2006) as well as in a large study of 16,961 Swedish twin pairs which found the heritability of IBS to be 25% (Svedberg, Johansson, Wallander, & Pedersen, 2008). To date, only a single twin study using 888 MZ and 982 DZ twin pairs from the U.K. Adult Twin Registry found no evidence for heritability of IBS, with similar concordance rate among MZ (17%) and DZ twins (16%; Mohammed, Cherkas, Riley, Spector, & Trudgill, 2005).

Thus far, more than 100 genetic variants in more than 60 genes from various pathways have been investigated in IBS association studies (Saito et al., 2007, 2011). In a recent review, Saito (2011) highlighted a conceptual model (Figure 13.1) illustrating the linkages among IBS phenotypes and interaction

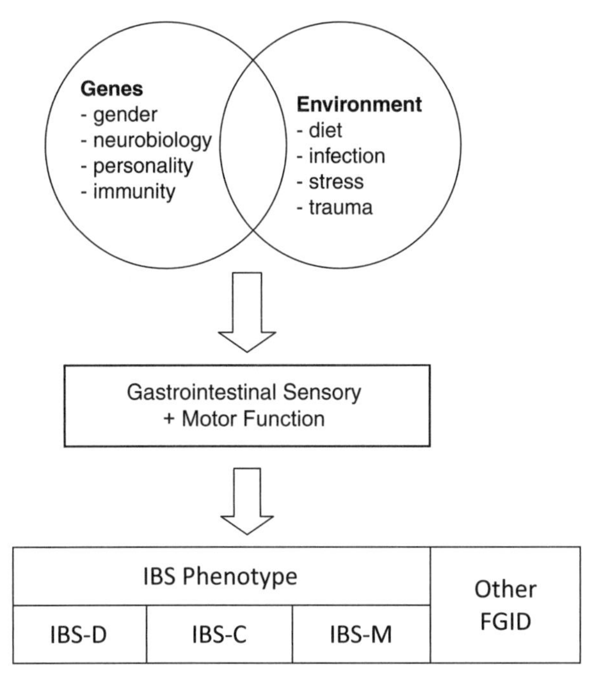

FIGURE 13.1 Gene-environment paradigm in IBS development. IBS-D = IBS-diarrhea; IBS-C = IBS-constipation; IBS-M = IBS-mixed diarrhea and constipation. FGID = functional gastrointestinal disorders. With permission from Saito, Y. A. (2011). The role of genetics in IBS. *Gastroenterology Clinics of North America*, *40*(1), 45–67.

of genes and environment. Clinical endophenotypes including motility and visceral sensitivity markers may serve as important biomarkers linking genetic and environmental influences with symptom expression including bowel pattern and pain. This current review focuses on four genes involved in the regulation of neurotransmission that have been shown to be associated with at least subgroups of patients with IBS.

Serotonin Transporter

Serotonin (5-HT) is an important neurochemical mediator in the CNS and the GI tract (Mawe, Coates, & Moses, 2006; Spiller, 2008). It plays a crucial role in the regulation of mood and behavior by the brain as well as the control of motility and sensory signaling in the gut (Mawe et al., 2006). For the past 3 decades, investigators have examined the relationship of serotonin with IBS symptoms. Evidence for a causal role for serotonin signaling alterations in IBS comes from studies that have found evidence for altered serotonin signaling in the gut and in peripheral blood of IBS patients (Atkinson, Lockhart, Whorwell, Keevil, &

Houghton, 2006; Bellini et al., 2003; Dunlop et al., 2005; Houghton, Atkinson, Lockhard, Whorwell, & Keevil, 2007; Miwa, Echizen, Matsueda, & Umeda, 2001; Torres, Gainetdinov, & Caron, 2003; Zuo et al., 2007). Animal studies support the presence of multiple serotonin receptors on enterochromaffin cells as well as enteric neurons in the GI tract (Chen et al., 2001; Liu, Rayport, Jiang, Murphy, & Gershon, 2002). Alterations in SERT transcript levels have been observed in the intestine of patients with IBS (Kerckhoffs, Ter Linde, Akkermans, & Samson, 2008). Drug therapies employing serotonin agonists and/or antagonists modulate GI motility and intestinal pain/discomfort sensitivity (Saad, 2011; Spiller, 2011). However, these initial drug therapies were later either removed from the market (tegaserod) or restricted in their use (alosetron) because of the adverse side effects found post marketing. The investigation of serotonergic agents for IBS continues with more selectively targeted agents.

Alterations in serotonin biosynthesis or reuptake change its availability in the gut and the brain. The SERT protein is localized on the presynaptic membrane of serotonergic neurons and on enterochromaffin cells in the gut. Serotonin transporter (SERT) acts as the major regulator of serotonergic neurotransmission throughout the body by controlling intensity and duration of serotonergic signaling via re-uptake of 5-HT into the synapse. The SERT gene is located on chromosome 17q11.1-17q12 and organized into 14 exons spanning approximately 38kb (Lesch et al., 1994; Ramamoorthy et al., 1993; See Figure 13.2). The most frequently studied genetic variant of SERT, 5-HTTLPR, is a 43 base pair (bp) insertion/deletion polymorphism located in the promoter region, resulting in a short (s) and a long (l) allele. The shorter (s) variant of the gene is associated with lower SERT expression compared to the long (l) allele (Greenberg, Tolliver, Huang, Li, Bengel, & Murphy

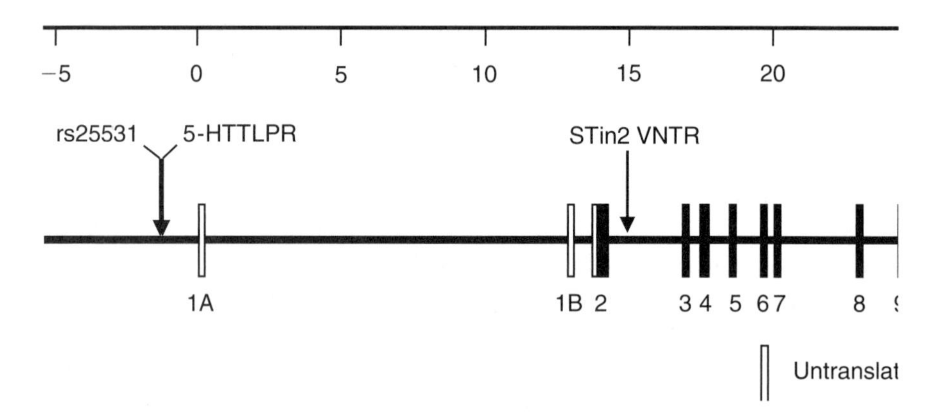

FIGURE 13.2 Structure of the human SERT gene. Distances (kb) are indicated by the ruler at the top. Arrows indicate the most frequently investigated SERT variants.

1999; Heils et al., 1996; Hu et al., 2006). The prevalence of the three alleles varies across the ethnic groups, with the s-allele being most frequent in Asians, and least frequent in African Americans compared to European American (Gelernter, Kranzler, & Cubells, 1997; Markoutsaki et al., 2011).

Previous studies in blood of IBS patients have shown a tight inverse correlation between GI symptoms and SERT binding, implicating loss of SERT function as a potential contributing factor in IBS (Bellini et al., 2003). Animal studies indicate that targeted deletion of the SERT gene alters intestinal motility and transit. Mice with congenitally lost or impaired SERT function resemble patients with IBS in that they suffer from IBS-like GI symptoms of alternating diarrhea and constipation in combination with behaviors suggesting depression and anxiety (Ansorge, Zhou, Lira, Hen, & Gingrich, 2004; Chen et al., 2001; Lira et al., 2003; Liu et al., 2002; Popa, Léna, Alexandre, & Adrien, 2008).

SERT polymorphisms are linked with mood state, most notably depression (Beevers, Wells, Ellis, & McGeary, 2009; Homberg & Lesch, 2011) and reduced emotional resilience (Stein, Campbell-Sills, & Gelernter, 2009). A number of studies have shown that the s-allele of 5-HTTLPR is associated with a higher vulnerability to psychosocial and medical stress (Caspi et al., 2003; Cervilla et al., 2007; Homberg, 2011; Xie et al., 2009; Zalsman et al., 2006; for review see Caspi, Hariri, Holmes, Uher, & Moffitt, 2010). Although an earlier meta-analysis refuted these associations (Risch et al., 2009), a more recent meta-analysis confirms them (Karg, Burmeister, Shledon, & Sen, 2011). Others have found an association between the 5-HTTLPR s-allele and anxiety symptoms (Jarrett et al., 2007). Evidence for a direct association of 5-HTTLPR with IBS is inconclusive, as some studies found an association between IBS and the 5-HTTLPR s/s genotype (Park et al., 2006; Pata et al., 2001; Yeo et al., 2004) whereas others did not (Fukudo et al., 2009; Kim et al., 2004; Lee et al., 2004; Niesler et al., 2010; Sikander et al., 2009).

As shown in Table 13.1, several studies have compared SERT polymorphisms among IBS subgroups based on predominant bowel pattern (constipation, diarrhea) history. A 2007 meta-analysis found that 5-HTTLPR genotype is not significantly associated with bowel (constipation, diarrhea) symptoms in IBS in European American or Asians (Van Kerkhoven, Laheij, & Jansen, 2007). However, using the barostat (inflatable balloon in rectum) in a study of 122 patients with IBS and 39 healthy controls, Camilleri et al. (2007) found that genetic variations in serotonergic systems were significantly associated with rectal compliance. Specifically, the investigators found that the *l/s* genotype was associated with increased rectal compliance and higher pain ratings during the barostat evaluation. A limited number of studies suggest that SERT polymorphisms are associated with the response to serotonergic agents for IBS (Camilleri et al., 2002; Li et al., 2007).

TABLE 13.1

Studies Examining Serotonin Reuptake Transporter Protein in Patients With Irritable Bowel Syndrome and Healthy Controls

Author and Population	Patients (number)	Results
Pata, C., Erdal, M. E., Derici, E. Yazar, A., Kanik A., & Ulu O. (2002). Serotonin transporter gene polymorphism in irritable bowel syndrome. *American Journal of Gastroenterology*, *97*(7), 1780–1784.	Turkey IBS-C (26); IBS-D (18); IBS-Alt (10); 91 Controls	NS IBS vs. Control s/s ↑ in IBS-C > IBS-D
Yeo, A., Boyd, P., Lumsden, S., Saunders, T., Handley, A., Stubbins, M., . . . Hicks, G. A. (2004). Association between a functional polymorphism in the serotonin transporter gene and diarrhoea predominant irritable bowel syndrome in women. *Gut*, *53*(10), 1452–1458.	US IBS (194) Controls (448)	s/s ↑ IBS-D *OR* = 2.23 (95% CI, 1.51–3.31)
Kim, H. J., Camilleri, M., Carlson, P. J., Cremonini, F., Ferber, I., Stephens, D., . . . Urrutia, R. (2004). Association of distinct alpha(2) adrenoceptor and serotonin transporter polymorphisms with constipation and somatic symptoms in functional gastrointestinal disorders. *Gut*, *53*(6), 829–837.	US Rome II: IBS-C (90) IBS-D (128) IBS-Alt (38)	IBS-C *OR* = 0.7 (95% CI, 0.3, 1.4) IBS-D *OR* = 1.0 (95% CI, 0.6, 1.7) IBS-Alt *OR* = 0.7 (95% CI, 0.3, 1.4)
Park, J. M., Choi, M. G., Park J. A., Oh, J. H., Cho, Y. K., Lee, I. S., . . . Chung, I. S. (2006). Serotonin transporter gene polymorphism and irritable bowel syndrome. *Neurogastroenterology Motility*, *18*(11), 995–1000.	Korea IBS (190) Controls (437)	NS IBS subgroup Trend: s/s IBS > Controls
Sikander, A., Rana, S. V., Sharma, S. K., Sinha, S. K., Arora, S. K., Prasad, K. K., & Singh, K. (2010). Association of alpha 2A adrenergic receptor gene (ADRAlpha2A) polymorphism with irritable bowel syndrome, microscopic and ulcerative colitis. *Clinica Chimica Acta*, *411*(1–2), 59–63.	India IBS (151) Controls (100)	s/s: IBS-C 52%, IBS-D 26%, IBS-M 33%

(Continued)

TABLE 13.1

Studies Examining Serotonin Reuptake Transporter Protein in Patients With Irritable Bowel Syndrome and Healthy Controls (Continued)

Author and Population	Patients (number)	Results
Niesler, B., Kapeller, J., Fell, C., Atkinson, W., Moller, D., Fischer, C., . . . Houghton, L. A. (2010). 5-HTTLPR and STin2 polymorphisms in the serotonin transporter gene and irritable bowel syndrome: Effect of bowel habit and sex. *European Journal of Gastroenterology Hepatology*, 22(7), 856–861.	UK Rome II: IBS-D (97); IBS-C (99); Controls (92)	s/s IBS-D 16%; IBS-C 14%; Controls 23.9% NS, Gender, $p < .05$
Markoutsaki, T., Karantanos, T., Gazoulli, M., Anagnou, N. P., Ladas, S. D., Karamanolis, D. G. (2011). Serotonin transporter and G protein beta 3 subunit gene polymorphisms in Greeks with irritable bowel syndrome. *Digestive Diseases and Sciences*, 56(11), 3276–3280.	Greece Rome III: IBS Controls (238)	s/s: 68.5% IBS; 56.7% Controls No association with phenotype

Note. IBS-C = irritable bowel syndrome constipation predominant; IBS-D = irritable bowel syndrome diarrhea predominant; IBS-Alt = irritable bowel syndrome alternating bowel pattern predominant. 95% CI = 95% Confidence Interval.

A second SERT polymorphism, STin2 VNTR, located in intron 2 and consisting of a variable number of nearly identical 17bp segments, has also been investigated in IBS, but no evidence for association with bowel symptoms was found (Kohen et al., 2009). However, in another study of IBS patients, participants who were homozygous for the short allele of 5-HTTLPR or carried a STin2.9 VNTR allele were significantly more likely to have a history of co-morbid depression (Jarrett et al., 2007). The three most frequent alleles are composed of 9, 10, and 12 repeat elements. These serve as transcriptional regulators with allele-specific enhancer-like properties. Kohen et al. (2009) also showed preliminary evidence that the single nucleotide polymorphism (SNP) rs25531, which is located immediately upstream of 5-HTTLPR, is associated with IBS. In this study, the carriers of the rare G allele of rs25531 had approximately 3-fold increased odds of IBS compared to carriers of the more common A-allele. However, the pilot nature of this study, limited to a northwestern U.S. European American sample, along with the small sample size, need to be considered in interpreting these results. Multiple as yet unknown rare SERT variants may also play an important role, as recent literature suggests that common diseases are most likely caused by a combination of rare and common variants (Bodmer & Bonilla, 2008; Nakamura, Ueno, Sano, & Tanabe, 2000; Schork, Murray, Frazer, & Topol, 2009). Additional studies including investigations of genetic associations with psychological factors, pain sensitivity, as well as motility in larger samples of IBS patients is warranted.

Tryptophan Hydroxylase

Tryptophan hydroxylase (TPH) is the rate-limiting enzyme of serotonin synthesis; therefore, *TPH* gene variants have been evaluated for possible associations with disorders whose underlying pathophysiology is related to serotonin. TPH has two isoforms, TPH1 and TPH2, with overall 71% identity in amino acid sequence in humans (Walther & Bader, 2003; Walther et al., 2003). TPH1 consists of 444 amino acids and is encoded on chromosome 11p15.3-p14 with a length of 29 kb and composed of 11 exons (Paoloni-Giacobino et al., 2000), whereas TPH2 consists of 490 amino acids and is encoded on chromosome 12q21.1 with a length of 93.6 kb and composed of 11 exons (Zill et al., 2004, 2007). TPH2 is mainly expressed in the brain, whereas TPH1 is expressed both in the brain and in the periphery such as enterochromaffin (EC) cells in the gut (Walther & Bader, 2003; Walther et al., 2003; Zill et al., 2007).

Because there is a high co-morbidity between IBS and depression, *TPH* gene variants might be possible candidates to influence IBS etiology (Jun, Kohen, Cain, Jarrett, & Heitkemper, 2011; Villani et al., 2010). Variants of the *TPH* gene have also been investigated for possible migraine without aura (Jung et al., 2010) and psychological disorders such as major depression (Gizatullin, Zaboli, Jönsson, Asberg, & Leopardi, 2006), suicidal behavior (Galfalvy, Huang, Oquendo, Currier, & Mann, 2009; Saetre et al., 2010), bipolar disorder (Chen, Glatt, & Tsuang, 2008), attention-deficit/hyperactivity disorder (Halmøy et al., 2010), and anger-related personality traits (Manuck et al., 1999; Rujescu et al., 2002).

Recently, a study reported possible associations between two TPH1 gene single nucleotide polymorphisms (SNPs; rs4537731 and rs211105) and daily reporting of GI symptoms including diarrhea, bloating, and loose stools in European-American women with IBS (n = 199; Jun et al., 2011). This study also showed possible associations between a TPH2 SNP in the promoter region (rs4590625) and stool characteristics, such as diarrhea and constipation. However, there were no associations between any of the tested TPH gene SNPs and a diagnosis of IBS. Villani and colleagues (2010) also reported no associations between TPH1 SNPs (rs1800532 and rs1799913) and TPH2 SNPs (rs11178997, rs4570625 and rs4565946) with the development of post-infectious (PI)-IBS. This study compared 228 subjects who developed gastroenteritis from microbial contamination of the municipal water supply in Walkerton, Canada, in 2000 and reported PI-IBS 2–3 years later, with 581 subjects who developed gastroenteritis but did not develop PI-IBS over the same time (Villani et al., 2010). However, the authors did not investigate rs453771 and rs211105 of theTPH1 gene. Further studies are needed to investigate associations between TPH gene variants and IBS symptoms in larger independent samples.

Studies investigating TPH1 expression in patients with IBS have given inconsistent results (Coates et al., 2004; Mawe et al., 2006). One study reported significantly reduced TPH1 mRNA levels in the colonic mucosa in both diarrhea-predominant and constipation-predominant IBS patients compared to healthy controls (Coates et al., 2004). Another study showed higher mucosal TPH1 mRNA levels in patients with chronic constipation compared to healthy controls (Costedio et al., 2010).

A recent study demonstrated that SNP rs453771 of the TPH1 gene is positively associated with cerebrospinal fluid (CSF) concentrations of the major serotonin metabolites 5-hydroxyindoleacetic acid (5-HIAA) in 132 healthy European Americans ($p = 0.003$; Andreou et al., 2010). The levels of 5-HIAA were higher in those with minor allele G/G and A/G as compared to those with the A/A genotype. However, another study did not show this association in 343 patients with major depression (European Americans, African American, Hispanic; Galfalvy et al., 2009). More functional studies are needed to explain the biological mechanism whereby TPH SNPs could contribute to various phenotypes of serotonin-related disorders as well as influence blood and CSF levels of serotonin.

Alpha2-Adrenoreceptor

The GI tract is extensively innervated by extrinsic noradrenergic neurons that release catecholamines that influence various GI functions including secretion, motility, and sensory. Polymorphic variation within alpha2-adrenoreceptor (ADR) gene contributes to inter-individual variability in cardiovascular responses to laboratory mental challenges implicating the autonomic nervous system (McCaffery, Pogue-Geile, Ferrell, Petro, & Manuck, 2002).

There is some preliminary evidence that pharmacologic agents targeted at alpha$_2$-adrenoreceptors (α_2-ADR) may be useful in the management of IBS (Camilleri et al., 2008). Alpha$_2$-ADRs are located on presynaptic membrane and modulate release of norepinephrine. Agents that block α_2-ADR influence gastric accommodation and transit. This has been shown in animal studies where stimulation of presynaptic alpha$_2$-adrenoreceptors produces decreases acetylcholine release from enteric neurons as well as serotonin release in the brain (Tack & Wood, 1992).

There are three α_2 adrenoreceptor subtypes (2_A, 2_B, and 2_C) that have been characterized. The possible significance of α_2-ADR in the IBS was explored by analyzing a point mutation at the nucleotide position–1291 (rs1800544) in the promoter of α_{2A}-adrenoreceptor gene (expected to reduce the receptor expression), and a four amino acid deletion in the coding region of α_{2C}-adrenoreceptor gene (322–325del) known to impair binding to G protein. Patients with IBS-C (n = 90), IBS–D (n = 128), IBS with alternating bowel function (n = 38), and patients with chronic abdominal pain (n = 20), were compared to healthy volunteers (n = 120; Kim et al., 2004). Logistic regression analysis showed independent associations of polymorphisms at α_{2A}- and α_{2C}-ADR with IBS-C. Additionally, an association between the α_{2C}-ADR polymorphism and high scores of severity and frequency for somatic pain symptoms was found. Another group of investigators found associations between variation in $\alpha_{2c\text{-}}$ADR and $\alpha_{2A\text{-}}$ADR and functional GI disorders using gastric emptying as an endophenotypic marker (Grudell et al., 2008). The results from an Indian study indicated that α_{2A}-1291C > G polymorphism was associated with IBS-D but not IBS-C (Sikander et al., 2010). However, the association of this polymorphism with pain was not explored. Together these studies suggest that genetic factors related to α_{2A}- and α_{2c}-ADR may interact with other genetic as well as environmental factors and contribute to clinical manifestations of IBS.

There are multiple points at which α_2-ADR can contribute to increased pain sensitivity. Central α_{2C}-ADR on spinal interneurons modulate the descending inhibitory pathways from brainstem that down-regulate dorsal horn neurons and peripheral sensation (Phillips & Corces, 2009). Winston and colleagues (2010) revealed that administration of norepinephrine to rats induced visceral hypersensitivity and increased expression of nerve growth factor in response to colonic distension. Furthermore, blockade of adrenergic receptors prevented this stress-induced hypersensitivity suggesting sensitization of primary afferents.

Catechol-O-Methyltransferase

Catechol-O-methyltransferase (COMT) is a key enzyme involved in the breakdown of the neurotransmitters dopamine, epinephrine, and norepinephrine. A common G to A substitution in the coding region of the gene results in a valine-to-methionine substitution (*val158met*). The valine-to-methionine substitution results in a threefold to fourfold reduction in the activity of the COMT enzyme. The val/val, val/met, and met/met genotypes predict high, intermediate, and low COMT enzyme activity. Because of its role in neurotransmission and pain, COMT may be important in the pathophysiology of IBS. Homozygosity for 158*met* leads to a threefold to fourfold reduction in COMT enzymatic activity compared with homozygosity for 158*val*.

In a study of 867 Swedish subjects (445 women), representative of the general population who were compared to 70 consecutively sampled IBS patients

(61 women) and genotyped for the val158met polymorphism, there was a significantly higher occurrence of the val/val genotype in patients compared to controls (30% vs. 20% $p = .046$; Karling et al., 2011). The IBS patients carrying the val/val genotype also had increased bowel frequency as measured on a daily diary (2.6 vs. 1.8 stools per day; $p = .03$) as compared to the val/met and met/met carriers. In another study, carriers of the lower-activity COMT version (met/met homozygotes) were found to have higher sensory and affective ratings of pain and a more negative internal affective state (Zubieta et al., 2003). Relating to mental health, val/val homozygotes were found to be more prone to persistent generalized anxiety in a large ($n = 962$) Australian population-based longitudinal study (Olsson et al., 2007). This propensity was greater in those who had both the COMT val/val and SERT s/s polymorphisms. Other investigators have concurrently examined both SERT and COMT variants. Lonsdorf showed that carriers of the 5-HTTLPR *s* allele exhibit conditioned startle potentiation, whereas carriers of the COMT *met/met* genotype failed to extinguish conditioned fear (Lonsdorf et al., 2011). Mandelli et al. (2007), in a large sample of patients with mood disorders, found an association between the 5-HTTLPR s-containing genotypes polymorphism and self report of adverse events. They also found the COMT *val158met* variant interacted in mediating the response to stressful life events. However, a possible epistatic interaction of these genes in IBS has not yet been fully investigated.

Variations in COMT have been examined for their potential roles in modulating pharmacologic pain management needs. In one study of patients with cancer pain ($N = 197$) those patients found to have the most frequent haplotype (34.5%) needed lower morphine doses (Rakvåg et al., 2008). The contribution of COMT to pain sensitivity may be influenced by ethnicity (Vargas-Alarcón et al., 2007). For example, in a study of Spanish ($n = 78$ fibromyalgia IBS; $n = 80$ Control) and Mexican ($n = 57$ fibromyalgia; $n = 33$ Control) women, an association between fibromyalgia and COMT haplotype associated with high pain was found in the Spanish patients but not observed in the Mexican group. In a European population of 867 subjects and 70 IBS patients, there was a higher occurrence of the val/val genotype in the IBS group. However, they found no significant differences in chronic pain (defined as pain the past 2 years) among the 3 genotypes. Given the small number of sample IBS patients and the heterogeneity of the condition additional examination of COMT is warranted.

SUMMARY

To understand the complexity of a heterogeneous condition such as IBS, mechanistic models that account for genetics, environmental factors such as early childhood adverse events, motility, inflammation, and daily and lifetime psychological distress and stress need to be considered. Unidimensional models that include only GI transit, peripheral cytokines, or stress hormone levels are not sufficient. Whether genetics contribute directly to the risk for IBS or indirectly through influencing vulnerability to environmental risk or response to protective factors is not known at this time. Furthermore, given the multifactorial nature of IBS, gene-gene interactions (epistasis) are likely at play indicating that it is unlikely that one single genetic factor is responsible for IBS.

ACKNOWLEDGMENTS

This effort was supported by NINR, NIH NR01094, and P30 NR04001.

REFERENCES

Andreou, D., Saetre, P., Werge, T., Andreassen, O. A., Agartz, I., Sedvall, G. C., . . . Jönsson E. G. (2010). Tryptophan hydroxylase gene 1 (TPH1) variants associated with cerebrospinal fluid 5-hydroxyindole acetic acid and homovanillic acid concentrations in healthy volunteers. *Psychiatry Research*, *180*(2–3), 63–67.

Ansorge, M. S., Zhou, M., Lira, A., Hen, R., & Gingrich, J. A. (2004). Early-life blockade of the 5-HT transporter alters emotional behavior in adult mice. *Science*, *306*(5697), 879–881.

Atkinson, W., Lockhart, S., Whorwell, P. J., Keevil, B., & Houghton, L. A. (2006). Altered 5-hydroxytryptamine signaling in patients with constipation- and diarrhea-predominant irritable bowel syndrome. *Gastroenterology*, *130*(1), 34–43.

Beevers, C. G., Wells, T. T., Ellis, A. J., & McGeary, J. E. (2009). Association of the serotonin transporter gene promoter region (5-HTTLPR) polymorphism with biased attention for emotional stimuli. *Journal of Abnormal Psychology*, *118*(3), 670–681.

Bellentani, S., Baldoni, P., Petrella, S., Tata, C., Armocida, C., Marchegiano, P., . . . Manenti F. (1990). A simple score for the identification of patients at high risk of organic diseases the colon in the family doctor consulting room. The Local IBS Study Group. *Family Practice*, *7*(4), 307–312.

Bellini, M., Rappelli, L., Blandizzi, C., Costa, F., Stasi, C., Colucci, R., . . . Del Tacca, M. (2003). Platelet serotonin transporter in patients with diarrhea-predominant irritable bowel syndrome both before and after treatment with alosetron. *American Journal of Gastroenterology*, *98*(12), 2705–2711.

Bengtson, M. B., Rønning, T., Vatn, M. H., & Harris, J. R. (2006). Irritable bowel syndrome in twins: Genes and environment. *Gut*, *55*(12), 1754–1759.

Bharucha, A. E., Seide, B. M., Zinsmeister, A. R., & Melton, L. J., III. (2008). Relation of bowel habits to fecal incontinence in women. *American Journal of Gastroenterology*, *103*(6), 1470–1475

Bodmer, W., & Bonilla, C. (2008). Common and rare variants in multifactorial susceptibility to common diseases. *Nature Genetics*, *40*(6), 695–701.

Cain, K. C., Jarrett, M. E., Burr, R., Rosen, S., Hertig, V. L., & Heitkemper, M. M. (2009). Gender differences in gastrointestinal, psychological, and somatic symptoms in irritable bowel syndrome. *Digestive Diseases and Sciences*, *54*(7), 1542–1549.

Camilleri, M., Andrews, C. N., Bharucha, A. E., Carlson, P. J., Ferber, I., . . . Coulie, B. (2007). Alterations in expression of p11 and SERT in mucosal biopsy specimens of patients with irritable bowel syndrome. *Gastroenterology*, *132*(1), 17–25.

Camilleri, M., Atanasova, E., Carolson, P. J., Ahmad, U., Kim, H. J., Viramontes, B. E., . . . Urrutia, R. (2002). Serotonin-transporter polymorphism pharmacogenetics in diarrhea-predominant irritable bowel syndrome. *Gastroenterology*, *123*(2), 425–432.

Camilleri, M., Busciglio, I., Carlson, P., McKinzie, S., Burton, D., Baxter, K., . . . Zinsmeister, A. R. (2008). Candidate genes and sensory functions in health and irritable bowel syndrome. *American Journal of Physiology. Gastrointestinal and Liver Physiology*, *295*(2), 219–225.

Caspi, A., Hariri, A. R., Holmes, A., Uher, R., & Moffitt, T. E. (2010). Genetic sensitivity to the environment: The case of the serotonin transporter gene and its implications for studying complex diseases and traits. *American Journal of Psychiatry*, *167*(5), 509–527.

Caspi, A., Sugden, K., Moffitt, T. E., Taylor, A., Craig, I. W., Harrington, H., . . . Poulton, R. (2003). Influence of life stress on depression: Moderation by a polymorphism in the 5-HTT gene. *Science*, *301*(5631), 386–389.

Cervilla, J. A., Molina, E., Rivera, M., Torres-González, F., Bellón, J. A., Moreno, B., . . . Guitérrez, B. (2007). The risk for depression conferred by stressful life events is modified by variation at the serotonin transporter 5HTTLPR genotype: Evidence from the Spanish PREDICT-Gene cohort. *Molecular Psychiatry*, *12*(8), 748–755.

Chen, C., Glatt, S. J., & Tsuang, M. T. (2008). The tryptophan hydroxylase gene influences risk for bipolar disorder but not major depressive disorder: Results of meta-analyses. *Bipolar Disorders*, *10*(7), 816–821.

Chen, J. J., Li, Z., Pan, H., Murphy, D. L., Tamir, H., Koepsell, H., & Gershon, M. D. (2001). Maintenance of serotonin in the intestinal mucosa and ganglia of mice that lack the high-affinity serotonin transporter: Abnormal intestinal motility and the expression of cation transporters. *Journal of Neuroscience*, *21*(16), 6348–6361.

Coates, M. D., Mahoney, C. R., Linden, D. R., Sampson, J. E., Chen, J., Blaszyk, H., . . . Moses, P. L. (2004). Molecular defects in mucosal serotonin content and decreased serotonin reuptake transporter in ulcerative colitis and irritable bowel syndrome. *Gastroenterology*, *126*(7), 1657–1664.

Costedio, M. M., Coates, M. D., Brooks, E. M., Glass, L. M., Ganguly, E. K., Blaszyk, H., . . . Mawe, G. M. (2010). Mucosal serotonin signaling is altered in chronic constipation but not in opiate-induced constipation. *American Journal of Gastroenterology*, *105*(5), 1173–1180.

Digesu, G. A., Panayi, D., Kundi, N., Tekkis, P., Fernando, R., & Khullar, V. (2010). Validity of the Rome III criteria in assessing constipation in women. *International Urogynecology Journal*, *21*(10), 1185–1193.

Drossman, D. A. (2006). The functional gastrointestinal disorders and the Rome III process. *Gastroenterology*, *130*(5), 1377–1390.

Drossman, D. A., Camilleri, M., Mayer, E. A., & Whitehead, W. E. (2002). AGA technical review on irritable bowel syndrome. *Gastroenterology*, *123*(6), 2108–2131.

Drossman, D. A., Morris, C. B., Schneck, S., Hu, Y. J., Norton, N. J., Norton, W. F., . . . Bangdiwala, S. I. (2009). International survey of patients with IBS: Symptom features and their severity, health status, treatments, and risk taking to achieve clinical benefit. *Journal of Clinical Gastroenterology*, *43*(6), 541–550.

Dunlop, S. P., Coleman, N. S., Blackshaw, E., Perkins, A. C., Singh, G., Marsden, C. A., & Spiller, R. C. (2005). Abnormalities of 5-hydroxytryptamine metabolism in irritable bowel syndrome. *Clinical Gastroenterology and Hepatology*, *3*(4), 349–357.

Folks, D. G. (2004). The interface of psychiatry and irritable bowel syndrome. *Current Psychiatry Reports*, *6*(3), 210–215.

Fukudo, S., Kanazawa, M., Mizuno, T., Hamaguchi, T., Kano, M., Watanabe, S., . . . Aoki, M. (2009). Impact of serotonin transporter gene polymorphism on brain activation by colorectal distention. *NeuroImage*, *47*(3), 946–951.

Galfalvy, H., Huang, Y. Y., Oquendo, M. A., Currier, D., & Mann, J. J. (2009). Increased risk of suicide attempt in mood disorders and TPH1 genotype. *Journal of Affective Disorders*, *115*(3), 331–338.

Gelernter, J., Kranzler, H., & Cubells, J. F. (1997). Serotonin transporter protein (SLC6A4) allele and haplotype frequencies and linkage disequilibria in African- and European-American and Japanese populations and in alcohol-dependent subjects. *Human Genetics*, *101*(2), 243–246.

Gizatullin, R., Zaboli, G., Jönsson, E. G., Asberg, M., & Leopardi, R. (2006). Haplotype analysis reveals tryptophan hydroxylase (TPH) 1 gene variants associated with major depression. *Biological Psychiatry*, *59*(4), 295–300.

Greenberg, B. D., Tolliver, T. J., Huang, S. J., Li, Q., Bengel, D., & Murphy, D. L. (1999). Genetic variation in the serotonin transporter promoter region affects serotonin uptake in human blood platelets. *American Journal of Medical Genetics*, *88*(1), 83–87.

Grudell, A. B., Camilleri, M., Carlson, P., Gorman, H., Ryks, M., Burton, D., . . . Zinsmeister, A. R. (2008). An exploratory study of the association of adrenergic and serotonergic genotype and gastrointestinal motor functions. *Neurogastroenterology and Motility*, *20*(3), 213–219.

Halder, S. L., Locke, G. R., III, Schleck, C. D., Zinsmeister, A. R., Melton, L. J. III, & Talley, N. J. (2007). Natural history of functional gastrointestinal disorders: A 12-year longitudinal population-based study. *Gastroenterology*, *133*(3), 799–807.

Halmøy, A., Johansson, S., Winge, I., McKinney, J. A., Knappskog, P. M., & Haavik, J. (2010). Attention-deficit/hyperactivity disorder symptoms in offspring of mothers with impaired serotonin production. *Archives of General Psychiatry*, *67*(10), 1033–1043.

Heils, A., Teufel, A., Petri, S., Stöber, G., Riederer, P., Bengel, D., & Lesch, K. P. (1996). Allelic variation of human serotonin transporter gene expression. *Journal of Neurochemistry*, *66*(6), 2621–2624.

Heitkemper, M., Carter, E., Ameen, V., Olden, K., & Cheng, L. (2002). Women with irritable bowel syndrome: Differences in patients' and physicians' perceptions. *Gastroenterology Nursing*, *25*(5), 192–200.

Homberg, J. R., & Lesch, K. P. (2011). Looking on the bright side of serotonin transporter gene variation. *Biological Psychiatry*, *69*(6), 513–519.

Houghton, L. A., Atkinson, W., Lockhart, C., Whorwell, P. J., & Keevil, B. (2007). Sigmoid-colonic motility in health and irritable bowel syndrome: A role for 5-hydroxytryptamine. *Neurogastroenterology and Motility*, *19*(9), 724–731.

Hu, X. Z., Lipsky, R. H., Zhu, G., Akhtar, L. A., Taubman, J., Greenberg, B. D., . . . Goldman, D. (2006). Serotonin transporter promoter gain-of-function genotypes are linked to obsessive-compulsive disorder. *American Journal of Human Genetics*, *78*(5), 815–826.

Hungin, A. P., Chang, L., Locke, G. R., Dennis, E. H., & Barghout, V. (2005). Irritable bowel syndrome in the United States: Prevalence, symptom patterns and impact. *Alimentary Pharmacology & Therapeutics*, *21*(11), 1365–1375.

Jarrett, M. E., Kohen, R., Cain, K. C., Burr, R. L., Poppe, A., Navaja, G. P., & Heitkemper, M. M. (2007). Relationship of SERT polymorphisms to depressive and anxiety symptoms in irritable bowel syndrome. *Biological Research for Nursing*, *9*(2), 161–169.

Jun, S., Kohen, R., Cain, K. C., Jarrett, M. E., & Heitkemper, M. M. (2011). Associations of tryptophan hydroxylase gene polymorphisms with irritable bowel syndrome. *Neurogastroenterology and Motility*, *23*(3), 233–239.

Jung, A., Huge, A., Kuhlenbäumer, G., Kempt, S., Seehafer, T., Evers, S., Berger, K., & Marziniak, M. (2010). Genetic TPH2 variants and the susceptibility for migraine: Association of a TPH2 haplotype with migraine without aura. *Journal of Neural Transmission*, *117*(11), 1253–1260.

Kanazawa, M., Endo, Y., Whitehead, W. E., Kano, M., Hongo, M., & Fukudo, S. (2004). Patients and nonconsulters with irritable bowel syndrome reporting a parental history of bowel problems have more impaired psychological distress. *Digestive Diseases and Sciences*, *49*(6), 1046–1053.

Karg, K., Burmeister, M., Shledon, K., & Sen, S. (2011). The serotonin transporter promoter variant (5-HTTLPR), stress, and depression meta-analysis revisited: Evidence of genetic moderation. *Archives of General Psychiatry*, *68*(5), 444–454.

Karling, P., Danielsson, Å., Wikgren, M., Söderström, I., Del-Favero, J., Adolfsson, R., & Norrback K. F. (2011). The relationship between the val158met catechol-O-methyltransferase (COMT) polymorphism and irritable bowel syndrome. *PLoS One*, *6*(3), e18035.

Kerckhoffs, A. P., Ter Linde, J. J., Akkermans, L. M., & Samsom, M. (2008). Trypsinogen IV, serotonin transporter transcript levels and serotonin content are increased in small intestine of irritable bowel syndrome patients. *Neurogastroenterology and Motility*, *20*(8), 900–907.

Kim, H. J., Camilleri, M., Carlson, P. J., Cremonini, F., Ferber, I., Stephens, D., . . . Urrutia, R. (2004). Association of distinct alpha(2) adrenoceptor and serotonin transporter polymorphisms with constipation and somatic symptoms in functional gastrointestinal disorders. *Gut*, *53*(6), 829–837.

Kohen, R., Jarrett, M. E., Cain, K. C., Jun, S. E., Navaja, G. P., Symonds, S., & Heitkemper, M. M. (2009). The serotonin transporter polymorphism rs25531 is associated with irritable bowel syndrome. *Digestive Diseases and Sciences*, *54*(12), 2663–2670.

Lee, D. Y., Park, H., Kim, W. H., Lee, S. I., Seo, Y. J., & Choi, Y. C. (2004). [Serotonin transporter gene polymorphism in healthy adults and patients with irritable bowel syndrome]. *Korean Journal of Gastroenterology*, *43*(1), 18–22.

Lembo, A., Zaman, M., Jones, M., & Talley, N. J. (2007). Influence of genetics on irritable bowel syndrome, gastro-oesophageal reflux and dyspepsia: A twin study. *Alimentary Pharmacology & Therapeutics*, *25*(11), 1343–1450.

Lesch, K. P., Balling, U., Gross, J., Strauss, K., Wolozin, B. L., Murphy, D. L, & Riederer, P. (1994). Organization of the human serotonin transporter gene. *Journal of Neural Transmissions. General Section*, *95*(2), 157–162.

Levy, R. L., Jones, K. R., Whitehead, W. E., Feld, S. I., Talley, N. J., & Corey, L. A. (2001). Irritable bowel syndrome in twins: Heredity and social learning both contribute to etiology. *Gastroenterology*, *121*(4), 799–804.

Levy, R. L., Whitehead, W. E., Von Korff, M. R., & Feld, A. D. (2000). Intergenerational transmission of gastrointestinal illness behavior. *American Journal of Gastroenterology*, *95*(2), 451–456.

Li, Y., Nie, Y., Xie, J., Tang, W., Liang, P., Sha, W., Yang, H., . . . Zhou, Y. (2007). The association of serotonin transporter genetic polymorphisms and irritable bowel syndrome and its influence on tegaserod treatment in Chinese patients. *Digestive Diseases and Sciences*, *52*(11), 2942–2949.

Lira, A., Zhou, M., Castanon, N., Ansorge, M. S., Gordon, J. A., Francis, J. H., . . . Gingrich, J. A. (2003). Altered depression-related behaviors and functional changes in the dorsal raphe nucleus of serotonin transporter-deficient mice. *Biological Psychiatry*, *54*(10), 960–971.

Liu, M. T., Rayport, S., Jiang, Y., Murphy, D. L., & Gershon, M. D. (2002). Expression and function of 5-HT3 receptors in the enteric neurons of mice lacking the serotonin transporter. *American Journal of Physiology Gastrointestinal and Liver Physiology*, *283*(6), 1398–1411.

Locke, G. R., III, Zinsmeister, A. R., Talley, N. J., Fett, S. L., & Melton, L. J., III. (2000). Familial association in adults with functional gastrointestinal disorders. *Mayo Clinic Proceedings*, *75*(9), 907–912.

Lonsdorf, T. B., Golkar, A., Lindstöm, K. M., Fransson, P., Schalling, M., Ohman, A., & Ingvar, M. (2011). 5-HTTLPR and COMTval158met genotype gate amygdala reactivity and habituation. *Biological Psychology*, *87*(1), 106–112.

Mandelli, L., Serretti, A., Marino, E., Pirovano, A., Calati, R., & Colombo, C. (2007). Interaction between serotonin transporter gene, catechol-O-methyltransferase gene and stressful life events in mood disorders. *International Journal of Neuropsychopharmacology*, *10*(4), 437–447.

Manuck, S. B., Flory, J. D., Ferrell, R. E., Dent, K. M., Mann, J. J., & Muldoon, M. F. (1999). Aggression and anger-related traits associated with a polymorphism of the tryptophan hydroxylase gene. *Biological Psychiatry*, *45*(5), 603–614.

Markoutsaki, T., Karantanos, T., Gazoulli, M., Anagnou, N. P., Ladas, S. D., & Karamanolis, D. G. (2011). Serotonin transporter and G protein beta 3 subunit gene polymorphisms in Greeks with irritable bowel syndrome. *Digestive Diseases and Sciences*, *56*(11), 3276–3280.

Mawe, G. M., Coates, M. D., & Moses, P. L. (2006). Review article: Intestinal serotonin signalling in irritable bowel syndrome. *Alimentary Pharmacology & Therapeutics*, *23*(8), 1067–1076.

Mayer, E. A. (2008). Clinical practice. Irritable bowel syndrome. *New England Journal of Medicine*, *358*(16), 1692–1699.

McCaffery, J. M., Pogue-Geile, M. F., Ferrell, R. E., Petro, N., & Manuck, S. B. (2002). Variability within alpha- and beta-adrenoreceptor genes as a predictor of cardiovascular function at rest and in response to mental challenge. *Journal of Hypertension*, *20*(6), 1105–1114.

Miwa, J., Echizen, H., Matsueda, K., & Umeda, N. (2001). Patients with constipation-predominant irritable bowel syndrome (IBS) may have elevated serotonin concentrations in colonic mucosa as compared with diarrhea-predominant patients and subjects with normal bowel habits. *Digestion*, *63*(3), 188–194.

Mohammed, I., Cherkas, L. F., Riley, S. A., Spector, T. D., & Trudgill, N. J. (2005). Genetic influences in irritable bowel syndrome: A twin study. *American Journal of Gastroenterology*, *100*(6), 1340–1344.

Morris-Yates, A., Talley, N. J., Boyce, P. M., Nandurkar, S., & Andrews, G. (1998). Evidence of a genetic contribution to functional bowel disorder. *American Journal of Gastroenterology*, *93*(8), 1311–1317.

Nakamura, M., Ueno, S., Sano, A., & Tanabe, H. (2000). The human serotonin transporter gene linked polymorphism (5-HTTLPR) shows ten novel allelic variants. *Molecular Psychiatry*, *5*(1), 32–38.

Niesler, B., Kapeller, J., Fell, C., Atkinson, W., Moller, D., Fischer, C., . . . Houghton, L. A. (2010). 5-HTTLPR and STin2 polymorphisms in the serotonin transporter gene and irritable bowel syndrome: Effect of bowel habit and sex. *European Journal of Gastroenterology & Hepatology*, *22*(7), 856–861.

Olsson, C. A., Byrnes, G. B., Anney, R. J., Collins, V., Hemphill, S. A., Williamson, R., & Patton, G. C. (2007). COMT Val(158)Met and 5HTTLPR functional loci interact to predict persistence of anxiety across adolescence: Results from the Victorian Adolescent Health Cohort Study. *Genes, Brain, and Behavior*, *6*(7), 647–652.

Pamuk, O. N., Pamuk, G. E., & Celik, A. F. (2003). Revalidation of description of constipation in terms of recall bias and visual scale analog questionnaire. *Journal of Gastroenterology and Hepatology*, *18*(12), 1417–1422.

Paoloni-Giacobino, A., Mouthon, D., Lambercy, C., Vessaz, M., Coutant-Zimmerli, S., Rudolph, W., . . . Buresi, C. (2000). Identification and analysis of new sequence variants in the human tryptophan hydroxylase (TpH) gene. *Molecular Psychiatry*, *5*(1), 49–55.

Park, J. M., Choi, M. G., Park, J. A., Oh, J. H., Cho, Y. K., Lee, I. S., . . . Chung, I. S. (2006). Serotonin transporter gene polymorphism and irritable bowel syndrome. *Neurogastroenterology and Motility*, *18*(11), 995–1000.

Pata, C., Erdal, M. E., Derici, E., Yazar, A., Kanik, A., & Ulu, O. (2002). Serotonin transporter gene polymorphism in irritable bowel syndrome. *American Journal of Gastroenterology*, *97*(7), 1780–1784.

Phillips, J. E., & Corces, V. G. (2009). CTCF: Master weaver of the genome. *Cell*, *137*(7), 1194–1211.

Popa, D., Léna, C., Alexandre, C., & Adrien, J. (2008). Lasting syndrome of depression produced by reduction in serotonin uptake during postnatal development: Evidence from sleep, stress, and behavior. *Journal of Neuroscience*, *28*(14), 3546–3554.

Rakvåg, T. T., Ross, J. R., Sato, H., Skorpen, F., Kaasa, S., & Klepstad, P. (2008). Genetic variation in the Catechol-OMethyltransferase (COMT) gene and morphine requirements in cancer patients with pain. *Molecular Pain*, *4*, 64.

Ramamoorthy, S., Bauman, A. L., Moore, K. R., Han, H., Yang-Feng, T., Chang, A. S., . . . Blakely, R. D. (1993). Antidepressant- and cocaine-sensitive human serotonin transporter: Molecular cloning, expression, and chromosomal localization. *Proceedings of the National Academy of Sciences of the United States of America*, *90*(6), 2542–2546.

Risch, N., Herrell, R., Lehner, T., Liang, K. Y., Eaves, L., Hoh, J., . . . Merikangas, K. R. (2009). Interaction between the serotonin transporter gene (5-HTTLPR), stressful life events, and risk of depression: A meta-analysis. *Journal of the American Medical Association*, *301*(23), 2462–2471.

Rujescu, D., Giegling, I., Bondy, B., Gietl, A., Zill, P., & Möller, H. J. (2002). Association of anger-related traits with SNPs in the TPH gene. *Molecular Psychiatry*, *7*(9), 1023–1029.

Russo, M. W., Gaynes, B. N., & Drossman, D. A. (1999). A national survey of practice patterns of gastroenterologists with comparison to the past two decades. *Journal of Clinical Gastroenterology*, *29*(4), 339–343.

Saad, R. J. (2011). Peripherally acting therapies for the treatment of irritable bowel syndrome. *Gastroenterology Clinics of North America*, *40*(1), 163–182.

Saetre, P., Lundmark, P., Wang, A., Hansen, T., Rasmussen, H. B., Djurovic, S., . . . Jönsson, E. G. (2010). The tryptophan hydroxylase 1 (TPH1) gene, schizophrenia susceptibility, and suicidal behavior: A multi-centre case-control study and meta-analysis. *American Journal of Medical Genetics. Part B, Neuropsychiatric Genetics*, *153B*(2), 387–396.

Saito, Y. A. (2011). The role of genetics in IBS. *Gastroenterology Clinics of North America*, *40*(1), 45–67.

Saito, Y. A., Locke, G. R., III, Zimmerman, J. M., Holtmann, G., Slusser, J. P., de Andrade, M., . . . Talley, N. J. (2007). A genetic association study of 5-HTT LPR and GNbeta3 C825T polymorphisms with irritable bowel syndrome. *Neurogastroenterology and Motility*, *19*(6), 465–470.

Saito, Y. A., Mitra, N., & Mayer, E. A. (2010). Genetic approaches to functional gastrointestinal disorders. *Gastroenterology*, *138*(4), 1276–1285.

Saito, Y. A., Petersen, G. M., Larson, J. J., Atkinson, E. J., Fridley, B. L., de Andrade, M., . . . Talley, N. J. (2010). Familial aggregation of irritable bowel syndrome: A family case-control study. *American Journal of Gastroenterology*, *105*(4), 833–841.

Saito, Y. A., Zimmerman, J. M., Harmsen, W. S., De Andrade, M., Locke, G. R., III, Petersen, G. M., & Talley, N. J. (2008). Irritable bowel syndrome aggregates strongly in families: A family-based case-control study. *Neurogastroenterology and Motility*, *20*(7), 790–797.

Schork, N. J., Murray, S. S., Frazer, K. A., & Topol, E. J. (2009). Common vs. rare allele hypotheses for complex diseases. *Current Opinion in Genetics and Development*, *19*(3), 212–219.

Sikander, A., Rana, S. V., Sharma, S. K., Sinha, S. K., Arora, S. K., Prasad, K. K., & Singh, K. (2010). Association of alpha 2A adrenergic receptor gene (ADRAlpha2A) polymorphism with irritable bowel syndrome, microscopic and ulcerative colitis. *Clinica Chimica Acta*, *411*(1–2), 59–63.

Spiller, R. (2008). Serotonergic agents and the irritable bowel syndrome: What goes wrong? *Current Opinion in Pharmacology*, *8*(6), 709–714.

Spiller, R. C. (2011). Targeting the 5-HT(3) receptor in the treatment of irritable bowel syndrome. *Current Opinion in Pharmacology*, *11*(1), 68–74.

Stein, M. B., Campbell-Sills, L., & Gelernter, J. (2009). Genetic variation in 5HTTLPR is associated with emotional resilience. *American Journal of Medical Genetics. Part B, Neuropsychiatric Genetics*, *150B*(7), 900–906.

Svedberg, P., Johansson, S., Wallander, M. A., & Pedersen, N. L. (2008). No evidence of sex differences in heritability of irritable bowel syndrome in Swedish twins. *Twin Research and Human Genetics*, *11*(2), 197–203.

Tack, J. F., & Wood, J. D. (1992). Actions of noradrenaline on myenteric neurons in the guinea pig gastric antrum. *Journal of the Autonomic Nervous System*, *41*(1–2), 67–77.

Talley, N. H., Boyce, P., & Jones, M. (1998). Identification of distinct upper and lower gastrointestinal symptom grouping in an urban population. *Gut*, *42*(5), 690–695.

Torres, G. E., Gainetdinov, R. R., & Caron, M. G. (2003). Plasma membrane monoamine transporters: Structure, regulation and function. *Nature Reviews. Neuroscience*, *4*(1), 13–25.

Van Kerkhoven, L. A., Laheij, R. J., & Jansen, J. B. (2007). Meta-analysis: A functional polymorphism in the gene encoding for activity of the serotonin transporter protein is not associated with the irritable bowel syndrome. *Alimentary Pharmacology & Therapeutics*, *26*(7), 979–986.

Vargas-Alarcón, G., Fragoso, J. M., Cruz-Robles, D., Vargas, A., Lao-Villadóniga, J. I., . . . Martínez-Lavín, M. (2007). Catechol-O-methyltransferase gene haplotypes in Mexican and Spanish patients with fibromyalgia. *Arthritis Research & Therapy*, *9*(5), R110.

Varma, M. G., Wang, J. Y., Berian, J. R., Patterson, T. R., McCrea, G. L., & Hart, S. L. (2008). The constipation severity instrument: A validated measure. *Diseases of the Colon and Rectum*, *51*(2), 162–172.

Villani, A. C., Lemire, M., Thabane, M., Belisle, A., Geneau, G., Garg, A. X., . . . Marshall, J. K. (2010). Genetic risk factors for post-infectious irritable bowel syndrome following a water-borne outbreak of gastroenteritis. *Gastroenterology*, *138*(4), 1502–1513.

Walther, D. J., & Bader, M. (2003). A unique central tryptophan hydroxylase isoform. *Biochemical Pharmacology*, *66*(9), 1673–1680.

Walther, D. J., Peter, J. U., Bashammakh, S., Hörtnagl, H., Voits, M., Fink, H., & Bader, M. (2003). Synthesis of serotonin by a second tryptophan hydroxylase isoform. *Science*, *299*(5603), 76.

Whorwell, P. J., McCallum, M., Creed, F. H., & Roberts, C. T. (1986). Non-colonic features of irritable bowel syndrome. *Gut*, *27*(1), 37–40.

Winston, J. H., Xu, G. Y., & Sarna, S. K. (2010). Adrenergic stimulation mediates visceral hypersensitivity to colorectal distension following heterotypic chronic stress. *Gastroenterology*, *138*(1), 294–304.

Wong, R. K., Palsson, O. S., Turner, M. J., Levy, R. L., Feld, A. D., von Korff, M., & Whitehead, W. E. (2010). Inability of the Rome III Criteria to distinguish functional constipation from constipation-subtype irritable bowel syndrome. *American Journal of Gastroenterology*, *105*(10), 2228–2234.

Xie, P., Kranzler, H. R., Poling, J., Stein, M. B., Anton, R. F., Brady K., . . . Gelernter, J. (2009). Interactive effect of stressful life events and the serotonin transporter 5-HTTLPR genotype on posttraumatic stress disorder diagnosis in 2 independent populations. *Archives of General Psychiatry*, *66*(11), 1201–1209.

Yeo, A., Boyd, P., Lumsden, S., Saunders, T., Handley, A., Stubbins, M., . . . Hicks, G. A. (2004). Association between a functional polymorphism in the serotonin transporter gene and diarrhoea predominant irritable bowel syndrome in women. *Gut*, *53*(10), 1452–1458.

Zalsman, G., Huang, Y. Y., Oquendo, M. A., Burke, A. K., Hu, X. Z., Brent, D. A., . . . Mann J. J. (2006). Association of a triallelic serotonin transporter gene promoter region (5-HTTLPR) polymorphism with stressful life events and severity of depression. *American Journal of Psychiatry*, *163*(9), 1588–1593.

Zill, P., Baghai, T. C., Zwanzger, P., Schüle, C., Eser, D., Rupprecht, R., . . . Ackenheil, M. (2004). SNP and haplotype analysis of a novel tryptophan hydroxylase isoform (TPH2) gene provide evidence for association with major depression. *Molecular Psychiatry*, *9*(11), 1030–1036.

Zill, P. A., Büttner, A., Eisenmenger, W., Möller, H. J., Ackenheil, M., & Bondy, B. (2007). Analysis of tryptophan hydroxylase I and II mRNA expression in the human brain: A post-mortem study. *Journal of Psychiatry Research*, *41*(1–2), 168–173.

Zubieta, J. K., Heitzeg, M. M., Smith, Y. R., Bueller, J. A., Xu, K., Xu, Y., . . . Goldman, D. (2003). COMT val158met genotype affects mu-opioid neurotransmitter responses to a pain stressor. *Science*, *299*(5610), 1240–1243.

Zuo, X. L., Li, Y. Q., Yang, X. Z., Guo, M., Guo, Y. T., Lu, X. F., . . . Desmond, P. V. (2007). Plasma and gastric mucosal 5-hydroxytryptamine concentrations following cold water intake in patients with diarrhea-predominant irritable bowel syndrome. *Journal of Gastroenterology and Hepatology*, *22*(12), 2330–2337.

CHAPTER 14

Type 2 Diabetes, Genomics, and Nursing

Necessary Next Steps to Advance the Science Into Improved, Personalized Care

Patricia C. Underwood

ABSTRACT

Type 2 diabetes mellitus (T2DM) is an inherited, chronic disorder with long-term complications; including cardiovascular disease the leading cause of mortality in the United States. The prevalence of T2DM and its complications are on the rise in the United States, highlighting the need for improved individualized prevention and treatment strategies. Exciting advancements in the field of genomics has led to the recent discovery of numerous genetic markers for T2DM; completing a promising first step toward improved, individualized prevention and treatment strategies for T2DM. These genomic markers, identified using genome-wide association studies (GWAS), candidate gene, and rare variant methodology, identify new physiologic pathways underlying the development of T2DM. Much more work is needed to successfully translate the identification of genetic markers for T2DM into improved, individualized prevention and treatment strategies. As front line providers and leaders of prevention and treatment strategies for chronic disease, nurses, nurse practitioners, and nurse scientists must contribute to this translational effort. Thus, it is important for

http://dx.doi.org/10.1891/0739-6686.29.281

nurses at all levels to (a) be aware of the current science of genetics and T2DM and (b) participate in the translation of this genetic information into improved, personalized patient care.

The aim of this review is to (a) provide an overview of the current state of the science of genetic markers and T2DM and (b) highlight essential next steps to successfully translate the identification of genetic markers for T2DM into improved prevention and treatment strategies; focusing particularly on the role of nursing in this process.

INTRODUCTION

Type 2 diabetes mellitus (T2DM) is a complex disease that affects millions of individuals internationally and is associated with increased risk for cardiovascular disease and mortality (Levitzky et al., 2008). The incidence of T2DM is increasing worldwide leading to poor health outcomes and increased health care costs (Lipscombe, 2007; Mbanya, Motala, Sobngwi, Assah, & Enoru, 2010). Family studies have demonstrated that approximately 30%–70% of T2DM can be attributed to heritable factors (Poulsen, Kyvik, Vaag, & Beck-Nielsen, 1999) and numerous studies have demonstrated an association with specific gene variants and T2DM. These genomic markers identified using genome-wide association study (GWAS), candidate gene, and rare variant methodology, highlight new physiologic pathways underlying the development of T2DM. This review will examine these studies providing an overview of the known genetic variants associated with T2DM and will introduce recently identified gene–environment interactions that influence the development of T2DM in humans. Further, the review will examine nursing's role in translating the genetic findings to improved prevention and treatment strategies for individuals with T2DM.

Because T2DM is the most prevalent form of diabetes and most likely to be seen by nurses and nurse practitioners, this review will focus only on T2DM. Readers are referred to extensive reviews on the genetics of other forms of diabetes including, type 1 diabetes mellitus (T1DM), mature onset of diabetes of the young (MODY), gestational diabetes mellitus, and neonatal diabetes for details on these topics (Bluestone, Herold, & Eisenbarth, 2010; Greeley, Tucker, Naylor, Bell, & Philipson, 2011; Steck & Rewers, 2011).

GENETIC MARKERS FOR TYPE 2 DIABETES MELLITUS

Heritability and Approaches to Identify Genomic Markers

Type 2 diabetes mellitus (T2DM) is a heritable condition with heritability estimates (proportion of T2DM attributed to genetics) close to 70%, even after accounting for age, body mass index (BMI), and gender (Almgren et al., 2011; Elbein, Sun, Scroggin, Teng, & Hasstedt, 2001). Further, family history is a known predictor of the development of T2DM (Hariri et al, 2006). The search for genomic markers that would identify individuals at risk for T2DM and potentially provide insight into the pathophysiology contributing to disease development is ongoing. Three approaches have been used to identify genomic factors of T2DM: (1) Genome-wide association studies (GWAS), (2) candidate gene studies, and (3) rare variant association studies. The GWAS approach scans the entire genome to determine if genetic variants occur more frequently in individuals with the disease versus healthy controls (Amos, 2007). The approach is nonhypothesis-based and provides insight into new physiologic pathways potentially contributing to T2DM. The research design requires extremely large sample sizes to account for the numerous multiple comparisons tested (Amos, 2007; Underwood & Read, 2008). In contrast, the candidate gene approach uses prior scientific evidence to examine whether genes within physiologic pathways, known to contribute to T2DM, associate with the disease. The candidate gene approach can be done in smaller sample sizes but prevents scientists from identifying new pathways not previously known to be associated with the disease (Underwood &Read, 2008). The most recent approach directly sequences genes of interest to determine if rare genetic variants associate with disease (Asimit & Zeggini, 2011). Together, these three approaches have identified close to 40 genetic variants that associate with T2DM and provide insight into the pathophysiology contributing to disease onset.

Pathophysiology: Beta Cell Dysfunction Versus Altered Glucose Metabolism

Studies clearly demonstrate that T2DM develops from a combination of problems with insulin secretion in the pancreas and/or insulin resistance in the liver, adipose tissue, or skeletal muscle. Understanding the physiologic pathways contributing to T2DM enable scientists to search for genes within these pathways and/or provide targeted pharmacological and behavioral interventions that address the underlying pathophysiology contributing to the disease.

Insulin resistance, a leading contributor to T2DM, may develop from alterations with either (a) intracellular insulin signaling or (b) altered glucose uptake into the muscle or adipose tissue (Kahn, Zraika, Utzschneider, & Hull, 2009). These pathways has been studied extensively and demonstrate a strong interplay between different signaling proteins and pathways (White, 1997). Disruption of these signaling pathways can result in altered insulin utilization and subsequent insulin resistance (Boura-Halfon & Zick, 2009); however, few genes involved with altered glucose metabolism and glucose uptake have been associated with

T2DM or insulin resistance except for the peroxisome proliferator activated receptor (PPARγ), insulin receptor subtrate 1 (IRS1), and caveolin-1 (CAV1; Altshuler et al., 2000; Kovacs et al., 2003; Pojoga et al., 2011).

Alternatively, many genes that affect either the production of insulin and/or insulin secretion have been found to contribute to T2DM. Individuals with T2DM have decreased beta cell mass with impaired insulin secretion (Talchai, Lin, Kitamura, & Accili, 2009). Although the mechanism is yet to be determined, genetic markers of T2DM provide insight that some of the physiologic pathways contributing to the development of T2DM are related to altered insulin production, altered incretin response, and/or alterations in insulin secretion signaling (Billings & Florez, 2010; See Table 14.1).

Genome-Wide Association Studies

In the past ten years, a plethora of studies have reported on the findings of GWAS for T2DM. Many genetic variants have been identified that associate with T2DM providing insight into some of the pathophysiology underlying this complex disease.

Genes Involved With Insulin Secretion

The most well-studied gene associated with T2DM and involved in processes of insulin secretion is the transcription factor 7-like 2 (TCF7L2) gene. This gene was first found to be associated with T2DM in three Caucasian cohorts where authors demonstrated that the microsatellite marker DG10S, a marker for the TCF7L2 gene, carried a population attributable risk of 28% for T2DM (Grant et al., 2006). These findings have been replicated in numerous other cohorts (Chauhan et al. 2010; Sale et al., 2007; Sladek et al., 2007; Wen et al., 2010). Initially, scientists proposed that the gene variants resulted in altered levels of glucagon like peptide-1 (GLP-1) and altered the GLP-1 response resulting in decreased insulin release. A subsequent study evaluated the expression of TCF7L2 in human islets and described that TCF7L2 gene expression has an effect on the enteroinsular axis and insulin secretion (Lyssenko et al., 2007).

Another well-studied gene that is associated with risk for T2DM is the hepatocyte nuclear-4a (HNF4a) gene. This gene is involved in regulating gene expression and insulin secretion in the beta cells. Variants in this gene are associated with T2DM and maturity onset diabetes of the young. The association of variants of this gene with T2DM has been identified in populations of individuals with Ashkenazi Jewish heritage (Silander et al., 2004), as well as from the United Kingdom (Weedon et al., 2004) and other European cohorts (Voight et al. 2010), providing insight into the numerous factors that affect insulin secretion in individuals.

TABLE 14.1
Large Cohorts Examined by Genome-Wide Association Studies of Type 2 Diabetes Mellitus

Name	Geographic Location	Description
Wellcome Trust Case Control Consortium (WTCCC): T2DM Genetics Consortium www.wtccc.org.uk	United Kingdom	2,000 individuals with T2DM and 3,000 healthy controls. Affymetrix Gene Chip used for genotyping
Diabetes Genetics Initiative (DGI) http://www.broadinstitute.org/diabetes	Sweden, Finland	Swedish and Finnish samples matched for body mass index, gender, geographic location. Sib-ship data available for a subset of individuals.
Finland-United States Investigation of Non-insulin dependent diabetes mellitus (FUSION) http://fusion.sph.umich.edu/	Finland, United States	Family triads available with and without diabetes. Data analysis supported by the National Human Genome Research Institute (NHGRI) and genotyped using Illumina platform.
Diabetes Epidemiology: Collaborative analysis of Diagnostic criteria in Europe (deCODE) http://www.decode.com/research/	Europe and Asia	Subjects had a 2-hour oral glucose tolerance test to evaluate T2DM. Data analyzed at the Diabetes Prevention Unit of the National Institute for Health and Welfare in Helsinki, Finland.
KORA http://epi.helmholtz-muenchen.de/kora-gen/index_e.php	Germany	Population-based health survey conducted in Augsburg, Germany and surrounding two counties
Nord Trøndelag Health Study (HUNT)	Norway	All inhabitants of Norway greater than 20 years invited to participate in General Health Examination performed 3 times over 22 years.

(Continued)

TABLE 14.1

Large Cohorts Examined by Genome-Wide Association Studies of Type 2 Diabetes Mellitus (Continued)

Name	Geographic Location	Description
Nurses Health Study (NHS) http://www.channing.harvard.edu/nhs/	United States: California, Connecticut, Florida, Maryland, Massachusetts, Michigan, New Jersey, New York, Ohio, Pennsylvania, and Texas.	Population based health survey conducted in United States registered nurses ages 30 years–55 years with follow up every 4 years
Framingham Heart Study (FHS) http://www.framingham heartstudy.org/	United States: Framingham, Massachusetts	Population based health survey conducted in the Unites States in individuals between the ages of 30 years and 62 years with follow up
Diabetes Genetics Replication and Meta-analysis (DIAGRAM) http://www.well.ox.ac.uk/DIAGRAM/	Europe and the United States	Meta-analysis from FUSION, DGI, WTCCC
African American Cohort	Southern United States: Wake Forest University	Small cohort of (approximately 1,000 individuals) of African American descent examined for T2DM and end-stage renal disease (ESRD)
Meta-analyses of Glucose and Insulin related traits (MAGIC) http://www.magic consortium.org/	Europe and the United States	Meta-analysis from FUSION, DGI, WTCCC that analyzes fasting insulin, fasting glucose, HbA1c levels and recently T2DM

Note. Website URLs were updated on November 15, 2011.

Two other genes have been identified as regulators of insulin secretion in T2DM. The SLC308A gene encodes a zinc transporter molecule involved in transporting zinc from the cytoplasm into insulin secretory vesicles and subsequently affects the biosynthesis of insulin (Florez, Manning et al., 2007; Sladek et al., 2007) and insulin secretion (Strawbridge et al. 2011). Variants in the gene, particularly the non-synonymous SNP rs13266634 (R325W) have been associated with T2DM is multiple populations including Pakistani (Rees et al., 2011), Chinese (Wu et al., 2008; Xiang et al., 2008); Japanese (Horikawa et al., 2008); Finnish (Hertel et al., 2008); British, (Zeggini et al., 2007); and French (Sladek et al., 2007). A second gene involved with insulin secretion is the potassium inwardly rectifying channel subfamily J member 11 (KCNJ11) gene. This gene encodes part of the potassium channel of the beta cell that controls insulin secretion (Gloyn et al., 2003). The association of genetic variants of the KCNJ11 gene and T2DM were first discovered in a large English cohort (Gloyn et al., 2003) and subsequently replicated in Indian (Chauhan et al., 2010; Chavali et al., 2011), East Asian (Yang et al., 2011), White (Florez, Manning et al., 2007), and Japanese populations (Tabara et al., 2009). The ATP-binding cassette subfamily C member 8 (ABCC8) gene, encoding the sulfonylurea receptor involved with insulin secretion from the beta cell, has also been associated with T2DM in some (Florez, Jablonski et al., 2007; Yokoi et al., 2006), but not other studies (Gloyn et al., 2003).

Other well-studied genes associated with T2DM and replicated in multiple populations include calpain 10 (Bodhini et al., 2011; Ezzidi et al., 2010; Horikawa et al., 2000; Kang et al., 2006; Tsuchiya et al., 2006), glucokinase regulator gene (GCKR; Bi et al., 2010; Dupuis et al., 2010; Ling et al., 2011; Onuma et al., 2010), and the melatonin receptor 1B (MTNR1B) gene (Chambers et al., 2009; Huth et al., 2009; Ling et al., 2011; Prokopenko et al., 2009; Rönn et al., 2009; Saxena et al., 2007). Calpain 10 is a protease involved in many calcium regulated cellular processes including signal transduction, cell proliferation, and cell differentiation (Carlsson et al., 2005). The calpain 10 gene is ubiquitously expressed in various tissues and variants in the gene have been associated with decreased calpain 10 expression in the skeletal muscle and adipose tissue with decreased insulin sensitivity (Ridderstråle, Parikh, & Groop, 2005). The GCKR gene encodes the glucokinase regulatory protein involved with glycolysis in liver hepatocytes; alterations in this gene are likely involved with hepatic insulin resistance (Bi et al., 2010). A surprising discovery occurred with the third gene, MTNR1B, with the realization that this gene is associated with T2DM (Lyssenko et al., 2009). MTNR1B is one of two melatonin receptors and a member of the G protein-coupled receptor family that mediates the effects of melatonin (Rönn et al., 2009). Because melatonin is known to be involved with the regulation of circadian rhythms, this discovery led to the investigation that altered circadian rhythms and sleep disturbances may affect glucose metabolism. Interestingly, MTNR1B has been shown to regulate insulin secretion as well (Prokopenko et al, 2009). Together, these genes highlight the multiple pathways (hepatic insulin resistance, insulin secretion) associated with T2DM.

Recent novel gene associations with T2DM in humans have been discovered, but need further evaluation and replication. Zeggini and colleagues (2007) conducted a GWAS in 10,128 individuals of European descent from nine cohorts and found associations with the JAZF zinc finger 1 (JAZF1), cell division cycle 123 homolog (CDC123/CAMK10), tetraspanin8/leucine rich repeat containing G protein-coupled receptor 5 (TSPAN8/LGR5), thyroid adenoma associated (THADA), ADAMS9, and NOTCH2 genes with T2DM. Further, Dupuis and colleagues (2010) identified associations with the adenylate cyclase 5 (ADCY5), prospero homeobox 1 (PROX1), diacylgycerole kinase beta (DGKB), and TMEM genes with T2DM. Additional analyses are warranted to examine the role of these genes with T2DM development and whether these genes continue to identify individuals with T2DM in multiple ethnicities.

In recent GWA studies researchers investigated the association of various T2DM genes with the underlying physiology of T2DM including fasting insulin, fasting glucose, hemoglobin A1C (HbA1c), and C-peptide levels providing a clear delineation between genes involved with insulin processing, insulin secretion, and hepatic glucose output. The Meta-analysis of Glucose and Insulin related traits Consortium (MAGIC) has provided great insight into the role of the TCF7L2, SCL30A8, MTNR1B, and GCKR genes in insulin secretion (Ingelsson et al., 2010). Further, new genes such as the fatty acid desaturase gene (FADS) has been associated with high density lipoproteins (HDL) and fasting glucose providing insight into mechanisms contributing to the co-occurrence of impaired glucose tolerance and dyslipidemia, a dyad often leading to T2DM (Billings & Florez, 2011; Dupuis et al., 2010).

Candidate Gene Studies

Although GWAS studies have identified many genetic markers for T2DM, they only explain approximately 10% of heritability of T2DM indicating that other genes are likely contributing to disease development. The failure of GWAS to identify other genetic markers for T2DM may be related to (1) heterogeneous phenotype of T2DM, (2) incomplete coverage of genes using GWAS genotyping platforms, and (3) environment–gene and gene–gene interactions that are not examined in the common GWAS approach. Another approach that has been successful is the candidate gene approach; a hypothesis-driven approach where genes known to be involved with a particular physiologic process are examined for an association with a disease.

One of the most well-known genetic markers of T2DM is a proline (Pro) to alanine (Ala) change (Pro12Ala) in the peroxisome proliferator-activated receptor gamma (PPARγ) gene. This SNP, first identified using a candidate gene approach and described by Deeb and colleagues (1998), is a missense mutation in codon 12 of Exon B of the PPARγ gene. PPARγ is a ligand activated transcription factor and member of the nuclear hormone receptor subfamily (Gurnell, 2003). Numerous studies demonstrate a role for PPARγ as a regulator of glucose metabolism and adipogenesis and it is well established that PPARγ agonists improve insulin sensitivity in type 2 diabetics (Nathan et al., 2009). It is hypothesized that the proline to alanine substitution results in decreased PPARγ stimulation resulting in the observed decreased risk of T2DM (Deeb et al., 1998; Stumvoll et al., 2001). This association has been observed in multiple cohorts and confirmed via meta-analysis (Altshuler et al., 2000; Lu et al., 2011; Muller, Bogardus, Beamer, Shuldiner, & Baier, 2003). The single nucleotide polymorphism's (SNP's) effect on the individual is relatively small; however, the population attributable risk (the difference between the incidence of a disease in a population exposed to the disease versus a population unexposed to the disease) is large because the more common allele (Pro) occurs approximately 85% of the time and is associated with a 25% increased risk of T2DM in the general public (Altshuler et al., 2000).

Another candidate gene found to be associated with T2DM is the insulin receptor substrate-1 (IRS1) gene, a gene directly involved in the intracellular insulin signaling pathway. Variants in IRS1 gene have been associated with decreased risk for T2DM, increased IRS1 receptor protein levels, decreased fasting insulin levels in 14,358 individuals of European descent (Rung et al., 2009). The association of variants of the IRS1 gene with T2DM has been identified in some populations of Mexican, Pima Indians, and British descent (Burguete-Garcia et al., 2010; Kovacs et al., 2003; Zeggini et al., 2004), but not other studies conducting in North American and Polish populations (Florez, et al., 2007).

Candidate gene approaches have also been used to identify genes associated with fasting insulin and *homeo*static *a*ssessment *m*odel of *i*nsulin *r*esistance levels (HOMA-IR); a measurement of insulin resistance in humans. These studies demonstrate a role for genes in the renin-angiotensin system (angiotensinogen) and the insulin signaling pathway (caveolin-1) with insulin resistance and hypertension; two prevalent characteristics of T2DM (Pojoga et al., 2011; Underwood et al., 2011).

Rare Variants

Recent efforts have been put forth to identify rare genetic variants that may be contributing to the development of complex disease. The hypothesis underlying this effort is that collectively, rare variants may contribute to the inheritance pattern of T2DM that is still unexplained by techniques analyzing common genetic variants. Techniques for sequencing rare genetic variants are becoming rapidly available although statistical approaches for assessing the data are still under developed (Lin & Tang, 2011). Interestingly, rare variants for T2DM have been identified in a population of individuals from France, United States, and Italy (Chiefari et al., 2011). This group identified that genetic variants in the high mobility group A1 gene (HMGA1) that occur in 7% of the studied population are associated with T2DM. The HMGA1 is a transcription factor involved in regulation of the insulin receptor and subsequent insulin signaling. Further studies are necessary to replicate this and other rare variant association studies; however, early data is promising.

Ethnic Differences

Most of the studies examining genetic associations with T2DM have been done in large cohorts of individuals of European descent (see Table 14.2). Few studies have examined the association of these genes with T2DM in other ethnicities. Waters and colleagues (2010) analyzed 19 common genetic variants associated with T2DM in individuals of African American, Latino, Japanese American, and Native Hawaiian descent, finding no difference in the genes positively associated with T2DM. However, other studies demonstrate strong differences in genetic markers for T2DM in Mexican Americans and Chinese cohorts (Below et al., 2011; Hayes et al., 2007; Hu et al., 2010). Further, few studies have examined the relationship between genetic variants and gender in T2DM development; however, some suggest gender may influence the genotype–phenotype effect (Ryoo, Woo, Kim, & Lee, 2011). It is essential that we begin to focus on the influence that different ethnicities and gender play on the association of genetic markers and T2DM.

Gene–Environment Interactions

Environmental factors such as lifestyle, behavior, and health status (exercise, diet, metabolic syndrome, and smoking) are known predictors of T2DM. It is becoming increasingly clear that these factors may influence the expression of certain genes related to T2DM; however, investigation of this phenomenon is in its early stages. Preliminary results demonstrate a relationship between BMI and genetic variants of the PPARγ and interleukin-6 (IL-6) gene where individuals with the greatest BMI and genetic risk allele have the greatest risk of developing T2DM or having insulin resistance (Fernández-Real et al., 2000; Grallert et al., 2006; Hamid et al., 2005; Herbert et al., 2006; Underwood et al., 2011). Early results from the Diabetes Prevention Program (DPP), an intensive lifestyle intervention program, have also demonstrated interactions between either lifestyle factors or medication treatment and genetic risk for T2DM. For example, lifestyle

TABLE 14.2

A Selection of Genes Associated With Insulin Resistance or Type 2 Diabetes Mellitus Using GWAS, Candidate Gene, and Rare Variant Study Designs

Gene	Year	Approach	Population
PPARγ	2010	Candidate Gene and GWAS	Caucasian (United States and European), Pima Indians, Mexicans, Asians, South Indian
TCF7L2	2006	Linkage analysis and GWAS	Caucasian (United States and European), Pima Indians, Mexicans, Asians, South Indian, African American
HNF4a	2004	GWAS	Caucasian (United States and European)
SLC30A8	2007	GWAS	Caucasian (United States and European), Pakistani, Chinese, Japanese
KCNJ11	2003	Candidate gene and GWAS	Caucasian (United States and European), Japanese, East Asian, Indian
CALP10	2000	GWAS	Caucasian (United States and European), Korean, Japanese, Indian
GCKR	2009	GWAS	Caucasian (United States and European), Chinese, Japanese
IRS1	2009	GWAS, candidate gene	Caucasian (United States and European)
FADS1	2010	GWAS	Caucasian: MAGIC Consortium
ADCY5	2010	GWAS	Caucasian: MAGIC Consortium
PROX1	2010	GWAS	Caucasian: MAGIC Consortium
DGKB	2010	GWAS	Caucasian: MAGIC Consortium
TMEM	2010	GWAS	Caucasian: MAGIC Consortium
JAZF1	2008	GWAS	DIAGRAM
CDC123/CAMK10	2008	GWAS	DIAGRAM
TSPAN8/LGR5	2008	GWAS	DIAGRAM

(*Continued*)

TABLE 14.2

A Selection of Genes Associated With Insulin Resistance or Type 2 Diabetes Mellitus Using GWAS, Candidate Gene, and Rare Variant Study Designs (Continued)

Gene	Year	Approach	Population
THADA	2008	GWAS	DIAGRAM
ADAMS9	2008	GWAS	DIAGRAM
NOTCH2	2008	GWAS	DIAGRAM
HMGA1	2011	Rare Variant	Caucasian: United States

Note. PPARγ = peroxisome proliferator-activated receptor gamma gene; IRS1 = insulin receptor substrate-1 gene; TCF7L2 = transcription factor 7-like 2; KCNJ11 = potassium inwardly-rectifying channel subfamily J member 11; CALP10 = calpain 10; GCKR = glucokinase regulator gene; MTNR1B = the melatonin receptor 1B gene; HNF4a = the hepatocyte nuclear-4a gene; KCNJ11 = potassium inwardly-rectifying channel subfamily J member 11 gene; FADS = fatty acid desaturase gene; JAZF1 = the JAZF zinc finger 1 gene; CDC123/CAMK10 = cell division cycle 123 homolog gene; TSPAN8/LGR5 = tetraspanin8/leucine rich repeat containing G protein-coupled receptor 5 genes; THADA = thyroid adenoma associated gene; ADCY5 = adenylate cyclase 5 gene; PROX1 = prospero homeobox 1 gene; DGKB = diacylgycerole kinase beta gene; HMGA1 = high mobility group A1 gene; DIAGRAM = Diabetes Genetics Replication and Meta-analysis; GWAS = genome-wide association study; MAGIC = Meta-analysis of Glucose- and Insulin-related-traits Consortium.

intervention was found to modify the effect of genetics variants of GCKR on triglyceride levels in individuals (Pollin et al., 2011). Further, lifestyle intervention was found to influence the effect of variants of the SLC30A8 gene on pro-insulin levels in individuals with pre-diabetes (Majithia & Florez, 2009). Thus, a role for environmental influence on genetic associations with T2DM is evident; however, more studies with larger samples sizes and greater power must be conducted to further elucidate this topic.

Role of Nursing in the Design and Implementation of Genomic-Based Prevention and Treatment Strategies

It is clear that much more work is needed to successfully translate the identification of genetic markers for T2DM into improved, individualized prevention and treatment strategies; however, the studies outlined thus far indicate that promising first steps have been. As front line providers and leaders of prevention and treatment strategies for chronic disease, nurses, nurse practitioners, and nurse scientists must contribute to this translational effort. Thus, it is important for nurses, at all levels, to be aware of the current science of genetics and T2DM and the necessary next steps for the translational of this information into improved, personalized patient care.

Efforts to translate the aforementioned findings have primarily focused on development of a genotype risk score for T2DM that would enable clinicians to identify individuals at greater risk for T2DM. Unfortunately, initial attempts to develop genetic risk evaluations have failed. For example, Meigs and colleagues (2008) analyzed the influence of a T2DM genetic risk score using 18 identified T2DM genetic markers. The genetic risk score provided no additional information than previously known predictors of T2DM such as family history, BMI, and age. However, an analysis of the study population younger than 50 years demonstrated that the genetic risk score did provide unique information to identify individuals at risk for T2DM, suggesting that evaluating genetic risk may be more valuable when attempting to identify individuals at a younger age (de Miguel-Yanes et al., 2011). Subsequent analyses in other cohorts using 18 SNPS for T2DM also failed to indicate that a genetic risk score provides information above and beyond previously identified risk factors for T2DM (Zeggini et al., 2007). Recent promising studies, using 34 genetic risk markers for T2DM, were more successful demonstrating that a higher genomic risk score was associated with increased risk for progression to T2DM even after accounting for age, gender, ethnicity, waist circumference, and treatment assignment (Hivert et al., 2011; Jablonski et al., 2010). Interestingly, the DPP lifestyle intervention was effective in individuals with the highest genomic risk profile suggesting that targeted intervention in individuals at greatest genetic risk may provide benefit (Hivert et al., 2011).

Because known genetic markers for T2DM explain a small proportion of the heritable risk for T2DM, much more work is necessary to identify individuals at greatest genetic risk to develop personalized prevention and treatment plans. Meigs et al. (2007) point out that failure to identify all T2DM genetic markers is likely related to the heterogeneous definition of T2DM, missed genetic variants using GWAS platforms, and underpowered studies. More work must be done to clarify the clinical subsets of T2DM and identify genetic markers associated with well-defined intermediate phenotypes that are more homogeneous. This approach has been successful for other complex illnesses including hypertension (Chamarthi et al., 2007; Williams, 1994; Williams et al., 1992; Williams & Fisher, 1997).

As we begin to move from the identification of genetic markers for T2DM to the implementation of genomic-guided, individualized prevention and treatment strategies for T2DM, nurses must now combine their proven skill set using complex disease prevention and treatment strategies (Dennison et al., 2007; Hill et al., 1999; Hill et al., 2003; Reichgott, Pearson, & Hill, 1983) with genetic knowledge to design and test more effective genomic-based intervention strategies. These intervention strategies will likely use specific pharmacologic therapy (particularly for nurse practitioners), induce behavioral change (similar to the

Diabetes Prevention Program), and education strategies targeted towards an individual's genotype that may improve health outcomes. Examples of specific genomic-based nurse led interventions that may lead to improved clinical outcomes include (1) a targeted weight loss intervention in individuals homozygous for the C allele of the IL-6 gene that improves insulin resistance and potentially prevents T2DM in this population and (2) providing a sulfonylurea agent as first line treatment in individuals harboring the KCNJ11 genetic variant that is known to alter insulin secretion. As first line providers for chronic disease education, particularly in the case of diabetes education by certified diabetes nurse educators, nurses are in a key position to design and implement studies that evaluate the effectiveness of genomic-based intervention strategies.

Use of family history in nursing interventions may be the first step in providing genomic-based intervention strategies. Because many clinicians are not always successful in assessing family history when establishing an individual's risk for chronic disease (Hariri, Yoon, Moonesinghe, Valdez, & Khoury, 2006; Hariri et al., 2006); research suggests that nurses are beginning to use family history to guide and implement interventions to improve outcomes for cardiovascular disease and T2DM (Mudd & Martinez, 2011; Pestka, Lim, & Png, 2010; Pestka et al., 2010). Studies evaluating clinician educational background on use of family history to evaluate chronic disease risk are lacking, but recognition that nurses are using family history to guide practice suggests that nurses are in a key position to work with genomic based interventions. More work must be done to determine how to increase the use of family history and genetic information for specific clinical intervention.

Much research is needed to translate the findings related to genetics and T2DM into improved clinical practice and better health outcomes. First, we must identify the remaining genetic contributions to T2DM. Second, we must continue to train and develop clinical scientists that will focus on genetics and clinical trial/intervention evaluation to design and evaluate the genomic-based intervention studies described. Third, these scientists must develop collaborative relationships with each other to develop large cohorts to continue to evaluate gene–environmental effects and replicate genetic findings in other ethnicities. Fourth, we must insure that research funding continues in this area for designing and implementing personalized, genetic-based health care that may lead to more effective interventions decreasing the incidence and prevalence of chronic disease. This will likely lead to decreased health care costs (related to disease prevention and patient compliance) and improved health outcomes.

Nurses will play a large role in the development and implementation of genetic research findings into clinical practice. Because of this important role, genetic education must continue at all levels of nursing to insure nurses are able to implement research findings into clinical practice. Nurses represent the largest group of health care providers in the health care system and are in a key position to insure personalized health care is provided to individuals. Focusing on the topics listed previously will insure we are participants in the exciting next steps of incorporating genomics into improved evaluation, treatment, and management of T2DM.

SUMMARY

Advances in the area of genomics and T2DM have led to increased knowledge of the underpinnings of T2DM; however, much more work is necessary to translate these findings into improved clinical care for patients. Initial studies found little success in using genotype to improve the identification of individuals at greatest risk for T2DM above and beyond conventional risk assessment tools. However, recent studies demonstrate that as more genomic variants are identified, use of genomic markers to identify individuals at risk and design targeted intervention strategies may be possible. As the largest group of health care providers that provide T2DM prevention programs to patients, nurses are in a key position to develop and implement genomic-based interventions for the prevention and treatment of T2DM. Success will lie in the profession's capacity to train qualified scientists to design and implement studies, educators to teach the importance of genetics to nursing students, and savvy clinicians that can effectively implement evidence based interventions to individuals at risk and with T2DM.

REFERENCES

Almgren, P., Lehtovirta, M., Isomaa, B., Sarelin, L., Taskinen, M. R., Lyssenko, V., . . . Botnia Study Group. (2011). Heritability and familiality of type 2 diabetes and related quantitative traits in the Botnia Study. *Diabetologia, 54*(11), 2811–2819.

Altshuler, D., Hirschhorn, J. N., Klannemark, M., Lindgren, C. M., Vohl, M. C., Nemesh, J., . . . Lander, E. S. (2000). The common PPARgamma Pro12Ala polymorphism is associated with decreased risk of type 2 diabetes. *Nature Genetics, 26*(1), 76–80.

Amos, C. I. (2007). Successful design and conduct of genome-wide association studies. *Human Molecular Genetics, 16 Spec. No. 2*, R220–R225.

Asimit, J., & Zeggini, E. (2011). Rare variant association analysis methods for complex traits. *Annual Review in Genetics, 44,* 293–308.

Below, J. E., Gamazon, E. R., Morrison, J. V., Konkashbaev, A., Pluzhnikov, A., McKeigue, P. M., . . . Hanis, C. L. (2011). Genome-wide association and meta-analysis in populations from Starr County, Texas, and Mexico City identify type 2 diabetes susceptibility loci and enrichment for expression quantitative trait loci in top signals. *Diabetologia, 54*(8), 2047–2055.

Bi, M., Kao, W. H., Boerwinkle, E., Hoogeveen, R. C., Rasmussen-Torvik, L. J., Astor, B. C., . . . Köttgen, A. (2010). Association of rs780094 in GCKR with metabolic traits and incident diabetes and cardiovascular disease: The ARIC Study. *PLoS One, 5*(7), e11690.

Billings, L. K., & Florez, J. C. (2010). The genetics of type 2 diabetes: What have we learned from GWAS? *Annals of the New York Academy of Science, 1212*, 59–77.

Bluestone, J. A., Herold, K., & Eisenbarth, G. (2010). Genetics, pathogenesis and clinical interventions in type 1 diabetes. *Nature, 464*(7293), 1293–1300.

Bodhini, D., Radha, V., Ghosh, S., Sanapala, K. R., Majumder, P. P., Rao, M. R., & Mohan, V. (2011). Association of calpain 10 gene polymorphisms with type 2 diabetes mellitus in Southern Indians. *Metabolism, 60*(5), 681–688.

Boura-Halfon, S., & Zick, Y. (2009). Phosphorylation of IRS proteins, insulin action, and insulin resistance. *American Journal of Physiology. Endocrinology and Metabolism, 296*(4), E581–E591.

Burguete-Garcia, A. I., Cruz-Lopez, M., Madrid-Marina, V., Lopez-Ridaura, R., Hernández-Avila, M., Cortina, B., . . . Velasco-Mondragón, E. (2010). Association of Gly972Arg polymorphism of IRS1 gene with type 2 diabetes mellitus in lean participants of a national health survey in Mexico: A candidate gene study. *Metabolism, 59*(1), 38–45.

Carlsson, E., Poulsen, P., Storgaard, H., Almgren, P., Ling, C., Jensen, C. B., . . . Ridderstråle, M. (2005). Genetic and nongenetic regulation of CAPN10 mRNA expression in skeletal muscle. *Diabetes, 54*(10), 3015–3020.

Chamarthi, B., Kolatkar, N. S., Hunt, S. C., Williams, J. S., Seely, E. W., Brown, N. J., . . . Williams, G. H. (2007). Urinary free cortisol: An intermediate phenotype and a potential genetic marker for a salt-resistant subset of essential hypertension. *Journal of Clinical Endocrinology and Metabolism, 92*(4), 1340–1346.

Chambers, J. C., Zhang, W., Zabaneh, D., Sehmi, J., Jain, P., McCarthy, M. I., . . . Kooner, J. S. (2009). Common genetic variation near melatonin receptor MTNR1B contributes to raised plasma glucose and increased risk of type 2 diabetes among Indian Asians and European Caucasians. *Diabetes, 58*(11), 2703–2708.

Chauhan, G., Spurgeon, C. J., Tabassum, R., Bhaskar, S., Kulkarni, S. R., Mahajan, A., . . . Chandak, G. R. (2010). Impact of common variants of PPARg, KCNJ11, TCF7L2, SLC30A8, HHEX, CDKN2A, IGF2BP2, and CDKAL1 on the risk of type 2 diabetes in 5,164 Indians. *Diabetes, 59*(8), 2068–2074.

Chavali, S., Mahajan, A., Tabassum, R., Dwivedi, O. P., Chauhan, G., Ghosh, S., . . . Bharadwaj, D. (2011). Association of variants in genes involved in pancreatic β-cell development and function with type 2 diabetes in North Indians. *Journal of Human Genetics, 56*(10), 695–700.

Chiefari, E., Tanyolaç, S., Paonessa, F., Pullinger, C. R., Capula, C., Iiritano, S., . . . Brunetti, A. (2011). Functional variants of the HMGA1 gene and type 2 diabetes mellitus. *Journal of the American Medical Association, 305*(9), 903–912.

Deeb, S. S., Fajas, L., Nemoto, M., Pihlajamäki, J., Mykkänen, L., Kuusisto, J., . . . Auwerx, J. (1998). A Pro12Ala substitution in PPARgamma2 associated with decreased receptor activity, lower body mass index and improved insulin sensitivity. *Nature Genetics, 20*(3), 284–287.

de Miguel-Yanes, J. M., Shrader, P., Pencina, M. J., Fox, C. S., Manning, A. K., Grant, R. W., . . . DIAGRAM (+) Investigators. (2011). Genetic risk reclassification for type 2 diabetes by age below or above 50 years using 40 type 2 diabetes risk single nucleotide polymorphisms. *Diabetes Care, 34*(1), 121–125.

Dennison, C. R., Post, W. S., Kim, M. T., Bone, L. R., Cohen, D., Blumenthal, R. S., . . . Hill, M. N. (2007). Underserved urban african american men: Hypertension trial outcomes and mortality during 5 years. *American Journal of Hypertension, 20*(2), 164–171.

Dupuis, J., Langenberg, C., Prokopenko, I., Saxena, R., Soranzo, N., Jackson, A. U., . . . Barroso, I. (2010). New genetic loci implicated in fasting glucose homeostasis and their impact on type 2 diabetes risk. *Nature Genetics, 42*(2), 105–116.

Elbein, S. C., Sun, J., Scroggin, E., Teng, K., & Hasstedt, S. J. (2001). Role of common sequence variants in insulin secretion in familial type 2 diabetic kindreds: The sulfonylurea receptor, glucokinase, and hepatocyte nuclear factor 1alpha genes. *Diabetes Care, 24*(3), 472–478.

Ezzidi, I., Turki, A., Messaoudi, S., Chaieb, M., Kacem, M., Al-Khateeb, G. M., . . . Mtiraouri, N. (2010). Common polymorphisms of calpain-10 and the risk of Type 2 Diabetes in a Tunisian Arab population: A case-control study. *BMC Medical Genetics, 11*, 75.

Fernández-Real, J. M., Broch, M., Vendrell, J., Gutiérrez, C., Casamitjana, R., Pugeat, M., . . . Ricart, W. (2000). Interleukin-6 gene polymorphism and insulin sensitivity. *Diabetes, 49*(3), 517–520.

Florez, J. C., Jablonski, K. A., Kahn, S. E., Franks, P. W., Dabelea, D., Hamman, R. F., . . . Altshuler, D. (2007). Type 2 diabetes-associated missense polymorphisms KCNJ11 E23K and ABCC8 A1369S influence progression to diabetes and response to interventions in the Diabetes Prevention Program. *Diabetes, 56*(2), 531–536.

Florez, J. C., Manning, A. K., Dupuis, J., McAteer, J., Irenze, K., Gianniny, L., . . . Meigs, J. B. (2007). A 100K genome-wide association scan for diabetes and related traits in the Framingham Heart Study: Replication and integration with other genome-wide datasets. *Diabetes, 56*(12), 3063–3074.

Florez, J. C., Sjögren, M., Agapakis, C. M., Burtt, N. P., Almgren, P., Lindblad, U., . . . Groop, L. (2007). Association testing of common variants in the insulin receptor substrate-1 gene (IRS1) with type 2 diabetes. *Diabetologia, 50*(6), 1209–1217.

Gloyn, A. L., Weedon, M. N., Owen, K. R., Turner, M. J., Knight, B. A., Hitman, G., . . . Frayling, T. M. (2003). Large-scale association studies of variants in genes encoding the pancreatic beta-cell KATP channel subunits Kir6.2 (KCNJ11) and SUR1 (ABCC8) confirm that the KCNJ11 E23K variant is associated with type 2 diabetes. *Diabetes, 52*(2), 568–572.

Grallert, H., Huth, C., Kolz, M., Meisinger, C., Herder, C., Strassburger, K., . . . Rathmann, W. (2006). IL-6 promoter polymorphisms and quantitative traits related to the metabolic syndrome in KORA S4. *Experimental Gerontolology, 41*(8), 737–745.

Grant, S. F., Thorleifsson, G., Reynisdottir, I., Benediktsson, R., Manolescu, A., Sainz, J., . . . Stefansson, K. (2006). Variant of transcription factor 7-like 2 (TCF7L2) gene confers risk of type 2 diabetes. *Nature Genetics, 38*(3), 320–323.

Greeley, S. A., Tucker, S. E., Naylor, R. N., Bell, G. I., & Philipson, L. H. (2011). Neonatal diabetes mellitus: A model for personalized medicine. *Trends in Endocrinology and Metabolism, 21*(8), 464–472.

Gurnell, M. (2003). PPARgamma and metabolism: Insights from the study of human genetic variants. *Clinical Endocrinology, 59*(3), 267–277.

Hamid, Y. H., Rose, C. S., Urhammer, S. A., Glümer, C., Nolsøe, R., Kristiansen, O. P., . . . Pedersen, O. (2005). Variations of the interleukin-6 promoter are associated with features of the metabolic syndrome in Caucasian Danes. *Diabetologia, 48*(2), 251–260.

Hariri, S., Yoon, P. W., Moonesinghe, R., Valdez, R., & Khoury, M. J. (2006). Evaluation of family history as a risk factor and screening tool for detecting undiagnosed diabetes in a nationally representative survey population. *Genetics in Medicine, 8*(12), 752–759.

Hariri, S., Yoon, P. W., Qureshi, N., Valdez, R., Scheuner, M. T., & Khoury, M. J. (2006). Family history of type 2 diabetes: A population-based screening tool for prevention? *Genetic Medicine, 8*(2), 102–108.

Hayes, M. G., Pluzhnikov, A., Miyake, K., Sun, Y., Ng, M. C., Roe, C. A., . . . Hanis, C. L. (2007). Identification of type 2 diabetes genes in Mexican Americans through genome-wide association studies. *Diabetes, 56*(12), 3033–3044.

Herbert, A., Liu, C., Karamohamed, S., Liu, J., Manning, A., Fox, C. S., . . . Cupples, L. A. (2006). BMI modifies associations of IL-6 genotypes with insulin resistance: The Framingham Study. *Obesity, 14*(8), 1454–1461.

Hertel, J. K., Johansson, S., Raeder, H., Midthjell, K., Lyssenko, V., Groop, L., . . . Njølstad, P. R. (2008). Genetic analysis of recently identified type 2 diabetes loci in 1,638 unselected patients with type 2 diabetes and 1,858 control participants from a Norwegian population-based cohort (the HUNT study). *Diabetologia, 51*(6), 971–977.

Hill, M. N., Bone, L. R., Hilton, S. C., Roary, M. C., Kelen, G. D., & Levine, D. M. (1999). A clinical trial to improve high blood pressure care in young urban black men: Recruitment, follow-up, and outcomes. *American Journal of Hypertension, 12*(6), 548–554.

Hill, M. N., Han, H. R., Dennison, C. R., Kim, M. T., Roary, M. C., Blumenthal, R. S., . . . Post, W. S. (2003). Hypertension care and control in underserved urban African American men: Behavioral and physiologic outcomes at 36 months. *American Journal of Hypertension, 16*(11, Pt. 1), 906–913.

Hivert, M. F., Jablonski, K. A., Perreault, L., Saxena, R., McAteer, J. B., Franks, P. W., . . . Diabetes Prevention Program Research Group. (2011). Updated genetic score based on 34 confirmed type 2 diabetes Loci is associated with diabetes incidence and regression to normoglycemia in the diabetes prevention program. *Diabetes, 60*(4), 1340–1348.

Horikawa, Y., Miyake, K., Yasuda, K., Enya, M., Hirota, Y., Yamagata, K., . . . Kasuga, M. (2008). Replication of genome-wide association studies of type 2 diabetes susceptibility in Japan. *Journal of Clinical Endocrinology and Metabolism, 93*(8), 3136–3141.

Horikawa, Y., Oda, N., Cox, N. J., Li, X., Orho-Melander, M., Hara, M., . . . Bell, G. I. (2000). Genetic variation in the gene encoding calpain-10 is associated with type 2 diabetes mellitus. *Nature Genetics, 26*(2), 163–175.

Hu, C., Zhang, R., Wang, C., Wang, J., Ma, X., Hou, X., . . . Jia, W. (2010). Variants from GIPR, TCF7L2, DGKB, MADD, CRY2, GLIS3, PROX1, SLC30A8 and IGF1 are associated with glucose metabolism in the Chinese. *PLoS One, 5*(11), e15542.

Huth, C., Illig, T., Herder, C., Gieger, C., Grallert, H., Vollmert, C., . . . Heid, I. M. (2009). Joint analysis of individual participants' data from 17 studies on the association of the IL6 variant -174G>C with circulating glucose levels, interleukin-6 levels, and body mass index. Annals of Medicine, 41(2), 128–138.

Ingelsson, E., Langenberg, C., Hivert, M. F., Prokopenko, I., Lyssenko, V., Dupuis, J., . . . MAGIC Investigators. (2010). Detailed physiologic characterization reveals diverse mechanisms for novel genetic Loci regulating glucose and insulin metabolism in humans. *Diabetes, 59*(5), 1266–1275.

Jablonski, K. A., McAteer, J. B., de Bakker, P. I., Franks, P. W., Pollin, T. I., Hanson, R. L., . . . Diabetes Prevention Program Research Group (2010). Common variants in 40 genes assessed for diabetes incidence and response to metformin and lifestyle intervention in the diabetes prevention program. *Diabetes, 59*(10), 2672–2681.

Kahn, S. E., Zraika, S., Utzschneider, K. M., & Hull, R. L. (2009). The beta cell lesion in type 2 diabetes: There has to be a primary functional abnormality. *Diabetologia, 52*(6), 1003–1012.

Kang, E. S., Kim, H. J., Nam, M., Nam, C. M., Ahn, C. W., Cha, B. S., & Lee, H. C. (2006). A novel 111/121 diplotype in the Calpain-10 gene is associated with type 2 diabetes. *Journal of Human Genetics, 51*(7), 629–633.

Kovacs, P., Hanson, R. L., Lee, Y. H., Yang, X., Kobes, S., Permana, P. A., . . . Baier, L. J. (2003). The role of insulin receptor substrate-1 gene (IRS1) in type 2 diabetes in Pima Indians. *Diabetes, 52*(12), 3005–3009.

Levitzky, Y. S., Pencina, M. J., D'Agostino, R. B., Meigs, J. B., Murabito, J. M., Vasan, R. S., & Fox, C. S. (2008). Impact of impaired fasting glucose on cardiovascular disease: The Framingham Heart Study. *Journal of the American College of Cardiology, 51*(3), 264–270.

Lin, D. Y., & Tang, Z. Z. (2011). A general framework for detecting disease associations with rare variants in sequencing studies. *American Journal of Human Genetics, 89*(3), 354–367.

Ling, Y., Li, X., Gu, Q., Chen, H., Lu, D., & Gao, X. (2011a). Associations of common polymorphisms in GCKR with type 2 diabetes and related traits in a Han Chinese population: A case-control study. *BMC Medical Genetics, 12*, 66.

Ling, Y., Li, X., Gu, Q., Chen, H., Lu, D., & Gao, X. (2011b). A common polymorphism rs3781637 in MTNR1B is associated with type 2 diabetes and lipids levels in Han Chinese individuals. *Cardiovascular Diabetology, 10*, 27.

Lipscombe, L. L. (2007). The growing prevalence of diabetes in Ontario: Are we prepared? *Healthcare Quarterly, 10*(3), 23–25.

Lu, Y., Ye, X., Cao, Y., Li, Q., Yu, X., Cheng, J., . . . Zhou, L. (2011). Genetic variants in peroxisome proliferator-activated receptor γ and retinoid X receptor-α gene and type 2 diabetes risk: A case-control study of a Chinese Han population. *Diabetes Technology & Therapeutics, 13*(2), 157–164.

Lyssenko, V., Lupi, R., Marchetti, P., Del Guerra, S., Orho-Melander, M., Almgren, P., . . . Groop, L. (2007). Mechanism by which common variants in the TCF7L2 gene increase risk of type 2 diabetes. *Journal of Clinical Investigation, 117*(8), 2155–2163.

Lyssenko, V., Nagorny, C. L., Erdos, M. R., Wierup, N., Jonsson, A., Spégel, A., . . . Groop, L. (2009). Common variant in MTNR1B associated with increased risk for type 2 diabetes and impaired early insulin secretion. *Nature Genetics, 11*(1), 82–88.

Nathan, D. M., Bose, J. B., Davidson, M. B., Ferrannini, E., Holman, R. R., Sherwin, R., & Zinman, B. (2009). Medical management of hyperglycemica in type 2 diabetes: A consensus algorithm for the initiation and adjustment of therapy: A consensus statement of the American Diabetes Association and the European Association for the Study of Diabetes. *Diabetes Care, 32*(1), 193–203.

Majithia, A. R., & Florez, J. C. (2009). Clinical translation of genetic predictors for type 2 diabetes. *Current Opinion in Endocrinology, Diabetes, and Obesity, 16*(2), 100–106.

Mbanya, J. C., Motala, A. A., Sobngwi, E., Assah, F. K., & Enoru, S. T. (2010). Diabetes in sub-Saharan Africa. *Lancet, 375*(9733), 2254–2266.

Meigs, J. B., Manning, A. K., Fox, C. S., Florez, J. C., Liu, C., Cupples, L. A., & Dupuis, J. (2007). Genome-wide association with diabetes-related traits in the Framingham Heart Study. *BMC Medical Genetics, 8*(Suppl. 1), S16.

Meigs, J. C., Shrader, P., Sullivan, L. M., McAteer, J. B., Fox, C. S., Dupuis, J., . . . Cupples, L. A. (2008). Genotype score in addition to common risk factors for prediction of type 2 diabetes. *New England Journal of Medicine, 359*(21), 2208–2219.

Mudd, G. T., & Martinez, M. C. (2011). Translation of family health history questions on cardiovascular disease and type 2 diabetes with implications for Latina health and nursing practice. *Nursing Clinics of North America, 46*(2), 207–218.

Muller, Y. L., Bogardus, C., Beamer, B. A., Shuldiner, A. R., & Baier, L. J. (2003). A functional variant in the peroxisome proliferator-activated receptor gamma2 promoter is associated with predictors of obesity and type 2 diabetes in Pima Indians. *Diabetes, 52*(7), 1864–1871.

Onuma, H., Tabara, Y., Kawamoto, R., Shimizu, I., Kawamura, R., Takata, Y., . . . Osawa, H. (2010). The GCKR rs780094 polymorphism is associated with susceptibility of type 2 diabetes, reduced fasting plasma glucose levels, increased triglycerides levels and lower HOMA-IR in Japanese population. *Journal of Human Genetics, 55*(9), 600–604.

Pestka, E., Lim, S. H., & Png, H. H. (2010). Education outcomes related to including genomics activities in nursing practice in Singapore. *International Journal of Nursing Practice, 16*(3), 282–288.

Pestka, E. L., Derscheid, D. J., Ellenbecker, S. M., Schmid, P. J., O'Neil, M. L., Ray-Mihm, R. J., & Cox, D. L. (2010). Use of genomic assessments and interventions in psychiatric nursing practice. *Issues in Mental Health Nursing, 31*(10), 623–630.

Pojoga, L. H., Underwood, P. C., Goodarzi, M. O., Williams, J. S., Adler, G. K., Jeunemaitre, X., . . . Williams, G. H. (2011). Variants of the caveolin-1 gene: A translational investigation linking insulin resistance and hypertension. Journal of Clinical Endocrinology and Metabolism, 96(8), E1288–E1292.

Pollin, T. I., Jablonski, K. A., McAteer, J. B., Saxena, R., Kathiresan, S., Kahn, S. E., . . . Diabetes Prevention Program Research Group (2011). Triglyceride response to an intensive lifestyle intervention is enhanced in carriers of the GCKR Pro446Leu polymorphism. *Journal of Clinical Endocrinology and Metabolism, 96*(7), E1142–E1147.

Poulsen, P., Kyvik, K. O., Vaag, A., & Beck-Nielsen, H. (1999). Heritability of type II (non-insulin-dependent) diabetes mellitus and abnormal glucose tolerance—A population-based twin study. *Diabetologia, 42*(2), 139–145.

Prokopenko, I., Langenberg, C., Florez, J. C., Saxena, R., Soranzo, N., Thorleifsson, G., . . . Abecasis, G. R. (2009). Variants in MTNR1B influence fasting glucose levels. *Nature Genetics, 41*(1), 77–81.

Rees, S. D., Hydrie, M. Z., Shera, A. S., Kumar, S., O'Hare, J. P., Barnett, A. H., . . . Kelly, M. A. (2011). Replication of 13 genome-wide association (GWA)-validated risk variants for type 2 diabetes in Pakistani populations. *Diabetologia, 54*(6), 1368–1374.

Reichgott, M. J., Pearson, S., & Hill, M. N. (1983). The nurse practitioner's role in complex patient management: Hypertension. *Journal of the National Medical Association, 75*(12), 1197–1204.

Ridderstråle, M., Parikh, H., & Groop, L. (2005). Calpain 10 and type 2 diabetes: Are we getting closer to an explanation? *Current Opinion in Clinical Nutrition and Metabolism Care, 8*(4), 361–366.

Rönn, T., Wen, J., Yang, Z., Lu, B., Du, Y., Groop, L., . . . Ling, C. (2009). A common variant in MTNR1B, encoding melatonin receptor 1B, is associated with type 2 diabetes and fasting plasma glucose in Han Chinese individuals. *Diabetologia, 52*(5), 830–833.

Rung, J., Cauchi, S., Albrechtsen, A., Shen, L., Rocheleau, G., Cavalcanti-Proença, C., . . . Sladek, R. (2009). Genetic variant near IRS1 is associated with type 2 diabetes, insulin resistance and hyperinsulinemia. *Nature Genetics, 41*(10), 1110–1115.

Ryoo, H., Woo, J., Kim, Y., & Lee, C. (2011). Heterogeneity of genetic associations of CDKAL1 and HHEX with susceptibility of type 2 diabetes mellitus by gender. *European Journal of Human Genetics, 19*(6), 672–675.

Sale, M. M., Smith, S. G., Mychaleckyj, J. C., Keene, K. L., Langefeld, C. D., Leak, T. S., . . . Freedman, B. I. (2007). Variants of the transcription factor 7-like 2 (TCF7L2) gene are associated with type 2 diabetes in an African-American population enriched for nephropathy. *Diabetes, 56*(10), 2638–2642.

Saxena, R., Voight, B. F., Lyssenko, V., Burtt, N. P., de Bakker, P. I., Chen, H., . . . Purcell, S. (2007). Genome-wide association analysis identifies loci for type 2 diabetes and triglyceride levels. *Science, 316*(5829), 1331–1336.

Silander, K., Mohlke, K. L., Scott, L. J., Peck, E. C., Hollstein, P., Skol, A. D., . . . Collins, F. S. (2004). Genetic variation near the hepatocyte nuclear factor-4 alpha gene predicts susceptibility to type 2 diabetes. *Diabetes, 53*(4), 1141–1149.

Sladek, R., Rocheleau, G., Rung, J., Dina, C., Shen, L., Serre, D., . . . Froguel, P. (2007). A genome-wide association study identifies novel risk loci for type 2 diabetes. *Nature, 445*(7130), 881–885.

Steck, A. K., & Rewers, M. J. (2011). Genetics of type 1 diabetes. *Clinical Chemistry, 57*(2), 176–185.

Strawbridge, R. J., Dupuis, J., Prokopenko, I., Barker, A., Ahlqvist, E., Rybin, D., . . . Florez, J. C. (2011). Genome-wide association identifies nine common variants associated with fasting proinsulin levels and provides new insights into the pathophysiology of type 2 diabetes. *Diabetes, 60*(10), 2624–2634.

Stumvoll, M., Wahl, H. G., Löblein, K., Becker, R., Machicao, F., Jacob, S., & Haring, H. (2001). Pro12Ala polymorphism in the peroxisome proliferator-activated receptor-gamma2 gene is associated with increased antilipolytic insulin sensitivity. *Diabetes, 50*(4), 876–881.

Tabara, Y., Osawa, H., Kawamoto, R., Onuma, H., Shimizu, I., Miki, T., . . . Makino, H. (2009). Replication study of candidate genes associated with type 2 diabetes based on genome-wide screening. *Diabetes, 58*(2), 493–498.

Talchai, C., Lin, H. V., Kitamura, T., & Accili, D. (2009). Genetic and biochemical pathways of beta-cell failure in type 2 diabetes. *Diabetes, Obesity & Metabolism, 11*(Suppl. 4), 38–45.

Tsuchiya, T., Schwarz, P. E., Bosque-Plata, L. D., Geoffrey Hayes, M., Dina, C., Froguel, P., . . . Bell, G. I. (2006). Association of the calpain-10 gene with type 2 diabetes in Europeans: Results of pooled and meta-analyses. *Molecular Genetics and Metabolism, 89*(1–2), 174–184.

Underwood, P. C., Chamarthi, B., Williams, J. S., Sun, B., Vaidya, A., Raby, B. A., . . . Williams, G. H. (2011). Replication and meta-analysis of the gene-environment interaction between body mass index and the interleukin-6 promoter polymorphism with higher insulin resistance. *Metabolism.*

Underwood, P. C., & Read, C. Y. (2008). Genetic association studies in nursing practice and scholarship. *Journal of Nursing Scholarship, 40*(3), 212–218.

Underwood, P. C., Sun, B., Williams, J. S., Pojoga, L. H., Raby, B., Lasky-Su, J., . . . Williams, G. H. (2011). The association of the angiotensinogen gene with insulin sensitivity in humans: A tagging single nucleotide polymorphism and haplotype approach. *Metabolism, 60*(8), 1150–1157.

Voight, B. F., Scott, L. J., Steinthorsdottir, V., Morris, A. P., Dina, C., Welch, R. P., . . . GIANT Consortium. (2010). Twelve type 2 diabetes susceptibility loci identified through large-scale association analysis. *Nature Genetics, 42*(7), 579–589.

Waters, K. M., Stram, D. O., Hassanein, M. T., Le Marchand, L., Wilkens, L. R., Maskarinec, G., . . . Haiman, C. A. (2010). Consistent association of type 2 diabetes risk variants found in europeans in diverse racial and ethnic groups. *PLoS Genetics, 6*(8), pii.

Weedon, M. N., Owen, K. R., Shields, B., Hitman, G., Walker, M., McCarthy, M. I., . . . Frayling, T. M. (2004). Common variants of the hepatocyte nuclear factor-4alpha P2 promoter are associated with type 2 diabetes in the U.K. population. *Diabetes, 53*(11), 3002–3006.

Wen, J., Ronn, T., Olsson, A., Yang, Z., Lu, B., Du, Y., . . . Hu, R. (2010). Investigation of type 2 diabetes risk alleles support CDKN2A/B, CDKAL1, and TCF7L2 as susceptibility genes in a Han Chinese cohort. *PLoS One, 5*(2), e9153.

White, M. F. (1997). The insulin signalling system and the IRS proteins. *Diabetologia, 40*(Suppl. 2), S2–S17.

Williams, G. H. (1994). Genetic approaches to understanding the pathophysiology of complex human traits. *Kidney International, 46*(6), 1550–1553.

Williams, G. H., Dluhy, R. G., Lifton, R. P., Moore, T. J., Gleason, R., Williams, R., . . . Hollenberg, N. K. (1992). Non-modulation as an intermediate phenotype in essential hypertension. *Hypertension, 20*(6), 788–796.

Williams, G. H., & Fisher, N. D. (1997). Genetic approach to diagnostic and therapeutic decisions in human hypertension. *Current Opinion in Nephrology and Hypertensions, 6*(2), 199–204.

Wu, Y., Li, H., Loos, R. J., Yu, Z., Ye, X., Chen, L., . . . Lin, X. (2008). Common variants in CDKAL1, CDKN2A/B, IGF2BP2, SLC30A8, and HHEX/IDE genes are associated with type 2 diabetes and impaired fasting glucose in a Chinese Han population. *Diabetes, 57*(10), 2834–2842.

Xiang, J., Li, X. Y., Xu, M., Hong, J., Huang, Y., Tan, J. R., . . . Ning, G. (2008). Zinc transporter-8 gene (SLC30A8) is associated with type 2 diabetes in Chinese. *Journal of Clinical Endocrinology and Metabolism, 93*(10), 4107–4112.

Yang, L., Zhou, X., Luo, Y., Sun, X., Tang, Y., Guo, W., . . . Ji, L. (2011). Association between KCNJ11 gene polymorphisms and risk of type 2 diabetes mellitus in East Asian populations: A meta-analysis in 42,573 individuals. *Molecular Biology Reports, 39*(1), 645–659.

Yokoi, N., Kanamori, M., Horikawa, Y., Takeda, J., Sanke, T., Furuta, H., . . . Seino, S. (2006). Association studies of variants in the genes involved in pancreatic beta-cell function in type 2 diabetes in Japanese subjects. *Diabetes, 55*(8), 2379–2386.

Zeggini, E., Parkinson, J., Halford, S., Owen, K. R., Frayling, T. M., Walker, M., . . . McCarthy, M. I. (2004). Association studies of insulin receptor substrate 1 gene (IRS1) variants in type 2 diabetes samples enriched for family history and early age of onset. *Diabetes, 53*(12), 3319–3322.

Zeggini, E., Weedon, M. N., Lindgren, C. M., Frayling, T. M., Elliott, K. S., Lango, H., . . . Hattersley, A. T. (2007). Replication of genome-wide association signals in UK samples reveals risk loci for type 2 diabetes. *Science, 316*(5829), 1336–1341.

CHAPTER 15

Patient and Family Issues Regarding Genetic Testing for Cystic Fibrosis

A Review of Prenatal Carrier Testing and Newborn Screening

Kathleen J. H. Sparbel and Audrey Tluczek

ABSTRACT

Cystic fibrosis (CF) is a potentially life-shortening autosomal recessive genetic condition resulting in chronic progressive respiratory involvement, malnutrition, electrolyte imbalance, and male infertility. It is the most common autosomal inherited condition in the White population, and its presence is recorded with varying prevalence across ethnicities. Since the 1989 discovery of the genetic variant F508del, the most common cystic fibrosis transmembrane conductance regulator (CFTR) mutation, more than 1,900 CF mutations have been identified. The 1997 National Institutes of Health (NIH) Consensus Statement on Cystic Fibrosis, along with 2001 and 2005 recommendations from the American College of Obstetricians and Gynecologists (ACOG), provide the basis for population CF carrier screening in the prenatal setting. Recommendations for newborn screening (NBS) for cystic fibrosis were released in 2004, with NBS programs in the United States initiated thereafter.

http://dx.doi.org/10.1891/0739-6686.29.303

With the wide variety of CFTR mutations and mutation combinations, there is not a clear understanding of the genotype–phenotype correlations or of the anticipated clinical trajectory for an individual who has identified CFTR mutations. This ambiguity creates challenges for patients and families in decision making related to CFTR carrier screening during the prenatal period, understanding the results of newborn screening for CF, or coping with the new genetic knowledge obtained. This literature review examines research regarding genetic testing for CF as it related to population screening. Patient and family issues from both the prenatal period and newborn testing are reviewed. Opportunities for future nursing research and implication for nursing practice are discussed.

INTRODUCTION

Cystic fibrosis (CF) is a potentially life-shortening genetic condition that typically affects multiple body systems. Defects in electrolyte exchange at the cellular level lead to abnormally thick mucus production in exocrine-secreting organs, most notably the lungs, gastrointestinal track, reproductive systems, and sweat glands. This thick mucus contributes to exacerbation of chronic pulmonary bacterial infections resulting in chronic cough and progressive obstructive lung disease. The thickened mucus obstructs the pancreas, preventing enzymes from entering the intestines to digest food. This pancreatic insufficiency produces malabsorption, malnutrition, and growth failure. The vas deferens in males are similarly affected by the thick mucus causing male infertility. Abnormally high chloride and sodium loss through sweat can lead to electrolyte imbalance and characteristic salty sweat. A CF diagnosis is confirmed by the presence of CF symptoms, family history of CF, and/or a sweat test documenting chloride levels ≥ 60 mmol/L. There is no cure for CF; therefore, treatment is directed toward managing symptoms and preventing complications (Boucher, 2012).

CF is most common in White populations, and its presence is recorded with varying prevalence across ethnicities. The incidence is about 1 in 2,500 births among Whites; 1 in 4,000 among Hispanics; 1 in 15,000 among African Americans; and 1 in 100,000 among Asian Americans (Voter & Ren, 2008). CF affects about 30,000 children and adults in the United States (US) and 70,000 worldwide. About 1,000 new cases of cystic fibrosis are diagnosed each year in the US primarily through newborn screening. An additional 10 million Americans are asymptomatic carriers of one CF gene mutation (Cystic Fibrosis Foundation, 2011). In this chapter, mutation refers to symptom-causing changes in an allele. For individuals with CF, the prognosis is largely determined by the extent of pulmonary disease, which can vary based on genetic and environmental factors. Although historically CF has been considered a fatal childhood condition, improved treatment has increased the life expectancy to a median survival of 36.8 years (Cystic Fibrosis Foundation, 2006) with increasing numbers of individuals with CF living well into their 40s, 50s, and in rare cases 70s. Adults now account for more than 45% of people with CF in the CF Patient Registry (Cystic Fibrosis Foundation, 2011).

CF Genetics

CF symptoms result from mutations in the cystic fibrosis transmembrane conductance regulator (CFTR) gene located on chromosome 7 (Farrell & Fost, 2002; Kulczycki, Kostuch, & Bellanti, 2003). The CFTR gene encodes for a super protein, the CFTR protein, which controls sodium and chloride ion passage through the chloride channel within epithelial cell membranes (Lashley, 1998; Skatch, 2000).

Given the autosomal recessive nature of CF, an individual must receive a copy of a CFTR mutation from each parent in order to develop CF symptoms. Males and females are equally likely to inherit CFTR mutations. When two CF carriers procreate, each pregnancy carries a 25% chance the child will inherit two copies of CFTR mutations and develop CF symptoms, a 50% chance the child will inherit one CFTR mutation and be an asymptomatic CF carrier and a 25% chance the child with not inherit any CFTR mutations and remain asymptomatic (Cystic Fibrosis Foundation et al., 2009).

Genotype-Phenotype Variability

Genotype is the sequence of individual DNA bases that encode for amino acid and proteins that regulate physiologic functions. Phenotype is the physiological effect produced by the genetic code. Phenotypic expression can be altered by variable gene expressivity, modifier genes, mutation combinations, and environmental influences (Witt, 2003).

Since the discovery of the F508del mutation in 1989, more than 1,900 different changes have been found in the CFTR gene (Cystic Fibrosis Mutation Database, 2011). Most of these changes are of unknown clinical significance. CFTR mutations can be classified by the mechanism of cellular dysfunction associated with the mutation, for example, lack of CFTR protein production, abnormal protein function, or abnormal protein transportation within the cell. Almost all of the known CFTR mutations are point mutations or small deletions.

Class I, II, or III CFTR mutations are generally associated with more severe phenotypic CF presentations, whereas Class IV or V mutations have been associated with respiratory symptoms and pancreatic sufficiency (Luder, 2003).

Mutations can occur in various combinations, with the resulting clinical picture widely varying from mild to severe. The F508del, responsible for most CF cases worldwide, is a Class II mutation (Mendes et al., 2003). Persons with a homozygous F508del genotype tend to have more severe manifestations of CF (Kerem et al., 1990; Kerem & Kerem, 1996; Mickle & Cutting, 2000), whereas those who have a I, II, or III mutation paired with a IV or V mutation are likely to experience fewer, milder, or later onset symptoms (McKone, Emerson, Edwards, & Aitken, 2003). Given that the latter genotypes can have some CFTR protein production, the phenotype can vary from asymptomatic, to male infertility, to severe CF symptoms (Kerem & Kerem, 1996). Thus, the variation in disease severity has not been well correlated with the CF genotype (Cutting, 2010; Zielenski, 2000). Research continues to explore which contributions of environmental and genetic factors produce or mitigate symptoms (Mickle & Cutting, 1998). For example, genes encoding for inflammation cytokines (Hull & Thomson, 1998) and modifier genes have been associated with variation in CF phenotype (Cutting, 2010; Luder, 2003; Witt, 2003).

This variability in CF phenotype has led many to view CF is a spectrum condition and nomenclature has been developed to better depict the range in symptom severity. In addition to a categorical diagnosis of CF, there are now two new diagnostic classifications: CFTR protein-related disorder and CFTR-related metabolic syndrome (Castellani et al., 2008; Cystic Fibrosis Foundation et al., 2009). These new classifications are associated with some CFTR function and less severe CF symptoms—previously referred to as nonclassic CF. Although guidelines have been developed for the treatment of patients within these classifications, there is much variability in sweat testing procedures, diagnostic labels, how prognosis is described, and the services offered to these families (Nelson, Adamski, & Tluczek, 2011).

Cystic Fibrosis Carrier Testing

Advances in genetic technologies capable of identifying CFTR mutations responsible for 90% of CF cases led to the development of population-based screening tests, both for screening adults for their carrier status and for screening newborns for CF. In 1997, the National Institutes of Health developed a consensus statement recommending that cystic fibrosis carrier testing (CFCT) be offered to adults with a positive CF family history, those partnered with individuals who have CF, couples currently planning a pregnancy, and couples seeking prenatal care ("Genetic testing for cystic fibrosis," 1999). However, screening panels of the most common mutations result in carrier detection rates that vary by ethnic background. Carrier screening guidelines published in 2001 indicated that the 25 mutation panel are most sensitive for individuals of Ashkenazi Jewish descent (97%), followed by the following ethnicities: Northern Europeans (85%–90%), Southern Europeans (70%), African Americans (72%), Hispanics (57%), and Asians (30%; March of Dimes Defects Foundation, 2001). Based on the sensitivity of mutation screening panels for identifying CFTR mutations, the American College of Obstetricians and Gynecologists (ACOG) further refined the NIH Consensus Development Conference recommendations. They concluded cystic fibrosis carrier testing *should be offered* to the following groups: "couples with a family history of CF, partners of individuals with CF, and Caucasian couples of European or Ashkenazi Jewish descent planning a pregnancy or seeking prenatal care" (American College of Obstetricians and Gynecologists, 2001). They also state "information about CF screening *should be provided* to patients in other ethnic and racial groups" with additional counseling and screening available *upon patient request* (American College of Obstetricians and Gynecologists, 2001). Thus, ACOG differentiated between offering CF screening and making screening available dependent on risk and sensitivity of the recommended CF mutation screen. In addition to testing recommendations, the NIH panel, ACOG, and the Secretary's Advisory Committee on Genetic Testing (SACGT) specify all individuals and families should have access to appropriate genetic education, testing, and counseling services to ensure an informed decision is being made regarding genetic testing (American College of Obstetricians and Gynecologists, 2001; "Genetic testing for cystic fibrosis," 1999; SACGT, 2000).

After the screening implementation guidelines were released by ACOG in 2001, cystic fibrosis carrier testing (CFCT) began being routinely offered in prenatal primary care settings (ACOG, 2001). Although the American College of Medical Genetics (ACMG) proposed a 25-mutation basic panel to serve as the minimum standard for population-based CF carrier screening (Grody et al., 2001), the number of mutations tested and the resulting test sensitivity varies by the laboratory used (Tait et al., 2003). More recently, ACMG advised that two of the mutations, 1078delT and 1148T, be removed from the panel, as the incidence is either below the 0.1% standard or the allele has not been associated with CF symptoms (Committee on Genetics & American College of Obstetricians and Gynecologists, 2005; Watson et al., 2004). The ACMG/ACOG guidelines are noted to be a *screening* procedure, not a diagnostic test. Although some mutation carriers will not be detected, the mutations tested represent an acceptable compromise between cost, sensitivity, and phenotypic predictive value (Tsongalis, Belloni, & Grody, 2004). Thus, CF became the first genetic condition for which genetic screening tests of carrier status using DNA analysis have been implemented on a population-wide basis, regardless of race/ethnicity, or socioeconomic status.

Newborn Screening

Newborn screening (NBS) programs have been identifying presymptomatic infants with metabolic and/or genetic conditions as a public health initiative since the 1960s (American Academy of Pediatrics Newborn Screening Task Force, 2000). In 1991, Wisconsin became the first state to use DNA analysis in NBS for CF (Farrell et al., 1991; Gregg et al., 1993). CF remains the only condition for which DNA analysis is used in NBS to identify mutations responsible for the condition. With mounting evidence of nutritional benefits associated with early diagnosis, NBS programs rapidly added CF to the panels. By January 2010, all NBS programs in the United States included CF. Although the purpose of NBS is to identify infants with CF, the use of DNA analysis also detects infants who are carriers of only one CFTR mutation. Protocols vary from state to state, but commonly test for the most common mutations and can identify about 85% of CF cases in infants of Northern European descent. Not all carriers or all mutations that cause CF will be identified through routine NBS.

Most CF NBS programs incorporate a two-tiered algorithm, in which first blood specimens obtained in the first 24–48 hours of life are examined for immunoreactive trypsinogen (IRT). Levels in the top percentiles or above a particular level then undergo DNA analysis to detect the most common CFTR mutations. If one or two mutations are identified, a quantitative sweat test is performed to rule out or confirm a CF diagnosis. Sweat chloride levels can be classified as positive (indicative of CF), intermediate (possible CF), or negative (CF unlikely; Farrell et al., 2008).

Population screening for CF, both in carrier testing (preconceptionally or during pregnancy), and in NBS, have implications for the patient and their families. The vast array of genes, modifiers, and environmental influences affecting CF presentation complicates prediction of phenotypic presentation based on a specific genetic profile. Consequently, genetic counseling of the patient may be complex and fraught with uncertainty.

There are several notable differences between CFCT and NBS. CFCT focuses on whether a couple is at risk of having a child with CF. The decision to obtain genetic testing is voluntarily made. The results allow couples to make informed decisions about family planning. Expectant parents can make decisions regarding prenatal fetal diagnosis and whether to continue the pregnancy. By contrast, NBS is mandated by law with some provisions for parent refusal (e.g., for religious reasons) but parents rarely refuse. Abnormal screens require additional diagnostic sweat testing and most positive NBS results for CF ultimately prove to be false-positive. NBS procedures that involve DNA analysis also identify infants who are CF carriers for whom NBS offers no health benefits. However, there are reproductive implications for the parents, siblings, extended family, and child later in life. Uncovering nonpaternity is also possible.

PATIENT AND FAMILY ISSUES

The implementation of CFCT and NBS programs led to a proliferation of research regarding (a) patient perceptions about the acceptability of CFCT and NBS, (b) individuals' and couples' decision making regarding CFCT, (c) patient education guidelines, (d) patient education strategies, (e) provider–patient communication, (f) patient interpretation of genetic information, (g) psychosocial consequences of CFCT and CF NBS, and (h) consequences of CFCT and NBS on future testing and reproductive decisions.

Acceptability of CFCT and NBS

Studies have documented wide variability in the acceptability of CFCT ranging from 11% to 99% (Hartley et al., 1997; Henneman et al., 2001; Honnor, Zubrick, Walpole, Bower, & Goldblatt, 2000; Loader et al., 1996; Sorenson et al., 1997; Tambor et al., 1994; Witt et al., 1996). Factors associated with CFCT acceptability included whether the test was offered during pregnancy (Clayton et al., 1995; Witt et al., 1996), a proactive approach to offering immediate testing (Bekker et al., 1993), high patient education level and income (Fang et al., 1997; Honnor et al., 2000), having a choice to refuse (Hartley et al., 1997), perceptions of increased risk (Chen & Goodson, 2007; Markens, Browner, & Press, 1999; Tambor et al., 1994), and perceptions that testing is beneficial (Chen & Goodson, 2007; Fang et al., 1997; Sorenson et al., 1997). Unacceptability was associated with perceptions that screening is not beneficial (Chen & Goodson, 2007) and avoidance coping styles (Fang et al., 1997). Participants have also found CFCT acceptable just to know their own carrier status or for "confirmation of a healthy child" (Delvaux et al., 2001). One study of CF adult patients and parents did not support carrier testing for all persons planning pregnancy, but were more supportive of CF carrier testing within CF families (Henneman et al., 2001). Barriers to sibling testing of CF patients include perceived changes in family relationships and self-concept (Fanos & Johnson, 1995). Refusal of CFCT testing has been related to opposition to pregnancy termination, disapproval or nonparticipation of the partner, perceived low risk, error rate of the test, unacceptable levels of anxiety (Mennie, Gilfillan, Compton, Liston, & Brock, 1993), and offering a screening appointment by letter (Bekker et al., 1993; Watson et al., 1991).

Families who have had a positive family history of CF may be more likely to request CFCT or be receptive to testing (Callahan, Bloom, Sorenson, DeVellis, & Cheuvront, 1995). However, for the families of patients with CF who would be considered high risk for having a positive CF carrier status, that familiarity with both the disease and genetic implications has not necessarily resulted in decisions to have CFCT. In a Canadian study, researchers were surprised that screening participation, or uptake, of CFCT was less than 10% for blood relatives of CF patients despite the families receiving genetic counseling and free genetic testing (Surh, Cappelli, MacDonald, Mettler, & Dales, 1994). The researchers hypothesized that limited dissemination of information through families about the testing or the family members not being in the geographic area of the study may have been factors. However, they also speculated that a lack of self-perceived risk for being a carrier or family members not valuing the testing and knowledge enough to learn their own carrier status accounted for the low uptake rate.

Experience with CF may actually deter women from finding out CF carrier status. For siblings of persons with CF, not learning their carrier status may serve significant psychological functions for individuals at risk regarding personal guilt, desirability, family communication, and relationships (Fanos & Johnson, 1995). This research was reported prior to the availability of carrier testing, and no studies were found of findings in this area since CFCT implementation. Familiarity with CF has been related to higher comprehension of cystic fibrosis information and retention of testing results 6 months post testing in a preconception CFCT couple screening population (Henneman, Poppelaars, & ten Kate, 2002).

Even when persons found the test acceptable and were satisfied, they did not necessarily recommend CFCT to others (Delvaux et al., 2001). Many of these reports precede the development and release of the 1997 screening guidelines. This is consistent with most of the CF screening literature that reflects the pilot studies that serve as a basis for subsequent guideline development. There is little research on how CFCT is conceptualized and screening decisions are made in routine clinical practice. Sparbel and Williams (2009) submitted that the context of "being pregnant" caused pregnant women in a primary care setting to not consider carrier testing, interpreting this potential knowledge to be inconsistent with their intent to continue the pregnancy. Subsequent narrative analysis added insights regarding rejecting prenatal testing as outside of the "storyline" of a "perfect" pregnancy (Sparbel & Ayres, 2009).

Newborn screening for treatable conditions has been universally accepted by parents as well as professionals. Additionally, most parents and prospective parents tend to support NBS for less treatable or untreatable conditions (Detmar et al., 2008; Plass, van El, Pieters, & Cornel, 2010). Most parents of children with CF and those with false-positive NBS results tend to support NBS for CF (al-Jader, Goodchild, Ryley, & Harper, 1990; Helton, Harmon, Robinson, & Accurso, 1991; Lewis, Curnow, Ross, & Massiel, 2006; Mérelle et al., 2003; Tluczek, Orland, & Cavanagh, 2011). Preventing parents from embarking on stressful "diagnostic odysseys" is the most commonly cited justification. Other reasons include offering affected children early access treatment, informing parents about their child's and their own carrier status, gaining information about CF carriers' health, and preparing parents psychologically for symptom onset in children who are asymptomatic at the time of diagnosis. One report described an unintended benefit for CF NBS is the serendipitous diagnosis of CF in older siblings of infants with positive CF NBS. These older siblings had symptoms that were typical but not specific for CF (Sands, Zybert, & Nowakowska, 2010). However, some parents of infants with false-positive CF NBS results reportedly are less supportive for CF NBS because of the high rate of false-positive results and related parent anxiety (Tluczek, Orland, & Cavanagh, 2011).

CFCT Decision Making

Many factors affect how individuals and couples make decisions to undergo CFCT. Although patients often want information and discussion about the condition, many do not wish to make the decisions themselves (Kahn et al., 1997). Prenatal genetic testing decisions can either be viewed as an individual decision by the pregnant woman or a couple decision related to pregnancy management. If the sequential method of carrier testing is employed, the pregnant woman can make the initial testing decision herself. Whether she needs to enlist partner participation in the testing process itself, not only the decision about having testing, may influence her subsequent communication with family members about the CFCT decision. In some circumstances regarding prenatal diagnosis and possible pregnancy termination, the pregnant woman may decide to make decisions without family communication if she anticipates opposition with her decision (Atkin, Ahmad, & Anionwu, 1998).

CFCT studies have noted that more women than men agree to be screened. In one study, twice as many women as men decided to be tested (Bekker et al., 1993). After CFCT, women have also been shown to think more about the test than their partners, even when results are negative (53% versus 36%; Delvaux et al., 2001). This may be because of the attitudes about carrier testing being associated with reproduction and more "the woman's responsibility." The partner is not a benign factor, however, as partner nonparticipation or disapproval factored into CFCT testing decisions to decline for 10% of the population in a United Kingdom (UK) study (Mennie et al., 1993). The quality of the pregnant woman's relationship with her partner may influence whether the CFCT decision is viewed as an individual versus a couple process (Sparbel, 2008).

A process of joint consensus decision-making has been described among married couples (Godwin & Scanzoni, 1989). The context of spousal emotional interdependence can influence both coerciveness and degree of individual control maintained in this process. Couples whose relationships were characterized by economic equality, a history of high cooperativeness, and intrinsic gratifications tend to report a high degree of shared consensus in the decision making. The extent to which the views of each partner are respected or taken into account by the other partner can also be integral to the decision making (Daniels, Lewis, & Gillett, 1995). Other factors associated with couple decision making regarding CFCT include gender role preferences (Haber & Austin, 1992) and cultural influences (Webster, 2000).

Couple decision-making about prenatal genetic testing and diagnosis involves additional invasive procedures, such as amniocentesis. Pregnant women have been found to make the majority of decisions about amniocentesis. Partner presence during the genetic consultation during which amniocentesis was offered has also been shown to be predictive of testing uptake (Browner & Preloran, 1999; Browner, Preloran, & Cox, 1999). Partners tended to assume active roles in decision making when they perceived such a role as an expectation (Kenen, Smith, Watkins, & Zuber-Pitore, 2000).

Patient Education Guidelines

ACOG guidelines specify that pretest counseling, which includes information about CFCT, be offered to patients verbally and through written materials such as ACOG's pamphlet, *Cystic Fibrosis Carrier Testing: The Decision is Yours* (ACOG, 2001). The nondirective informed consent process should assure patient comprehension on the testing procedure and implications of results as well as emphasize the autonomous nature of the decision. Like most consents, ACOG does not incorporate understanding of psychological or social implications into its informed consent model. Whether the consent process results in increased patient autonomy in decision making is controversial (Atkin et al., 1998; Freda, DeVore, Valentine-Adams, Bombard, & Merkatz, 1998). Information offered to participants in research studies has been noted to meet some, but not all, recommended criteria for informed consent. Comprehension of the meaning and implications of positive test results is often missing. Despite a lack of meaningful comprehension of the test implications, more than 80 % of the study participants for maternal serum alpha-fetoprotein screening agreed to the test and signed informed consent documentation (Freda et al., 1998). This finding suggests some incongruity between genetic knowledge and acceptability or uptake of prenatal testing.

Guidelines are available on how genetic risk is established and when genetic testing should be offered. Little guidance exists about the content and process of genetic testing discussions. The information offered to pregnant women about prenatal genetic testing for chromosomal abnormalities has been described as inadequate for ensuring informed autonomous decision-making (Bernhardt et al., 1998). In the Bernhardt study of 169 pregnant women aged 21 and older with 21 obstetricians and 19 certified nurse-midwives, topics discussed most often in the initial prenatal visit regarding prenatal genetic testing were the practical details about the purpose of testing, and that testing is voluntary. Rarely were discussions comprehensive.

Guidelines for implementation of CF NBS programs advocated for the use of standardized educational materials for parents and providers (Comeau et al., 2007). Educational materials are often developed on a state-by-state basis. Common topics in written materials include CF fact sheets, CF carrier fact sheet, explanation of an initial positive result, inconclusive result, and repeat positive and inconclusive results (Comeau et al., 2007). Routine counseling for NBS results to enhance comprehension of results has been advocated. A tailored evidence-based approach to family centered genetic counseling following abnormal NBS for CF has been developed in Wisconsin. The approach uses both an educational and counseling approach to facilitate understanding of genetic information and a more positive experience with the NBS process (Tluczek, Zaleski, et al., 2011). The CF Foundation recently added new pages to its website (www.CFF.org) that are devoted specifically to NBS.

Patient Education Strategies

Educational strategies for increasing patient understanding of CFCT have included videotaped education (Witt et al., 1996), discussion between the health care provider and patient (Bernhardt et al., 1998), written material (Loeben, Marteau, & Wilfond, 1998), and computer-based instructional resources (Gason, Aitken, Delatycki, Sheffield, & Metcalfe, 2004). Clayton and colleagues (1995) found written and videotaped information as being equally effective in conveying information, with respondents answering an average of 86% of information correctly with either method. Computer-based learning has been found to be equally effective as oral educational presentation with the potential benefits of a consistent educational message, short delivery time, and limited resource commitments (Gason, et al., 2004). These various educational strategies have not resulted in a consensus about the best method for facilitating patient understanding and retention of genetic information, either on the short- or long-term basis.

Written information may offer a standardized approach but it can also provide underlying biased messages to the patient regarding both the condition to be tested and the consequences of testing. In one study, information about CF and reproductive options found in 28 pamphlets about CFCT for prenatal and other populations originating in the United States and the United Kingdom was assessed (Loeben et al., 1998). The depth of discussion about CF, reproductive options, and balance of positive/negative information of CF varied widely among the pamphlets available for CF education. Abortion was mentioned in about half the pamphlets, more often in the United Kingdom (UK) than in the United States. The wide variance in educational tone of the written information raises concerns about what content should be included. In a study conducted in the United Kingdom, the color of information literature (black and white versus glossy colored) affected the attitude of research participants toward genetic testing. Additionally, more positive attitudes toward genetic testing were observed if participants were asked questions about what they had read, than if they passively read the material (Michie, di Lorenzo, Lane, Armstrong, & Sanderson, 2004). Researchers call for the establishment of effective educational content and strategies for genetic counseling as a research priority (Marteau, 2000).

Provider–Patient Communication

Health care providers' knowledge and attitude toward testing and treatments frames the clinic interaction (Kahneman & Tversky, 1984; O'Connor, Pennie, & Dales, 1996; Sullivan, Hébert, Logan, O'Connor, & McNeely, 1996). Information about test probabilities and consequences are typically given during the clinical encounter. The same information can result in different patient decisions depending on how the information is labeled, known as "framing effects" (Levin, Gaeth, Schreiber, & Lauriola, 2002). Framing effects are difficult to avoid because there is no one optimal way to present information to all patients (Redelmeier, Rozin, & Kahneman, 1993). Provider knowledge and attitudes about a genetic condition and genetic testing can affect whether that testing is offered to the patient, the type and amount of risk information the provider presents to the patient, the advisability of genetic testing (Abramsky & Fletcher, 2002; Press & Browner, 1993; Wilkins-Haug, Horton, Cruess, & Frigoletto, 1996), and how patients interpret the results (Reyna, Lloyd, & Whalen, 2001). Provider perceptions of client personal characteristics can also influence the practitioner's view of the applicability of genetic testing and/or the educational strategies used when presenting genetic testing as an option (Press & Browner, 1993).

Considerable debate about population-based CFCT among health professionals has occurred since the identification of the CF gene and development of carrier testing. In 1993, most (86%) U.S. physicians favored offering CFCT if the testing were free. In one study, less than half (43%) of surveyed physicians and genetic counselors favored routinely offering CFCT to White couples, whereas even fewer obstetricians (35.5%) and genetic counselors (26.3%) favored routine screening (Faden et al., 1994). Physician practice patterns for offering CFCT to patients with no CF family history increased from 20% to 43% after CFCT was endorsed by the NIH (Doksum, Bernhardt, & Holtzman, 2001). This change in physician practice was not always accompanied by an increase in provider knowledge about CF. Professional practitioners often did not have accurate knowledge about key components of CF, such as inheritability, genetic risk, and meaning of testing results (Bekker et al., 1993; Bernhardt et al., 1998; Doksum et al., 2001; Rowley, Loader, Levenkron, & Phelps, 1993). Such inaccurate knowledge can adversely affect the way providers frame the amount, type, and tone of information delivered during clinical patient encounters.

The terms health professionals use can influence patients' interpretations of risk. Even when the intent of genetic counseling is nondirective, words are rarely value-free and have connotations that may communicate varying degrees of risk that can affect patient decision making (Abramsky & Fletcher, 2002; Wroe & Salkovskis, 2000). In a study conducted in the United Kingdom, expectant couples who were given negative information about sex-linked chromosomal abnormalities were more likely to terminate the pregnancy than couples received a mix of negative, positive, and neutral information (Hall, Abramsky, & Marteau, 2003). In NBS, misleading communication of carrier results were found in 41 out of 59 transcribed clinical interviews and associated with parents' inaccurate understanding of infant carrier status results (La Pean & Farrell, 2005). In another study, the large number of jargon words coupled with limited subsequent explanation suggested physician counseling may be too complex for some parents (Farrell, Duester, Donovan, & Christopher, 2008). The language used in the clinical encounter can also restrict the kinds of questions patients feel comfortable asking or information they volunteer (Anderson, 1999).

In the primary care setting, various professionals are involved in genetic education and counseling, including general practice and advanced practice nurses, physician assistants, social workers, physicians, as well as genetic counselors. Even if providers are knowledgeable, primary care settings seldom allow sufficient time for them to conveying complicated genetic risk information (Clayton et al., 1995). Primary care counseling sessions also tend to be more directive regarding prenatal testing than sessions with genetic professionals (Bartels, LeRoy, McCarthy, & Caplan, 1997). These suboptimal counseling conditions in the primary care setting and related framing effects can adversely influence patients' subsequent health care decisions.

Primary care providers are generally the gatekeepers for patient referrals regarding genetic testing and/or genetic counseling. In a U.S. study, prenatal carrier screening for several genetic conditions including CF carrier testing was only addressed in 12% of the visits in a primary care practice, disproportionately addressed with White patients, not addressed with African American patients, and addressed in only 30% of Jewish clients (Bernhardt et al., 1998). Given the high mutation rate for several genetic conditions in the Ashkenazi Jewish population, mutation panels should be routinely offered to that population. A survey of ACOG members showed that about half of the physicians reported that they do not include CF when taking a family history or offer CFCT to nonpregnant patients. One third of the 88.7% who offer CFCT to pregnant patients offer it only on a selective basis. Most do not offer CFCT in accordance with ACOG selection criteria guidelines for CFCT (Morgan, Driscoll, Mennuti, & Schulkin, 2004). A decade after the guidelines were released, most prenatal providers were aware of the CFCT guidelines, but 43% still lacked information regarding carrier rates, screening sensitivity, and residual risk (Darcy, Tian, Taylor, & Schrijver, 2011).

High screening participation, or uptake, rates for CFCT have been associated with providers offering immediate testing. When screening is invited by letter, thus requiring the patient to make an appointment, uptake rates tend to be lower, ranging from 4% to 12% (Bekker et al., 1993; Watson et al., 1991). However, when health providers offer testing in the clinical setting, the range of uptake increases from 70% in a U.S. study (Bekker et al., 1993) to 98% in a U.K. study (Harris et al., 1996). A study conducted in the United Kingdom showed that offering serum screening Down syndrome as part of a routine clinic appointment increased uptake rates as compared to requiring a separate appointment (Dormandy, Michie, Weinman, & Marteau, 2002).

Patient Interpretation of Genetic Information

Research repeatedly shows a significant proportion of individuals misunderstand or misinterpret CFCT testing results (Bekker et al., 1993; Denayer, Welkenhuysen, Evers-Kiebooms, Cassimann, & Van den Berghe, 1997; Gordon, Walpole, Zubrick, & Bower, 2003; Honnor et al., 2000; Levenkron, Loader, & Rowley, 1997). One study found that individuals who tested negative in carrier testing had up to 57% inaccurate recall (Axworthy, Brock, Bobrow, & Marteau, 1996). Individuals who test negative as CF carriers are also less likely to have an accurate understanding of their residual risk for having a child with CF than those found to be CF carriers (Botkin & Alemagno, 1992; Delvaux et al., 2001; Gordon et al., 2003; Hartley et al., 1997; Honnor et al., 2000). Risk perception of having a child post CFCT has been studied in Belgium in a population in which 91% of the participants had a family member or whose partner had a family member with cystic fibrosis (Denayer et al., 1997). Although most participants remembered whether the results were positive or negative, only one fourth could correctly report the estimated risk for having a child with CF. Most participants tended to underestimate the risk. In another study, all the couples who had one partner test positive for CF carrier status chose not to have genetic counseling and most did not know their risk for having a child with CF (Delvaux et al., 2001). Cheuvront and colleagues (1998) found that nonpregnant relatives of patients with CF had a 90% accurate recall rate regarding knowledge of CFCT regardless of whether they had received genetic counseling or received a mailed pamphlet with similar content. Although the two study groups had comparable knowledge, those who independently reviewed the information in their homes had more questions than those who received genetic counseling. Findings suggest that they would have liked more information and point to the benefits of patient-centered genetic counseling.

The application of genetic testing to CF NBS prompted the inclusion of genetic counseling in follow-up diagnostic sweat test appointments. Parents who received genetic counseling at specialized centers had better and more accurate recall of test results than those who had no counseling or received counseling in community settings (Ciske, Haavisto, Laxova, Rock, & Farrell, 2001; Mischler et al., 1998). Even a decade after parents received genetic counseling related to their infants' false-positive NBS results and heterozygote CF carrier status, parents who received genetic counseling had significantly higher levels of genetic knowledge than those who had not received genetic counseling (Cavanagh, Compton, Tluczek, Brown, & Farrell, 2010). A recent qualitative study examined parents' understanding of their children's false-positive CF NBS results and carrier status (Tluczek, Orland, & Cavanagh, 2011). Results showed that all parents accurately understood the meaning of negative diagnostic sweat test results. Most also believed that being a CF carrier was not associated with CF symptoms.

Despite recall of selected CF NBS information, parents continue to have inaccurate understanding of important implications of test results. Parents demonstrate misunderstandings about whether the child's carrier status could confer disease manifestations and the reproductive risk associated with being a CF carrier (Cavanagh et al., 2010; Ciske et al., 2001). Interruptions during the genetic counseling sessions adversely affected parents' retention of genetic information (Dillard, Shen, Tluczek, Modaff, & Farrell, 2007). A minority of parents in one study reported concerns about the accuracy of the sweat test, misconceptions about reproductive risk associated with one or two members of a couple being CF carriers, and misattribution about the CF carrier status (Tluczek, Orland, & Cavanagh, 2011). For parents questioned a decade after genetic counseling, most had already informed their children of the child's carrier status at a mean age of 9.2 years. Situational prompts were the most common reasons for disclosing this information to their children (Cavanagh et al., 2010; Ciske et al., 2001). The accuracy and adequacy of information parents give to children regarding their carrier status has implications for their genetic risk understanding and future decision making. Access to ongoing genetic counseling services for patients and families may be valuable to enhance understanding across generations. A system-wide analysis for quality improvement of CF NBS implementation implicates parents' misunderstanding of genetic counseling information as the greatest risk for system failure (Groose, Reynolds, Li, & Farrell, 2010).

Psychosocial Implications of CFCT and NBS

Psychosocial consequences of genetic testing and results can affect individuals and their families (Williams & Schutte, 1997). The testing process can give rise to feelings of anxiety associated with the potential of learning that one has a CFTR mutation and the related implication. Sixteen percent of CF carriers identified by population screening in the United Kingdom reported feeling worried about their test results and carriers had a poorer perception of their own health than noncarriers (Axworthy et al., 1996). Some women have indicated that the test generated unacceptable levels of anxiety for them and declined testing (Mennie et al., 1993; Sparbel & Williams, 2009; Witt et al., 1996). Individuals found to be CF carriers report less positive feelings about their CF test results compared to noncarriers, but not as negative as they anticipated they would feel (Gordon, Walpole, Zubrick, & Bower, 2003). Kenen (1996) suggests that the consequences of changes in self-perceptions following such testing and related decision-making behavior motivated by perceived "risk" for passing on genetically inherited diseases has not been fully explained and warrant additional investigation.

The literature shows a mix of positive and negative individual and family psychosocial consequences associated with CFCT (Callanan, Cheuvront, & Sorenson, 1999; Cheuvront et al., 1998; Fanos & Johnson, 1995; Henneman et al., 2001). Psychological effects on the family related to CFCT have generally been examined in high-risk populations (Callanan et al., 1999; Cheuvront et al., 1998; Fanos & Johnson, 1995; Henneman et al., 2001). Siblings of patients with CF identified concerns about the effects on family dynamics and relationships as a barrier for CFCT (Fanos & Johnson, 1995). One study conducted in Australia examined the effect of CFCT in a low-risk population (Gordon et al., 2003). Those found to be CF carriers tended to minimize adverse psychological effects on themselves. Both those who tested positive and those with negative results thought persons who tested positive would feel worse about their results than people who tested negative. Interestingly, carriers personally reported feeling more positive than they thought other carriers would feel regarding CF positive test results.

Cystic fibrosis NBS researchers have focused on the psychosocial effects on parents throughout the screening and diagnostic process as well as the impact on parents and their relationships with their children. Parents of infants with abnormal NBS results for CF have reported clinical levels of depressive symptoms during the days or weeks they await their infants' diagnostic test results (Tluczek, Koscik, Farrell, & Rock, 2005). This emotional distress was reportedly fueled by parents' lack of knowledge about CF and NBS as well as the normative adjustment to having a newborn. Such psychological effects have included parental anxiety, depression, hypervigilance, misunderstanding of the genetic results, and continuing concerns about the infant after a confirmatory negative CF result (Kharrazi & Kharrazi, 2005; Mérelle et al., 2003). Recommendations include limiting the time between the initial screen and confirmatory testing to decrease anxiety (Kharrazi & Kharrazi, 2005; Mérelle et al., 2003). Additionally, parents have described two dimensions of counseling that reduced their distress while enhancing their understanding: the provision of factual information (e.g., probability of CF diagnosis, sweat test procedure, and CF genetics), and emotional support (e.g., offering choices about timing/amount CF information, showing empathy, instilling hope, personalizing counseling, and providing hospitality; Tluczek et al., 2006). Despite the initial adverse psychological effects of false-positive CF NBS results, some studies suggest that there are minimal long-term adverse effects on parents' perceptions of their children's vulnerability (Beucher et al., 2010; Cavanagh et al., 2010). However, other studies raise questions about the effects of a CF diagnosis or CF carrier status on parents, their perceptions of their children's vulnerability, and the parent–child relationship. One study found that infants with CF and those found to be CF carriers had significantly more documented illnesses during the first year of life than a comparison group of infants with normal NBS results (Tluczek, McKechnie, & Brown, 2011). In another analysis of the same sample, mother–infant feeding interactions were compared across four groups based on infant diagnostic status, which included CF, CF carrier, congenital hypothyroidism, and normal NBS results (Tluczek, Clark, McKechnie, Orland, & Brown, 2010). Mothers of infants with CF were most likely to bottle feed, which was also associated with less sensitive and task-oriented interactions with their infants. Mothers of infants with CF were significantly more likely to have clinical levels of anxiety and depression than the other groups. Findings suggest that a CF diagnosis during the neonatal period may potentially pose a risk to maternal mental health and the quality of the newly developing parent–infant relationship. Interviews with parents of infants with false-positive CF NBS results revealed positive and negative psychosocial consequences associated with the receipt of genetic information through NBS (Tluczek, Orland, & Cavanagh, 2011). Positive consequences were described as an increased appreciation for the child's good health, strengthened family relationships, and feeling empathy for parents of affected children. Negative consequences included lingering concerns about their child's health and potentially limited future reproductive choices, as well as questioning the child's paternity. Parents also wondered whether their other children or relatives might have CF or be CF carriers. The newfound genetic information was viewed as an opportunity for some to obtain additional CFCT and to inform relatives about this option. Other parents felt burdened by a sense of responsibility to inform relatives with whom they had strained relationships. These findings suggest that a

CF diagnosis or identification of CF carrier status through NBS affects parents, their relationships with their children, and interactions with extended family members. Additional investigation of long-term psychosocial sequelae of a CF diagnosis or false-positive results are merited to reduce adverse iatrogenic psychosocial complications and inform clinical interventions.

Consequences of CFCT on Future Testing and Reproductive Decisions

Positive CFCT results precipitate subsequent decisions regarding further confirmatory carrier testing, prenatal diagnosis by amniocentesis, and possible decisions about the future of the pregnancy. The full scope of factors that influence such decisions is not clearly understood. Health professionals and patients may have divergent views from patients. Pregnancy termination has not been found to be the primary motivation for CFCT (Atkin et al., 1998).

Research indicates a wide discrepancy between persons accepting or willing to have CFCT and those who choose abortion of an affected fetus. In one study, 84% of pregnant women had a strong interest in being tested for CF carrier status, whereas only 29% were willing to terminate a fetus that had CF (Botkin & Alemagno, 1992). Some of those who declined initial testing cited opposition to abortion as the main reason for their refusal (Mennie et al., 1993; Sparbel & Williams, 2009). In another study, only 46% would consider termination, and just 30% of those whose results did not find a CFTR mutation indicated they would have considered termination if they had been found to have an affected fetus (Witt et al., 1996). Yet another study showed the use of prenatal diagnosis resulted in the termination of all pregnancies in which the fetus was found to have CF (Henneman, Kooij, Bouman, & ten Kate, 2002). Research continues to document discrepancies between testing and intent to abort or actually aborting, or seek further genetic counseling (Delvaux et al., 2001), including individuals who have a known family history of CF (Henneman et al., 2001; Wertz, Janes, Rosenfield, & Erbe, 1992). In the prenatal testing literature there is evidence that Maternal Serum Alpha-Fetoprotein (MSAFP) initial screening was not predictive of eventual uptake of amniocentesis and possible pregnancy termination. In some clinic settings, high MSAFP acceptance rates result from the provider de-emphasizing the association between screening and possible future prenatal diagnostic testing. Women accepting screening thus consider the MSAFP test as "normal prenatal care" (Markens et al., 1999). Sizeable proportions of women who accept screening report they would refuse an abortion or were uncertain of their decision for further testing.

Although the primary intent of NBS for CF is the identification of infants with the condition, the genetic information affords parent opportunities to consider additional genetic testing. A two-year post NBS study of families whose infants had positive NBS for CF found that parents indicate that despite the initial anxiety over testing, they would choose to have CF NBS again with a subsequent child (Beucher et al., 2010). Parents of children with CF diagnosed through NBS were asked about their attitudes regarding prenatal diagnosis and pregnancy termination. A comparison of their hypothetical versus actual reproductive behaviors showed a change over time. In most (67%) mothers, the planned behavior matched their actual behavior. However, one third of mothers changed their minds regarding the number of children they wanted, nearly half changed their views on prenatal diagnosis, and one third changed their views on pregnancy termination (for or against). The importance of continued genetic counseling after initial diagnosis and education was emphasized (Sawyer et al., 2006).

FUTURE DIRECTIONS

Cystic fibrosis illustrates how advances in genetic technologies and genomic health care are transforming our understanding of and approach to population screening for complex genetic conditions. The ability to identify mutations in the CFTR gene has resulted in offering genetic testing at three critical times in the reproductive cycle: when couples are planning to conceive, during pregnancy, and during the neonatal period. Thus, prospective parents can make informed family planning decisions and affected infants can receive prompt treatment. However, counseling such families has become increasingly complex and challenging because of the ever-increasing number of detectable CFTR mutations, lack of known clinical implications for most, wide range in phenotypic presentations, and inconsistent correlation between CFTR genetic profile and clinical manifestations. Research is needed to determine the most effective ways to help patients understand, retain, and make personal use of genetic information. We need to gain a better understanding of how genetic testing decisions are made and how best to facilitate informed decision-making consistent with patients' values and beliefs. Evidence-based clinical aids are needed to facilitate CFCT decisions and enhance NBS parent education and counseling. There is also a need for additional research to increase our understanding of the long-term effects of genetic information on individuals, couples, and their families regarding future reproductive decision making related to CFCT, and parent-child relationships related to NBS for CF. Individual's conceptions about health and genetic risk for offspring following CFCT or NBS is another fertile area for further inquiry.

Genetic testing for CF is available across the lifespan. Therefore, nurses in clinical practice need to remain informed about CFCT and NBS procedures and related laws within their jurisdictions as well as advances in CF research. Genetic literacy is central to competent nursing care. Additionally, nurses can use existing empirical evidence to inform educational and supportive interventions for families at the multiple points of care that can include prenatal care, the hospital stay at the time of delivery, primary care appointments for parents or infants, and through CF centers or genetic specialty services. Nurses can also nurture collaborative relationships with genetic counselors to enhance the continuity of care across setting.

Use of comprehensive tailored family-centered models for genetic counseling, such as the Wisconsin model (Tluczek, Zaleski et al., 2011), can inform future intervention development, research, and clinical practice. It is essential for nurses to understand couple and family dynamics associated with CFCT and positive CF NBS results to support family relationships and autonomous decision-making. As genomic medicine expands, nursing needs to be involved in addressing system capacity and policy issues to meet the increased demand for genetic patient education. Finally, the breadth of nursing roles in research, education, and practice offers countless opportunities for the integration of genetic information into personalized health care.

REFERENCES

Abramsky, L., & Fletcher, O. (2002). Interpreting information: What is said, what is heard—A questionnaire study of health professionals and members of the public. *Prenatal Diagnosis*, *22*(13), 1188–1194.

al-Jader, L. N., Goodchild, M. C., Ryley, H. C., & Harper, P. S. (1990). Attitudes of parents of cystic fibrosis children towards neonatal screening and antenatal diagnosis. *Clinical Genetics*, *38*(6), 460–465.

American Academy of Pediatrics Newborn Screening Task Force. (2000). Serving the family from birth to the medical home. Newborn screening: A blueprint for the future—a call for a national agenda on state newborn screening programs. *Pediatrics*, *106*(Pt. 2), 383–427.

American College of Obstetricians and Gynecologists (2001). *Preconception and prenatal carrier screening for cystic fibrosis: Clinical and laboratory guidelines*. Washington, DC: ACOG Publication.

American Congress of Obstetricians and Gynecologists. (2001). Cystic fibrosis carrier testing: The decision is yours [Brochure].

Anderson, G. (1999). Nondirectiveness in prenatal genetics: Patients read between the lines. *Nursing Ethics*, *6*(2), 126–136.

Atkin, K., Ahmad, W. I., & Anionwu, E. N. (1998). Screening and counselling for sickle cell disorders and thalassaemia: The experience of parents and health professionals. *Social Science & Medicine*, *47*(11), 1639–1651.

Axworthy, D., Brock, D. J. H., Bobrow, M., & Marteau, T. M. (1996). Psychological impact of population-based carrier testing for cystic fibrosis: 3-year follow-up. UK Cystic Fibrosis Follow-Up Study Group. *Lancet*, *347*(9013), 1443–1446.

Bartels, D. M., LeRoy, B. S., McCarthy, P., & Caplan, A. L. (1997). Nondirectiveness in genetic counseling: A survey of practitioners. *American Journal of Medical Genetics*, *72*, 172–179.

Bekker, H., Modell, M., Denniss, G., Silver, A., Mathew, C., Bobrow, M., & Marteau, T. (1993). Uptake of cystic fibrosis testing in primary care: Supply push or demand pull? *British Medical Journal*, *306*(6892), 1584–1586.

Bernhardt, B. A., Geller, G., Doksum, T., Larson, S. M., Roter, D., & Holtzman, N. A. (1998). Prenatal genetic testing: Content of discussions between obstetric providers and pregnant women. *Obstetrics and Gynecology*, *91*(5, Pt. 1), 648–655.

Beucher, J., Leray, E., Deneuville, E., Roblin, M., Pin, I., Bremont, F., . . . Roussey, M. (2010). Psychological effects of false-positive results in cystic fibrosis newborn screening: A two-year follow-up. *The Journal of Pediatrics*, *156*(5), 771–776.

Botkin, J. R., & Alemagno, S. (1992). Carrier screening for cystic fibrosis: A pilot study of the attitudes of pregnant women. *American Journal of Public Health*, *82*(5), 723–725.

Boucher, R. C. (2012). Chapter 259: Cystic fibrosis. In D. L. Longo, A. S. Fauci, D. L. Kasper, S. L. Hauser, J. L. Jameson, & J. Loscalzo (Eds.), *Harrison's principles of internal medicine* (Vol. 18e). Retrieved April 2, 2012, from http://www.accessmedicine.com.proxy.cc.uic.edu/content.aspx?aID=9128393.

Browner, C. H., & Preloran, H. M. (1999). Male partners' role in Latinas' amniocentesis decisions. *Journal of Genetic Counseling*, *8*(2), 85–108.

Browner, C. H., Preloran, H. M., & Cox, S. J. (1999). Ethnicity, bioethics, and prenatal diagnosis: The amniocentesis decisions of Mexican-origin women and their partners. *American Journal of Public Health*, *89*(11), 1658–1666.

Callahan, N. P., Bloom, D., Sorenson, J. R., DeVellis, B., & Cheuvront, B. (1995). CF carrier testing: Experience of relatives. *Journal of Genetic Counseling*, *4*(2), 83–95.

Callanan, N. P., Cheuvront, B. J., & Sorenson, J. R. (1999). CF carrier testing in a high risk population: Anxiety, risk perceptions, and reproductive plans of carrier by "non-carrier" couples. *Genetics in Medicine*, *1*(7), 323–327.

Castellani, C., Cuppens, H., Macek, M. J., Cassiman, J. J., Kerem, E., Durie, P., . . . Elborn, J. S. (2008). Consensus on the use and interpretation of cystic fibrosis mutation analysis in clinical practice. *Journal of Cystic Fibrosis*, *7*(3), 179–196.

Cavanagh, L., Compton, C. J., Tluczek, A., Brown, R. L., & Farrell, P. M. (2010). Long-term evaluation of genetic counseling following false-positive newborn screen for cystic fibrosis. *Journal of Genetic Counseling*, *19*(2), 199–210.

Chen, L. S., & Goodson, P. (2007). Factors affecting decisions to accept or decline cystic fibrosis carrier testing/screening: A theory-guided systematic review. *Genetics in Medicine*, *9*(7), 442–450.

Cheuvront, B., Sorensen, J. R., Callanan, N. P., Stearns, S. C., & DeVellis, B. M. (1998). Psychosocial and educational outcomes associated with home- and clinic-based pretest education and cystic fibrosis carrier testing among a population of at-risk relatives. *American Journal of Medical Genetics*, *75*(5), 461–469.

Ciske, D. J., Haavisto, A., Laxova, A., Rock, L. Z., & Farrell, P. M. (2001). Genetic counseling and neonatal screening for cystic fibrosis: An assessment of the communication process. *Pediatrics*, *107*(4), 699–705.

Clayton, E. W., Hannig, V. L., Pfotenhauer, J. P., Parker, R. A., Campbell, P. W., III, & Phillips, J. A., III. (1995). Teaching about cystic fibrosis carrier screening by using written and video information. *American Journal of Human Genetics*, *57*(1), 171–181.

Comeau, A. M., Accurso, F. J., White, T. B., Campbell, P. W., III, Hoffman, G., Parad, R. B., . . . O'Sullivan, B. P. (2007). Guidelines for implementation of cystic fibrosis newborn screening programs: Cystic fibrosis foundation workshop report. *Pediatrics*, *119*(2), e495–e518.

Committee on Genetics, & American College of Obstetricians and Gynecologists. (2005). ACOG committee opinion. Number 325, December 2005. Update on carrier screening for cystic fibrosis. *Obstetrics and Gynecology*, *106*(6), 1465–1468.

Cutting, G. R. (2010). Modifier genes in Mendelian disorders: The example of cystic fibrosis. *Annals of the New York Academy of Sciences, 1214*, 57–69.

Cystic Fibrosis Foundation. (2006). *New statistics show CF patients living longer*. Retrieved from http://www.cff.org/news/2006_news_archive/index.cfm?ID=6532&blnShowBack=False&idContentType=2561

Cystic Fibrosis Foundation. (2011). *About cystic fibrosis: What you need to know.* Retrieved from www.cff.org//AboutCF/

Cystic Fibrosis Foundation, Borowitz, D., Parad, R. B., Sharp, J. K., Sabadosa, K., Robinson, K. A., . . . Accurso, F. J. (2009). Cystic Fibrosis Foundation practice guidelines for the management of infants with cystic fibrosis transmembrane conductance regulator-related metabolic syndrome during the first two years of life and beyond. *Journal of Pediatrics*, *155*(6, Suppl.), S106–S116.

Cystic Fibrosis Mutation Database. (2011). *CFMBD statistics*. Retrieved from http://www.genet.sickkids.on.ca/StatisticsPage.html

Daniels, K. R., Lewis, G. M., & Gillett, W. (1995). Telling donor insemination offspring about their conception: The nature of couples' decision-making. *Social Science & Medicine*, *40*(9), 1213–1220.

Darcy, D., Tian, L., Taylor, J., & Schrijver, I. (2011). Cystic fibrosis carrier screening in obstetric clinical practice: Knowledge, practices, and barriers, a decade after publication of screening guidelines. *Genetic Testing and Molecular Biomarkers*, *15*(7–8), 517–523.

Delvaux, I., van Tongerloo, A., Messiaen, L., Van Loon, C., De Bie, S., Mortier, G., & De Paepe, A. (2001). Carrier screening for cystic fibrosis in a prenatal setting. *Genetic Testing*, *5*(2), 117–125.

Denayer, L., Welkenhuysen, M., Evers-Kiebooms, G., Cassimann, J. J., & Van den Berghe, H. (1997). Risk perception after CF carrier testing and impact of the test result on reproductive decision making. *American Journal of Medical Genetics*, *69*(4), 422–426.

Detmar, S., Dijkstra, N., Nijsingh, N., Rijnders, M., Verweij, M., & Hosli, E. (2008). Parental Opinions about the Expansion of the Neonatal Screening Programme. *Community Genetics*, *11*(1), 11–17.

Dillard, J. P., Shen, L., Tluczek, A., Modaff, P., & Farrell, P. M. (2007). The effect of disruptions during counseling on recall of genetic risk information: The case of cystic fibrosis. *Journal of Genetic Counseling*, *16*(2), 179–190.

Doksum, T., Bernhardt, B. A., & Holtzman, N. A. (2001). Carrier screening for cystic fibrosis among Maryland obstetricians before and after the 1997 NIH Consensus Conference. *Genetic Testing*, *5*(2), 111–116.

Dormandy, E., Michie, S., Weinman, J., & Marteau, T. M. (2002). Variation in uptake of serum screening: The role of service delivery. *Prenatal Diagnosis*, *22*(1), 67–69.

Faden, R. R., Tambor, E. S., Chase, G. A., Geller, G., Hofman, K. J., & Holtzman, N. A. (1994). Attitudes of physicians and genetics professionals toward cystic fibrosis carrier screening. *American Journal of Medical Genetics*, *50*(1), 1–11.

Fang, C. Y., Dunkel-Schetter, C., Tatsugawa, Z. H., Fox, M. A., Bass, H. N., Crandall, B. F., & Grody, W. W. (1997). Attitudes toward genetic carrier screening for cystic fibrosis among pregnant women: The role of health beliefs and avoidant coping style. *Women's Health*, *3*(1), 31–51.

Fanos, J. H., & Johnson, J. P. (1995). Barriers to carrier testing for adult cystic fibrosis sibs: The importance of not knowing. *American Journal of Medical Genetics*, *59*(1), 5–91.

Farrell, M., Deuster, L., Donovan, J., & Christopher, S. (2008). Pediatric residents' use of jargon during counseling about newborn genetic screening results. *Pediatrics*, *122*(2), 243–249.

Farrell, P. M., & Fost, N. (2002). Prenatal screening for cystic fibrosis: Where are we now? *Journal of Pediatrics*, *141*(6), 758–763.

Farrell, P. M., Mischler, E. H., Fost, N. C., Wilfond, B. S., Tluczek, A., Gregg, R. G., . . . Laessig, R. H. (1991). Current issues in neonatal screening for cystic fibrosis and implications of the CF gene discovery. *Pediatric Pulmonology*, 7, 11–18.

Farrell, P. M., Rosenstein, B. J., White, T. B., Accurso, F. J., Castellani, C., Cutting, G. R., . . . Campbell, P. W., III. (2008). Guidelines for diagnosis of cystic fibrosis in newborns through older adults: Cystic Fibrosis Foundation consensus report. *Journal of Pediatrics, 153*(2), S4–S14.

Freda, M. C., DeVore, N., Valentine-Adams, N., Bombard, A., & Merkatz, I. R. (1998). Informed consent for maternal serum alpha-fetoprotein screening in an inner city population: How informed is it? *Journal of Obstetric, Gynecologic, and Neonatal Nursing, 27*(1), 99–106.

Gason, A. A., Aitken, M., Delatycki, M. B., Sheffield, E., & Metcalfe, S. A. (2004). Multimedia messages in genetics: Design, development, and evaluation of a computer-based instructional resource for secondary school students in a Tay Sachs disease carrier screening program. *Genetics in Medicine, 6*(4), 226–231.

Genetic testing for cystic fibrosis. National Institutes of Health. Consensus Development Conference Statement on genetic testing for cystic fibrosis. (1999). *Archives of Internal Medicine, 159*(14), 1529–1539.

Godwin, D. D., & Scanzoni, J. (1989). Couple consensus during marital joint decision-making: A context, process, outcome model. *Journal of Marriage & the Family, 51*(4), 943–956.

Gordon, C., Walpole, I., Zubrick, S. R., & Bower, C. (2003). Population screening for cystic fibrosis: Knowledge and emotional consequences 18 months later. *American Journal of Medical Genetics, 120A*(2), 199–208.

Gregg, R. G., Wilfond, B. S., Farrell, P. M., Laxova, A., Hassemer, D., & Mischler, E. H. (1993). The application of DNA analysis in a population-screening program for neonatal diagnosis of cystic fibrosis (CF): Comparison of screening protocols. *American Journal of Human Genetics, 52*(3), 616–626.

Grody, W. W., Cutting, G. R., Klinger, K. W., Richards, C. S., Watson, M. S., Desnick, R. J., . . . American College of Medical Genetics. (2001). Laboratory standards and guidelines for population-based cystic fibrosis carrier screening. *Genetics in Medicine, 3*(2), 149–154.

Groose, M. K., Reynolds, R., Li, Z., & Farrell, P. M. (2010). Opportunities for quality improvement in cystic fibrosis newborn screening. *Journal of Cystic Fibrosis, 9*(4), 284–287.

Haber, L. C., & Austin, J. K. (1992). How married couples make decisions. *Western Journal of Nursing Research, 14*(3), 322–335.

Hall, S., Abramsky, L., & Marteau, T. M. (2003). Health professionals' reports of information given to parents following the prenatal diagnosis of sex chromosome anomalies and outcomes of pregnancies: A pilot study. *Prenatal Diagnosis, 23*(7), 535–538.

Harris, H., Scotcher, D., Hartley, N. E., Wallace, A., Craufurd, D., & Harris, R. (1996). Pilot study of the acceptability of cystic fibrosis carrier testing during routine antenatal consultations in general practice. *British Journal of General Practice, 46*(405), 225–227.

Hartley, N. E., Scotcher, D., Harris, H., Williamson, P., Wallace, A., Craufurd, D., & Harris, R. (1997). The uptake and acceptability to patients of cystic fibrosis carrier testing offered in pregnancy by the GP. *Journal of Medical Genetics, 34*(6), 459–464.

Helton, J. L., Harmon, R. J., Robinson, N., & Accurso, F. J. (1991). Parental attitudes toward newborn screening for cystic fibrosis. *Pediatric Pulmonology. Supplement, 7*, 23–28.

Henneman, L., Bramsen, I., Van Os, T. A., Reuling, I. E., Heyerman, H. G., van der Laag, J., . . . ten Kate, L. P. (2001). Attitudes toward reproductive issues and carrier testing among adult patients and parents of children with cystic fibrosis (CF). *Prenatal Diagnosis, 21*, 1–9.

Henneman, L., Kooij, L., Bouman, K., & ten Kate, L. P. (2002). Personal experiences of cystic fibrosis (CF) carrier couples prospectively identified in CF families. *American Journal of Medical Genetics, 110*(4), 324–331.

Henneman, L., Poppelaars, F. A., & ten Kate, L. P. (2002). Evaluation of cystic fibrosis carrier screening programs according to genetic screening criteria. *Genetics in Medicine, 4*(4), 241–249.

Honnor, M., Zubrick, S. R., Walpole, I., Bower, C., & Goldblatt, J. (2000). Population screening for cystic fibrosis in Western Australia: Community response. *American Journal of Medical Genetics, 93*(3), 198–204.

Hull, J., & Thomson, A. H. (1998). Contribution of genetic factors other than CFTR to disease severity in cystic fibrosis. *Thorax*, *53*(12), 1018–1021.

Kahn, B. E., Greenleaf, E., Irwin, J. R., Isen, A., M., Levin, I. R., Luce, M. F., . . . Young, M. J. (1997). Examining medical decision making from a marketing perspective. *Marketing Letters*, *8*(3), 361–375.

Kahneman, D., & Tversky, A. (1984). Choices, values, and frames. *American Psychologist*, *39*(4), 341–350.

Kenen, R., Smith, A. C. M., Watkins, C., & Zuber-Pitore, C. (2000). To use or not to use: Male partners' perspectives on decision making about prenatal diagnosis. *Journal of Genetic Counseling*, *9*(1), 33–45.

Kerem, E., Corey, M., Kerem, B. S., Rommens, J., Markiewicz, D., Levison, H., . . . Durie, P. (1990). The relation between genotype and phenotype in cystic fibrosis—Analysis of the most common mutation (delta F508). *New England Journal of Medicine*, *323*(22), 1517–1522.

Kerem, E., & Kerem, B. (1996). Genotype-phenotype correlations in cystic fibrosis. *Pediatric Pulmonology*, *22*(6), 387–395.

Kharrazi, M., & Kharrazi, L. D. (2005). Delayed diagnosis of cystic fibrosis and the family perspective. *Journal of Pediatrics*, *147*(3, Suppl.), S21–S25.

Kulczycki, L. L., Kostuch, M., & Bellanti, J. A. (2003). A clinical perspective of cystic fibrosis and new genetic findings: Relationship of CFTR mutations to genotype-phenotype manifestations. *American Journal of Medical Genetics*, *116A*(3), 262–267.

La Pean, A., & Farrell, M. H. (2005). Initially misleading communication of carrier results after newborn genetic screening. *Pediatrics*, *116*(6), 1499–1505.

Lashley, F. R. (1998). *Clinical genetics in nursing practice* (2nd ed.). New York, NY: Springer Publishing.

Levenkron, J. C., Loader, S., & Rowley, P. T. (1997). Carrier screening for cystic fibrosis: Test acceptance and one year follow-up. *American Journal of Medical Genetics*, *73*(4), 378–386.

Levin, I. P., Gaeth, G. J., Schreiber, J., & Lauriola, M. (2002). A new look at framing effects: Distribution of effect sizes, individual differences, and interdependence of types of effects. *Organizational Behavior and Human Decision Processes*, *88*(1), 411–429.

Lewis, S., Curnow, L., Ross, M., & Massiel, J. (2006). Parental attitudes to the identification of their infants as carriers of cystic fibrosis by newborn screening. *Journal of Paediatrics and Child Health*, *42*(9), 533–537.

Loader, S., Caldwell, P., Kozyra, A., Levenkron, J. C., Boehm, C. D., Kazazian, H. H., Jr., & Rowley, P. T. (1996). Cystic fibrosis carrier population screening in the primary care setting. *American Journal of Human Genetics*, *59*(1), 234–247.

Loeben, G. L., Marteau, T. M., & Wilfond, B. S. (1998). Mixed messages: Presentation of information in cystic fibrosis-screening pamphlets. *American Journal of Human Genetics*, *63*(4), 1181–1189.

Luder, E. (2003). Cystic fibrosis: The influence of the genotype and phenotype relationship on nutritional status. *Topics in Clinical Nutrition*, *18*(2), 92–99.

March of Dimes Defects Foundation. (2001). *Genetic screening pocket facts* [Brochure]. Staten Island, NY: Comprehensive Genetic Disease Program at the NYS Institute for Basic Research in Developmental Disabilities.

Markens, S., Browner, C. H., & Press, N. (1999). 'Because of the risks': How US pregnant women account for refusing prenatal setting. *Social Science & Medicine*, *49*(3), 359–369.

Marteau, T. M. (2000). Population screening for cystic fibrosis: A research agenda for the next 10 years. *American Journal of Medical Genetics*, *93*(3), 205–206.

McKone, E. F., Emerson, S. S., Edwards, K. L., & Aitken, M. L. (2003). Effect of genotype on phenotype and mortality in cystic fibrosis: A retrospective cohort study. *Lancet*, *361*(9370), 1671–1676.

Mendes, F., Roxo Rosa, M., Dragomir, A., Farinha, C. M., Roomans, G. M., Amaral, M. D., & Penque, D. (2003). Unusually common cystic fibrosis mutation in Portugal encodes a misprocessed protein. *Biochemical and Biophysical Research Communications, 311*(3), 665–671.

Mennie, M. E., Gilfillan, A., Compton, M. E., Liston, W. A., & Brock, D. J. (1993). Prenatal cystic fibrosis carrier screening: Factors in a woman's decision to decline testing. *Prenatal Diagnosis, 13*(9), 807–814.

Mérelle, M. E., Huisman, J., Alderden-van der Vecht, A., Taat, F., Bezemer, D., Griffioen, R. W., . . . Dankert-Roelse, J. E. (2003). Early versus late diagnosis: Psychological impact on parents of children with cystic fibrosis. *Pediatrics, 111*(2), 346–350.

Michie, S., di Lorenzo, E., Lane, R., Armstrong, K., & Sanderson, S. (2004). Genetic information leaflets: Influencing attitudes towards genetic testing. *Genetics in Medicine, 6*(4), 219–225.

Mickle, J. E., & Cutting, G. R. (1998). Clinical implications of cystic fibrosis transmembrane conductance regulator mutations. *Clinics in Chest Medicine, 19*(3), 443–458.

Mickle, J. E., & Cutting, G. R. (2000). Genotype-phenotype relationships in cystic fibrosis. *Medical Clinics in North America, 84*(3), 597–607.

Mischler, E. H., Wilfond, B. S., Fost, N., Laxova, A., Reiser, C., Sauer, C. M., . . . Farrell, P. M. (1998). Cystic fibrosis newborn screening: Impact on reproductive behavior and implication for genetic counseling. *Pediatrics, 102*(1, Pt. 1), 44–52.

Morgan, M. M., Driscoll, D. A., Mennuti, M. T., & Schulkin, J. (2004). Practice patterns of obstetrician-gynecologists regarding preconception and prenatal screening for cystic fibrosis. *Genetics in Medicine, 6*(5), 450–455.

Nelson, M. R., Adamski, C. R., & Tluczek, A. (2011). Clinical practices for intermediate sweat tests following abnormal cystic fibrosis newborn screens. *Journal of Cystic Fibrosis, 10*(6), 460–465.

O'Connor, A. M., Pennie, R. A., & Dales, R. E. (1996). Framing effects on expectations, decision, and side effects experienced: The case of the influenza immunization. *Journal of Clinical Epidemiology, 49*(11), 1271–1276.

Plass, A. M., van El, C. G., Pieters, T., & Cornel, M. C. (2010). Neonatal screening for treatable and untreatable disorders: Prospective parents' opinions. *Pediatrics, 125*(1), e99–e106.

Press, N., & Browner, C. H. (1993). 'Collective fictions': Similarities in reasons for accepting maternal serum alpha-fetoprotein screening among women of diverse ethnic and social class backgrounds. *Fetal Diagnosis Therapeutics, 8*(Suppl. 1), 97–106.

Redelmeier, D. A., Rozin, P., & Kahneman, D. (1993). Understanding patients' decisions: Cognitive and emotional perspectives. *Journal of the Allied Medical Association, 270*(1), 72–76.

Reyna, V. F., Lloyd, F. J., & Whalen, P. (2001). Genetic testing and medical decision making. *Archives of Internal Medicine, 161*(20), 2406–2408.

Rowley, P. T., Loader, S., Levenkron, J. C., & Phelps, C. E. (1993). Cystic fibrosis carrier screening: Knowledge and attitudes of prenatal care providers. *American Journal of Preventive Medicine, 9*(5), 261–266.

Sands, D., Zybert, K., & Nowakowska, A. (2010). Cystic fibrosis newborn screening enables diagnosis of elder siblings of recalled infants—Additional benefit. *Folia Histochemica et Cytobiologica, 48*(1), 163–165.

Sawyer, S. M., Cerritelli, B., Carter, L. S., Cooke, M., Glazner, J. A., & Massie, J. (2006). Changing their minds with time: A comparison of hypothetical and actual reproductive behaviors in parents of children with cystic fibrosis. *Pediatrics, 118*(3), e649–e656.

Secretary's Advisory Committee on Genetic Testing. (2000). *Enhancing the oversight of genetic tests: Recommendations of the SACGT*. Retrieved from http://www4.od.nih.gov/oba/sacgt.htm

Skatch, W. R. (2000). Defects in processing and trafficking of the cystic fibrosis transmembrane conductance regulator. *Kidney International, 57*(3), 825–831.

Sorenson, J. R., Cheuvront, B., DeVellis, B., Callanan, N., Silverman, L., Koch, G., . . . Fernald, G. (1997). Acceptance of home and clinic-based cystic fibrosis carrier education and testing by first, second, and third degree relatives of cystic fibrosis patients. *American Journal of Medical Genetics*, *70*(2), 121–129.

Sparbel, K. J. H. (2008). Decision making process in prenatal Cystic Fibrosis Carrier Testing (CFCT) decisions in primary care. *Midwest Nursing Research Society*. Indianapolis, IN.

Sparbel, K. J. H., & Ayres, L. (2009). 'Happily ever after': Telling the story of Cystic Fibrosis Carrier Testing (CFCT) decisions during pregnancy. *2009 Midwest Nursing Research Society*. Minneapolis, MN.

Sparbel, K. J., & Williams, J. K. (2009). Pregnancy as foreground in cystic fibrosis carrier testing decisions in primary care. *Genetic Testing and Molecular Biomarkers*, *13*(1), 133–142.

Sullivan, K. E., Hébert, P. C., Logan, J., O'Connor, A. M., & McNeely, P. D. (1996). What do physicians tell patients with end-stage COPD about intubation and mechanical ventilation? *Chest*, *109*(1), 258–264.

Surh, L. C., Cappelli, M., MacDonald, N. E., Mettler, G., & Dales, R. E. (1994). Cystic fibrosis carrier screening in a high-risk population: Participation based on a traditional recruitment process. *Archives of Pediatric and Adolescent Medicine*, *148*(6), 632–637.

Tambor, E. S., Bernhardt, B. A., Chase, G. A., Faden, R. R., Geller, G., Hofman, K. J., & Holtzman, N. A. (1994). Offering cystic fibrosis carrier screening to an HMO population: Factors associated with utilization. *American Journal of Human Genetics*, *55*(4), 626–637.

Tluczek, A., Clark, R., McKechnie, A. C., Orland, K. M., & Brown, R. L. (2010). Task-oriented and bottle feeding adversely affect the quality of mother-infant interactions after abnormal newborn screens. *Journal of Developmental Behavior in Pediatrics*, *31*(5), 414–426.

Tluczek, A., Koscik, R. L., Farrell, P. M., & Rock, M. J. (2005). Psychosocial risk associated with newborn screening for cystic fibrosis: Parents' experience while awaiting for sweat test appointment. *Pediatrics*, *115*(6), 1692–1703.

Tluczek, A., Koscik, R. L., Modaff, P., Pfeil, D., Rock, M. J., Farrell, P. M., . . . Sullivan, B. (2006). Newborn screening for cystic fibrosis: Parents' preferences regarding counseling at the time of infants' sweat test. *Journal of Genetic Counseling*, *15*(4), 277–291.

Tluczek, A., McKechnie, A. C., & Brown, R. L. (2011). Factors associated with parental perception of child vulnerability 12 months after abnormal newborn screening results. *Research in Nursing & Health*, *34*(5), 389–400.

Tluczek, A., Orland, K. M., & Cavanagh, L. (2011). Psychosocial consequences of false-positive newborn screens for cystic fibrosis. *Qualitative Health Research*, *21*(2), 174–186.

Tluczek, A., Zaleski, C., Stachiw-Hietpas, D., Modaff, P., Adamski, C. R., Nelson, M. R., . . . Josephson, K. D. (2011). A tailored approach to family-centered genetic counseling for cystic fibrosis newborn screening: The Wisconsin model. *Journal of Genetic Counseling*, *20*(2), 115–128.

Tsongalis, G. J., Belloni, D. R., & Grody, W. W. (2004). Cystic fibrosis mutation analysis: How many is enough? *Genetics in Medicine*, *6*(5), 456–458.

Voter, K. Z., & Ren, C. L. (2008). Diagnosis of cystic fibrosis. *Clinical Reviews in Allergy & Immunology*, *35*(3), 100–106.

Watson, E. K., Mayall, E., Chapple, J., Dalziel, M., Harrington, K., Williams, C., & Williamson, R. (1991). Screening for carriers of cystic fibrosis through primary health care services. *British Medical Journal*, *303*(6801), 504–507.

Watson, M., Cutting, G. R., Desnick, R. J., Driscoll, D. A., Klinger, K., Mennuti, M. . . . Grody, W. W. (2004). Cystic fibrosis population carrier screening: 2004 revision of American College of Medical Genetics mutation panel. *Genetics in Medicine*, *6*(5), 387–391.

Webster, C. (2000). Is spousal decision making a culturally situated phenomenon? *Psychology and Marketing*, *17*(12), 1035–1058.

Wertz, D. C., Janes, S. R., Rosenfield, J. M., & Erbe, R. W. (1992). Attitude toward the prenatal diagnosis of cystic fibrosis: Factors in decision making among affected families. *American Journal of Human Genetics*, *50*(5), 1077–1085.

Wilkins-Haug, L., Horton, J. A., Cruess, D. F., & Frigoletto, F. D. (1996). Antepartum screening in the office-based practice: Findings from the collaborative Ambulatory Research Network. *Obstetrics and Gynecology*, *4*(1), 483–489.

Williams, J. K., & Schutte, D. L. (1997). Benefits and burdens of genetic carrier identification. *Western Journal of Nursing Research*, *19*(1), 71–81.

Witt, D. R., Schaefer, C., Hallam, P., Wi, S., Blumberg, B., Fishback, A., . . . Palmer, R. (1996). Cystic fibrosis heterozygote screening in 5,161 pregnant women. *American Journal of Human Genetics*, *58*(4), 823–835.

Witt, H. (2003). Chronic pancreatitis and cystic fibrosis. *Gut*, *52*(Suppl. 2), ii31–ii41.

Wroe, A. L., & Salkovskis, P. M. (2000). The effects of 'non-directive questioning on an anticipated decision whether to undergo predictive testing for heart disease: An experimental study. *Behaviour Research and Therapy*, *38*(4), 389–403.

Zielenski, J. (2000). Genotype and phenotype in cystic fibrosis. *Respiration*, *67*(2), 117–133.

CHAPTER 16

Perinatal Genomics

Current Research on Genetic Contributions to Preterm Birth and Placental Phenotype

Gwen Latendresse

ABSTRACT

Significant maternal, fetal, and newborn morbidity and mortality can be attributed to complications of pregnancy. There are direct links between perinatal complications and poor fetal/newborn development and impaired cognitive function, as well as fetal, newborn, and maternal death. Many perinatal complications have pathophysiologic mechanisms with a genetic basis. The objective of this chapter is to focus on perinatal genomics and the occurrence of two specific complications: preterm birth and dysfunctional placental phenotype. This chapter includes discussions of genetic variation, mutation and inheritance, gene expression, and genetic biomarkers in relation to preterm birth, in addition to the impact of maternal tobacco smoke exposure on placental phenotype. The concept of epigenetics is also addressed, specifically the regulation of gene expression in the placenta and fetal origins of adult health and disease. There is great potential for nurse-researchers to make valuable contributions to perinatal genomics investigations, but this requires perseverance, increased genetics-based understanding and skills, as well as multidisciplinary mentorship.

http://dx.doi.org/10.1891/0739-6686.29.331

INTRODUCTION

A quick literature search on the topic of genetics and pregnancy reveals a rapidly accumulating collection of studies. Indeed, a PubMed search using the simple terms "genetics and pregnancy" returns 71,452 articles; 10,354 when limiting the search to humans, English, and published during the last 5 years. Most of these studies have been dedicated to prenatal screening and diagnosis of genetic conditions that directly affect the fetus, newborn, or child (i.e., trisomies 13, 18, 21, cystic fibrosis, Tay–Sachs disease, hemophilia, etc.). However, the objective of this chapter is to focus on perinatal genomics related to two specific pregnancy-related complications: preterm birth and dysfunctional placental phenotype. The chapter section on genetics of preterm birth discusses the contributions of genetic variation, mutation and inheritance, gene expression, and genetic biomarkers in relation to preterm birth. The chapter section on placental phenotype addresses the interaction of a specific environmental exposure—tobacco smoke—with the ability of the placenta to function as a "protective barrier" against overexposure to glucocorticoids, specifically cortisol. The concept of epigenetics is also discussed, specifically in relation to regulation of gene expression in the placenta, and the implications for fetal origins of adult health and disease. Both preterm birth and dysfunctional placental phenotype have significant impact on maternal, newborn, and adult health, and thus have been (and will continue to be) an important focus in perinatal genetics research.

Very few nurse-researchers are engaged in maternal or perinatal genetic studies. A literature search for perinatal genetics articles having nurse authors was successful in finding only a handful of publications. Five articles were identified; four review articles and one research report on placental gene expression and preeclampsia. However, with adequate skills-building, knowledge base, and experienced mentoring, nurse-researchers could make substantial contributions in this very important field. Nurse-researchers have typically been clinically oriented, not "bench" oriented. Nurse-researchers, however, have the capacity for taking on the task of translational science by bringing together clinical perspectives and bench science. Perinatal genetics research is an area with potential for nurse-researchers to make valuable contributions. This is discussed in the last section on the role of nurse-researchers in future perinatal genetics research.

THE GENETICS OF PRETERM BIRTH

Epidemiology and Biological Basis of Preterm Birth

Birth that occurs prior to the completion of 37 weeks of gestation is considered preterm. After several decades of increasing rates for preterm birth, and an all-time high of 12.8% in 2006, the United States witnessed decreased rates in 2007

(12.7%) and 2008 (12.3%; Martin, Osterman, & Sutton, 2010). This reduction may likely reflect a change in obstetrical practices in very recent years; primarily the avoidance of elective induction of labor and scheduled cesarean section deliveries prior to 39 completed weeks of pregnancy. However, 12.3% is still too high given the significant newborn and infant morbidity, mortality, and long-term health issues associated with preterm birth. The cause of spontaneous preterm birth is complex and multifactorial evidenced by the wide variety of factors associated with its occurrence and the lack of success in identifying a singular mechanism. There are well-documented environmental contributors to preterm birth, such as poverty, nutrition, chronic maternal stress, and exposure to tobacco smoke, as well as racial/ethnic disparities (i.e., preterm birth occurs in 17.5% of newborns of non-Hispanic Black women vs. 11.1% of non-Hispanic White women; Behrman & Stith Butler, 2007; Martin et al., 2010; Moutquin, 2003). Genomics investigations of the etiology of preterm birth have been increasingly on the forefront of research efforts and focus on known or suspected pathways involved in the physiological processes of birth and birth initiation. These pathways include (a) infection, immunity, and inflammation; (b) uterine contractility; (c) placental dysfunction, including coagulation factors; (d) connective tissue remodeling, primarily degradation of the uterine cervix and amniotic membranes; and (e) activation of the maternal-fetal hypothalamic-pituitary-adrenal axis (i.e., stress response; Behrman & Stith Butler, 2007).

Genetic Variation, Mutation, and Inheritance of Preterm Birth

It has been documented that preterm birth has a heritable component (Plunkett & Muglia, 2008). A Norwegian study that used a large birth registry (191,282 mothers and 127,830 fathers; Wilcox, Skjaerven, & Lie, 2008) reported that mothers who are born preterm have a relative risk of 1.54 (95%; CI = 1.42, 1.67) for delivering preterm, and the effect was stronger for mothers who were born < 35 weeks gestation (RR = 1.85, 95%; CI = 1.52, 2.27). For fathers who were born preterm, the risk was weak to nonexistent for having preterm offspring. The authors state that "this argues against major contributions of fetal genes inherited from either parent" as a basis for initiation of the preterm birth process by the fetus itself. "The increased risk of preterm delivery among mothers born preterm is consistent with heritable maternal phenotypes that confer a propensity to deliver preterm (Wilcox et al., 2008)." This also supports the theory that the etiology of preterm birth falls primarily along maternal pathophysiological pathways (i.e., involving maternal physiology), and not fetal pathways.

The heritability of preterm birth is thought to be the result of a polygenic process, rather than a classical Mendelian inheritance (single gene mutations such as cystic fibrosis, Tay–Sachs, and sickle-cell anemia; Chaudhari et al., 2008).

High-throughput DNA genotyping of candidate genes (e.g., genes with plausible function along the known or suspected physiological pathways) has enabled researchers to evaluate a far greater number of gene polymorphisms, and this has been the bulk of most research efforts. Most of these candidate genes fall along the inflammatory, coagulation, and cervical remodeling pathways to preterm birth. However, it is thought that gene–environment and gene–gene interaction play the major role in the actual risk for preterm birth. For example, the occurrence of a bacterial vaginosis infection during pregnancy in women who also carry a specific genetic variant of an inflammatory-related cytokine (tumor necrosis factor alpha—TNFα) demonstrate a higher risk for preterm birth than do women who have either of these alone (Macones et al., 2004). Indeed, gene–environment interactions could explain some of the disparity in preterm birth between non-Hispanic Blacks and non-Hispanic Whites (Dolan, 2010). The TNFα variant mentioned earlier, for example, is much more common in non-Hispanic Blacks (Menon et al., 2006). Table 16.1 provides a sampling of some of the studies reporting candidate genes in which single-nucleotide polymorphisms (SNPs) have an association with preterm birth risk. Research results are generally inconsistent; the most consistent and well studied are the TNFα polymorphisms.

Although numerous studies of candidate genes and preterm birth associations could be found in PubMed, very little evidence could be found for genome-wide association studies. This is likely because of the expense and extremely large sample sizes required for adequate statistical analyses, as well as the need for sophisticated epidemiological statistical approaches (Khoury et al., 2009). However, use of a genome-wide association approach could provide a nonbiased approach to identifying DNA SNPs, haplotypes, and foci along previously unknown pathways, as well as assist in the evaluation of disparities (Anum, Springel, Shriver, & Strauss, 2009; Plunkett & Muglia, 2008).

Gene Expression and Preterm Birth

Over 300 genes are differentially expressed in the placentas obtained from laboring versus nonlaboring women at term delivery (Lee et al., 2010). Furthermore, there are dramatic changes in global gene expression between early and late gestation (Winn et al., 2007). However, it is not clear if these gene expression profiles reflect changes that occur because of labor or play a role in the initiation of labor (or both). Some studies suggest that there are "pro labor" genes that are expressed both similarly and differently in uterine tissues of women who deliver preterm compared to term (Tattersall et al., 2008). The expression of microRNAs and siRNA may also be contributors to the preterm birth phenotype because of gene regulation (Mayor-Lynn, Toloubeydokhti, Cruz, & Chegini, 2011).

TABLE 16.1

Examples of Polymorphisms Associated With Biological Pathways to Preterm Birth

Author Info (Year)	Study Details	Gene[a] SNPs Identified in Association With Preterm Birth[b]	Biological Pathway or Function
(Annells et al., 2004)	22 SNPs in candidate genes (cytokines, mediators of apoptosis, and host defense). 202 cases and 185 controls.	Interleukin (IL)10, IL4	Inflammation
		Tumor Necrosis Factor alpha (TNFα)	
		Manose Binding Lectin (MBL2)	Connective tissue remodeling
(Bodamer et al., 2006)	MBL2 genotypes for 102 preterm infants 102 term infants	MBL2 variants	Connective tissue remodeling
(Crider, Whitehead, & Buus, 2005)	Systematic review: 18 studies prior to 2004	TNFa, IL1b, IL4	Inflammation
		IL6 "protective" for Whites compared to Blacks	Inflammation
		Matrix Metalloproteinases (MMP)1, MMP9	Connective tissue remodeling
		Toll-like receptor (TLR)1	Immune response
		Adrenergic Beta-2 Receptor (ADRB2)	Uterine contractility
		Vascular Endothelial Growth Factor (VEGF)	Placentation
		Factor V	Coagulation
(Plunkett & Muglia, 2008)	Exhaustive systematic review	TNFa, IL1R, IL4, IL6, IL10, IL18, TLR2, TLR4	Inflammation
		Lymphotoxin Alpha (LTA)	
		MBL2	Connective tissue remodeling
		MMP1, MMP9	
		Serpin Peptidase Inhibitor, Clade H, Member 1 (SERPINH1)	
		Solute Carrier Family, Member 2 (SLC23A2)	
		Glutathione S-Transferase Theta 1 (GSTT1)	

(*Continued*)

TABLE 16.1

Examples of Polymorphisms Associated With Biological Pathways to Preterm Birth (Continued)

Author Info (Year)	Study Details	Gene[a] SNPs Identified in Association With Preterm Birth[b]	Biological Pathway or Function
		ADRB2	Uterine contractility
		Factor 5, Factor 7	Coagulation
		Coagulation Factor XIII, A1 Polypeptide (F13A1)	
		Thrombomodulin (THBD)	
		Nitric Oxide Synthase 3 (NOS3)	Placentation
		Nitric Oxide Synthase 2a, Inducible (NOS2A)	
		VEGF	
		Paraoxonase 1 (PON1)	Response to environmental toxins
		Paraoxonase 2 (PON2)	
		Opioid Receptor, Mu 1 (OPRM1)	
		Methyltetrahydrofolate-Homocysteine Methyltransferase Reductase (MTRR)	Folate metabolism
		Serine Hydroxymethyl-transferase 1 (SHMT1)	
		Peroxisome Proliferator-Activated Receptor Gamma (PPARG)	Adipose tissue differentiation & metabolism
(Romero et al., 2010)	223 mothers and 179 fetuses of preterm birth	Interleukin-6 Receptor 1 (IL6R)-fetal	Inflammation
		IL2 Haplotype-fetal	
	599 mothers and 628 fetuses of term birth	Insulin-like Growth Factor 2 (IGF2) Haplotype-fetal	Growth factor
	Chilean Hispanic Population	Tissue Inhibitor of Metalloproteinase-2 (TIMP2)-maternal	Extracellular matrix
	190 candidate genes and 775 SNPs	Collagen, Type IV, Alpha 3 (COL4A3) Haplotype-maternal	

(Continued)

TABLE 16.1
Examples of Polymorphisms Associated With Biological Pathways to Preterm Birth (Continued)

Author Info (Year)	Study Details	Gene[a] SNPs Identified in Association With Preterm Birth[b]	Biological Pathway or Function
(Sata et al., 2009)	Japanese population (n = 414). Inflammatory cytokines. IL-1a, IL-1b, IL-2	IL1a	Inflammation
(Velez et al., 2008)	Maternal & fetal DNA from 370 U.S. White birth-events (172 cases and 198 controls). 1,536 SNPs in 130 candidate genes.	Factor V, Factor VII, tissue plasminogen activator (PLAT)-maternal	Coagulation
		IL10 Receptor Antagonist (IL10RN)-fetal	Inflammation

[a]Official gene names with corresponding symbols are used on first introduction of each specific gene, followed by use of gene symbols only for subsequent references.
[b]Reports are generally inconsistent; the most consistent and well studied are the TNFα polymorphisms.

Genetic Biomarkers for Preterm Birth

There are currently no genetic biomarkers that have been developed for use in clinical screening and/or diagnostic testing for preterm birth risk. This is likely because of the multifactorial nature of preterm birth, and the inability to distinguish a single SNP or genetic marker with acceptable sensitivity or specificity across various populations (or even in any given population). However, the future holds promise for a combined testing approach that includes identification of genetic susceptibility (i.e., immune-related genetic markers) in conjunction with identification of relevant environmental exposures, such as infection, smoking, obesity, or stress.

There is also early development of global gene expression testing of maternal blood early in gestation. Enquobahrie et al. (2009) report accurate prediagnostic separation of preterm birth cases from controls via identification of 209 differentially expressed genes in the maternal blood of women at 16 weeks of gestation. In addition to identifying several candidate genes along preterm birth pathways (i.e., immune system, inflammation), they also identified gene expression with functions in organ development, metabolism (lipids, carbohydrates, amino acids), and cell signaling. However, progression of global gene expression testing has not yet reached clinical application, nor is a cost/benefit ratio known. The development of biomarkers certainly has the potential for increasing health care costs without a significant impact on morbidity and mortality.

Proteomics (the study of protein structure and function; Anderson & Anderson, 1998), also holds promise for the development of biomarkers. Protein products of the genes, not simply gene expression, cause disturbances at the cellular level and hold promise as a future screening and/or diagnostic tool (Buhimschi, Weiner, & Buhimschi, 2006). Furthermore, identification of proteins can lead investigators towards a better understanding of the "upstream" genes involved, thus increasing our knowledge of the pathophysiological mechanisms contributing to the occurrence of preterm birth. For example, corticotrophin-releasing hormone is a polypeptide normally produced by the placenta and circulated in the maternal blood. Elevated maternal plasma corticotrophin-releasing hormone levels are associated with a higher risk of preterm birth (Latendresse & Ruiz, 2011; Wadhwa et al., 2004). Although corticotrophin-releasing hormone (CRH) itself has not demonstrated adequate sensitivity or specificity to render it a clinically useful biomarker for identifying women at risk for preterm birth (McGrath & Smith, 2002), it has led to a deeper understanding of the upstream contributors to corticotrophin-releasing hormone gene expression (such as cortisol and catecholamines; Sandman et al., 2006).

In contrast, vaginal swab testing for fetal fibronectin (a glycoprotein involved in the formation of the extracellular matrix that supports structures within the pregnant uterus) has become a widely used clinical biomarker, particularly as a negative predictor for women with threatened preterm birth (Honest, Bachmann, Gupta, Kleijnen, & Khan, 2002). Like CRH, the identification of fibronectin as a biomarker could lead to a better understanding of the mechanisms involved in preterm birth by identifying specific genes, polymorphisms, and gene expression contributing to preterm birth pathways (i.e., degradation of the extracellular matrix).

Summary of Genetics and Preterm Birth

Spontaneous preterm birth is a multifactorial, polygenic condition for which research in genetics has attempted to find answers to its origins and pathophysiologic mechanisms. Genetics studies have established that there is a level of heritability in preterm birth, and have identified several genetic variants in association with an increased risk for preterm birth. However, most of these associations are weak and reports are inconsistent. The most consistent and well-studied are the TNFα polymorphisms. Less is known about the contributions of gene–gene and gene–environment interactions in the occurrence of preterm birth, but evidence suggests a significant role.

PLACENTAL PHENOTYPE, GENE EXPRESSION, AND MATERNAL EXPOSURE TO TOBACCO SMOKE

Alterations in placental phenotype can have a profound impact on pregnancy and the fetus, including intrauterine growth restriction, small for gestational age, preterm birth, preeclampsia, and even "fetal programming" for risk of child and adult diseases (Founds et al., 2011; Jansson & Powell, 2007; Romero, Kusanovic, Chaiworapongsa, & Hassan, 2011; Vedmedovska, Rezeberga, Teibe, Melderis, & Donders, 2011). The origins of many pregnancy complications, such as preeclampsia, could not be explained until more recent advances in science were able to adequately investigate and implicate the placenta (Founds et al., 2011; Kanasaki & Kalluri, 2009). It has long been understood that the placenta is essential for delivering nutrients and oxygen to the fetus, and for removing metabolic wastes. Furthermore, the role of the placenta in determining gestational length and maturation of the fetus has become much better understood. However, the placental role as a "protective barrier" and as a contributor to adaptive (albeit sometimes harmful) response to environmental insult (such as exposure to maternal tobacco smoke, poor air quality, inadequate nutrition, and chronic maternal stress) is much less known.

One known environmental contributor to placental phenotype is maternal tobacco smoking. Maternal tobacco smoke has a well-documented negative impact on the placenta, including degenerative and inflammatory changes, increased collagen synthesis, interference in pro-angiogenic factors (i.e., vascular endothelial growth factor—VEGF), disturbed physiologic transformation of uterine blood vessels, alteration in trophoblastic invasion of the uterus, decreased placental growth factor, decreased placental weight, changes in villous structure (i.e., thickening of the trophoblastic membrane that reduces the diffusion of nutrients, etc., across the placenta), and a host of structural and functional changes (Jauniaux & Burton, 2007; U.S. Department of Health and Human Services, 2010). It is clear that alterations in placental phenotype (function and structure) contribute significantly to the health of the fetus and newborn; fetal hypoxia is no longer considered the sole contributor to smoking-associated adverse pregnancy outcome.

Smoking, Adverse Pregnancy Outcomes, and Pathophysiologic Mechanisms

Maternal tobacco smoke, through the actions of nicotine, carbon monoxide, polycyclic aromatic hydrocarbons, and thousands of other chemicals and substances contained in tobacco smoke can affect the fetus directly or indirectly (via the placenta or maternal health; U.S. Department of Health and Human Services, 2010). Moreover, any disruption in normal placental development, function, or

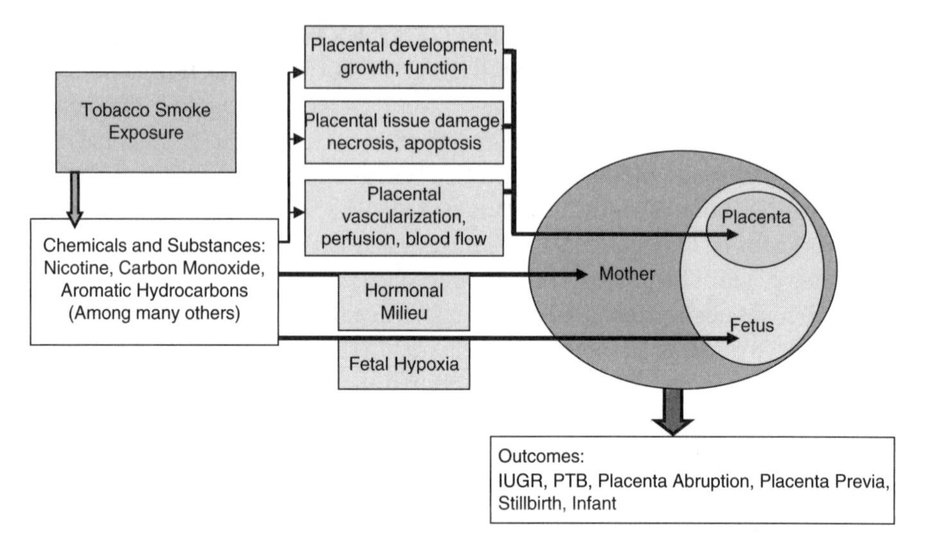

FIGURE 16.1 Pathophysiologic links between smoking and adverse pregnancy outcome.

blood flow has the potential to impair the exchange of oxygen, nutrients, and metabolic products between mother and fetus (Jauniaux & Burton, 2007; U.S. Department of Health and Human Services, 2010).

Figure 16.1 depicts a simplistic model of the pathophysiologic pathways that link maternal tobacco smoke with poor pregnancy outcome, including (a) interference in normal placental development (i.e., trophoblast invasion), growth, and function; (b) placental tissue damage, necrosis, and apoptosis; (c) poor placental vascularization (i.e., alteration in pro-angiogenic factors), perfusion, and blood flow; (d) maternal-fetal-placental hormonal milieu; and (e) chronic fetal hypoxia as a result of hemoglobin's high affinity to carbon monoxide (a byproduct of combustion, thus of cigarette smoke). The first three of this list are directly related to placental phenotype.

Maternal Tobacco Smoke, the Placental "Protective Barrier," and the Role of Glucocorticoids

Figure 16.1 exhibits the significant role of the placenta as a pathophysiologic link between maternal tobacco smoke and adverse pregnancy outcome. However, the impact of maternal tobacco smoke on the function of the placenta as a "protective barrier" against excessive fetal cortisol (an active form of glucocorticoid) exposure is not well understood. Excessive cortisol has a well-documented causative association with intrauterine growth restriction (Goedhart et al., 2009), and perhaps preterm birth (Giurgescu, 2009). More importantly, excessive fetal cortisol exposure may contribute significantly to the occurrence of chronic diseases later in life (i.e., hypertension, cardiovascular disease, type 2 diabetes, and mental health) via in utero "fetal programming" (Drake, Tang, & Nyirenda, 2007; Seckl & Holmes, 2007). Higher levels of cortisol are found in maternal serum and in umbilical cord blood of mothers who smoke or are exposed to nicotine during pregnancy (Lieberman et al., 1992; Varvarigou, Liatsis, Vassilakos, Decavalas, & Beratis, 2009; Varvarigou, Petsali, Vassilakos, & Beratis, 2006). The mechanisms are not well understood, but disruption of placental cortisol metabolism and protection against excessive fetal and placental cortisol exposure may be one pathway to adverse outcome (Mark, Augustus, Lewis, Hewitt, & Waddell, 2009; Yang, Julan, Rubio, Sharma, & Guan, 2006).

The Role of Placental Enzymes and Molecular Transporters as a "Protective Barrier" Against Excessive Placental and Fetal Exposure to Cortisol

Although cortisol is essential for normal development of the placenta, fetus, and in pregnancy/birth physiology (Challis et al., 2005; Sherwood, 2004), overexposure contributes to placental and fetal growth restriction, and an altered hormonal milieu that may contribute to preterm birth (Gennari-Moser et al., 2011; Giurgescu, 2009; Goedhart et al., 2009). Maternal cortisol levels are 10-fold higher in healthy, nonsmoking pregnant women than in the fetus and cortisol easily crosses from mother to the placenta and fetus (Ni et al., 2009). However, the placental enzyme 11beta-hydroxysteroid dehydrogenase type 2 (11β-HSD2) primarily produced in placental trophoblasts protects against excessive fetal cortisol levels by rapidly metabolizing cortisol into biologically inactive cortisone (Mark et al., 2009; Murphy & Clifton, 2003; Ni et al., 2009; Staud et al., 2006). Additionally, PGP-1, a glycoprotein transporter (also primarily in trophoblasts), limits cortisol access to the placenta and fetus by actively moving cortisol back into maternal circulation against the normally existing concentration gradient (Mark & Waddell, 2006; Sun et al., 2006).

Figure 16.2 depicts cortisol metabolism and the "protective barrier" provided by 11β-HSD2 and PGP-1 in the normal human placenta. Reduction in either 11β-HSD2 or PGP-1 disrupts the protective barrier resulting in increased cortisol levels delivered to the fetus. Furthermore, a second isoform of the 11β-HSD enzyme, type 1, catalyzes the conversion of cortisone back into bioavailable cortisol and contributes to overall levels, but to a much lesser degree (Alfaidy, Li, MacIntosh, Yang, & Challis, 2003).

Maternal tobacco smoke is associated with decreased 11β-HSD2 in the placenta (Jauniaux & Burton, 2007; Mark et al., 2009; Yang et al., 2006) but the impact on PGP-1 is not known. A reduction in 11β-HSD2 and PGP-1 in the

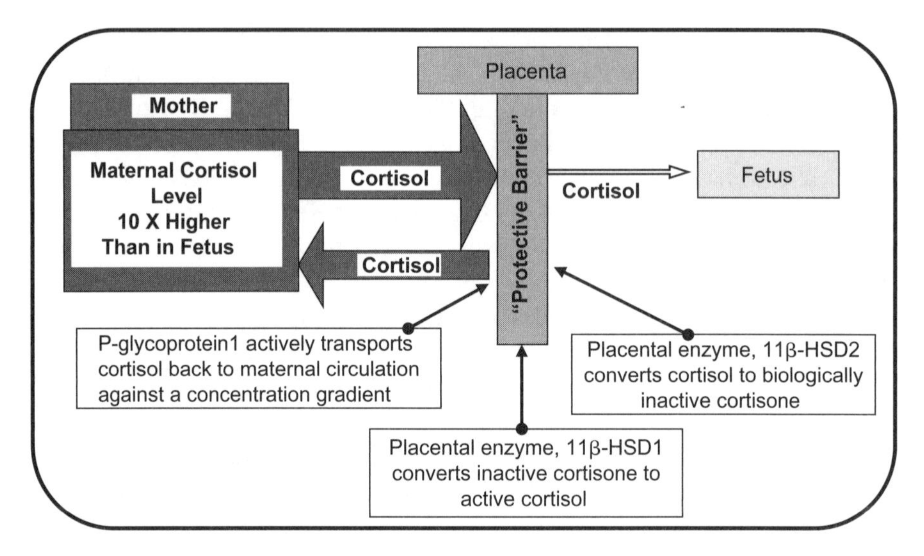

FIGURE 16.2 Placental 11β-HSD2 and PGP-1 create a "protective barrier" against placental and fetal cortisol overexposure.

placenta of a mother who smokes could be one explanation for the increased cortisol and subsequent increased risk of intrauterine growth restriction and preterm birth associated with maternal tobacco smoke, but this requires further study.

Placental HSD11β2 and ABCB1 Gene Expression

There is evidence of normal regulatory patterns of placental gene expression for the gene (official symbol HSD11β2) that codes for 11β-hydroxysteroid dehydrogenase 2, including significant decreases in mRNA, protein levels, and activity in the last several weeks of pregnancy (Murphy & Clifton, 2003). This is compatible with the normal increase in maternal and fetal serum cortisol levels observed late in pregnancy that is necessary for final maturation of the fetus and the initiation of parturition mechanisms. Reduction in placental HSD11β2 gene expression is observed in pregnancies complicated by intrauterine growth restriction (Mericq et al., 2009; Wächter, Masarik, Bürzle, Mallik, & von Mandach, 2009), as well as several environmental "insults," such as micronutrient (i.e., copper, zinc, vitamin E) deficiency (Rosario, Gomez, & Anbu, 2008), infection (Johnstone, Bocking, Unlugedik, & Challis, 2005), hypoxia (Homan, Guan, Hardy, Gratton, & Yang, 2006), and increased cadmium (one element of tobacco smoke) exposure (Yang et al., 2006), in addition to maternal tobacco smoke (Jauniaux & Burton, 2007; Mark et al., 2009; Yang et al., 2006). Indeed, maternal tobacco smoke is associated with alterations in the expression of well over 200 placental genes (Bruchova et al., 2010). Much less is known about placental ABCB1 (ATP-binding cassette, sub-family B1, the official gene name for PGP1 that encodes for P glycoprotein 1) gene expression, but there is some evidence that the normal expression of ABCB1 occurs in the same fashion as 11β-HSD2 expression, with decreases in mRNA and protein levels as pregnancy advances (Sun et al., 2006).

As previously mentioned, expression of these two specific placental genes, HSD11β2 and ABCB1, have relevance to pregnancy and newborn outcomes because of the function of their gene products, 11β-HSD2 and PGP-1 respectively, and the metabolism and transport of cortisol in the placenta. Downregulation of either HSD11β2 or ABCB1 can lead to increased cortisol exposure to the placenta and fetus.

Epigenetic Regulation of Gene Expression in the Placenta

Kouzarides (2007) provides a definition of epigenetics as "change in gene expression that occurs without a change in DNA sequence." Epigenetic regulation of gene expression occurs via biochemical structural modification of DNA via methylation, histone modification, and microRNA (miRNA) mediated control (among others). Such modifications render the DNA more or less accessible to transcription factors, thus, correspondingly increase or decrease gene expression "downstream" (Smith & Waddell, 2002). This can be viewed as a placental adaptive response to environmental change and insult (Jauniaux & Burton, 2007; Myatt, 2006) providing phenotypic "plasticity" in which the developing organism can adapt to specific environments early in life (without genetic mutation; Johnson & Tricker, 2010). Epigenetic modifications are often enduring (i.e., DNA methylation often persists and amenable covalent histone modifications take time; Aagaard-Tillery et al., 2008; MacLennan et al., 2004), perhaps heritable (Berger, Kouzarides, Shiekhattar, & Shilatifard, 2009), and with negative impact on health in exchange for improved survival in the short term (Joss-Moore, Metcalfe, Albertine, McKnight, & Lane, 2010).

Evidence is also mounting in support of the "Barker Hypothesis" and "fetal programming," which postulate that many adult diseases, such as cardiovascular and metabolic conditions (i.e., dyslipidemia, hypertension, diabetes, and obesity), have their origins in fetal/early life via epigenetic mechanisms (Armitage, Poston, & Taylor, 2008; Sinclair, Lea, Rees, & Young, 2007; Tamashiro & Moran, 2010). Furthermore, normal gene expression in the human placenta is highly regulated epigenetically throughout the gestational period, corresponding to the placental function appropriate to the changing needs of the fetus as pregnancy progresses. However, placental adaptive response to environmental challenge and insult (i.e., maternal tobacco smoke) can also occur via epigenetic mechanisms and could also reflect epigenetic alterations taking place in the fetus (Jauniaux & Burton, 2007; Myatt, 2006).

One example of how maternal tobacco smoke epigenetically modifies placental gene expression has been documented in relation to the placental cytochrome P450, family 1, subfamily A, polypeptide 1 (CYP1A1) gene, which encodes for an enzyme involved in the conversion of aromatic hydrocarbons into harmful hydrophilic DNA adducts (Suter et al., 2010). In mothers who smoke, the CYP1A1 gene has been found to be "hypomethylated" along the CpG sites immediately proximal to the xenobiotic response element transcription factor-binding element. This results in upregulation of placental CYP1A1 gene expression 4.4-fold in mothers who smoke compared to those who do not. Increased CYP1A1 gene expression ultimately results in increased harm to the fetus from aromatic hydrocarbon conversion. It is not known if this effect varies based on ethnicity/race, nor which specific compound in tobacco smoke leads to hypomethylation of the CYP1A1 gene. It is also not known if maternal tobacco smoke (or other environmental insults) affects the epigenetic regulation of HSD11β2 and ABCB1 gene expression in a similar fashion as for CYP1A1. However, differential methylation along the entire epigenome has been reported in the placentas of mothers who smoke compared to those who do not (Suter et al., 2011).

Summary of the Affect of Maternal Smoking on Placental Gene Expression

Maternal smoking is an environmental perinatal "insult" with well-documented associations with adverse perinatal outcomes, such as low birth weight, intrauterine growth restriction, and preterm birth. The expression of more than 200 placental genes is altered by maternal tobacco smoke (Bruchova et al., 2010), including HSD11β2 (Jauniaux & Burton, 2007; Mark et al., 2009; Yang et al., 2006) and perhaps ABCB1. Decreased placental HSD11β2 and ABCB1 gene expression can result in placental and fetal cortisol overexposure. For example, fetal overexposure to cortisol occurs during pregnancy in women who lack gene expression of HSD11β2 and thus cannot metabolize and inactivate cortisol in the placenta (Seckl & Holmes, 2007). Outcomes for these pregnancies include small for gestational age fetuses and newborns and altered newborn/child hypothalamic-pituitary-adrenal axis response. Overexposure to cortisol is a known contributor to placental and fetal growth restriction, preterm birth, and likely fetal programming of adult disease such as metabolic, cardiovascular, neuroendocrine, and psychiatric conditions (Seckl & Holmes, 2007). Epigenetic regulation of placental genes, in relation to environmental exposures such as maternal tobacco smoke, has received some research attention to date. However, epigenetics has rapidly become a research focus because of our increased understanding of the potentially profound impact of the environment on gene expression and health outcomes.

FUTURE RESEARCH AND A ROLE FOR NURSE RESEARCHERS

Future perinatal genetics research will predominantly focus on the following: (a) further understanding of the genetic origins that lead to adverse pregnancy outcome, including disparities; (b) identification of genetic biomarkers early in pregnancy (or preconception); (c) understanding the environmental impact on the genome and epigenome, and thus the sequence of events that leads to adverse pregnancy outcome (i.e., regulation of gene expression, post transcriptional and post translational mechanisms, protein synthesis, and protein function); and (d) therapies that prevent, ameliorate, or resolve the genetic contributions to pregnancy complications, including pharmacotherapeutics, nutritional, behavioral, and mental health interventions. Furthermore, there is a contemporary emphasis on "translational" or "bench to bedside" science, in which research questions will be addressed from within multidisciplinary research teams. There are compelling arguments that nurse-researchers are essential and ideally suited for participating in multidisciplinary teams, including perinatal genetics research teams (Woods & Magyary, 2010). Among the strongest of the arguments is that nurses have a foundation in "bedside" care, as well as understanding of the biologic basis of disease and injury. Furthermore, nursing has developed a solid evidence base for answering clinical questions and in delivering health care services. The combination of knowledge, expertise, and professional emphasis enables nurse researchers to understand "translational" science, and thus capable of making valuable contributions to perinatal genetics research with clinical relevancy.

Based on the topics discussed in this chapter, there are several areas in which nurse researchers who are interested in perinatal genomics can focus their attention. Most obvious are the environmental impacts on placental phenotype and adverse pregnancy outcomes, such as preterm birth, preeclampsia, and intrauterine growth restriction. For example, gaps in the research indicate a need for placental gene expression studies in relation to environmental insults, such as poor air quality, chronic maternal stress, inadequate nutrition, and exposures to toxins, heavy metals, and endocrine disruptors. Gene expression studies could focus attention on regulation, such as epigenetic regulation "upstream," as well as post transcriptional and posttranslational mechanisms that lead to protein synthesis and function "downstream." Moreover, the concerning disparities in the occurrence of pregnancy complications such as preterm birth may be explained by a gene–environment interaction and requires much further study (Dolan, 2010).

The identification and development of biomarkers to detect potential pregnancy complications early enough in gestation (or even preconception) for prevention, treatment, and/or education purposes could be quite valuable. It is important to keep in mind, however, that our ultimate goal in identification of any disease or condition is to reduce morbidity and mortality, not simply to predict or confirm a diagnosis. This idea is frequently lost in our quest to find biomarkers for risk identification for pregnancy complications, such as preterm birth or preeclampsia. For example, currently there are no therapies for treating preeclampsia, aside from delivery of the baby. Therefore, the use of biomarkers is not likely to affect preeclampsia-related morbidity and mortality. However, understanding biomarkers often leads researchers to an explanation for the pathophysiology of a disease or condition. Furthermore, should effective therapies be developed in the future, the value of biomarkers may increase substantially, particularly if they demonstrate high sensitivity and specificity in early pregnancy (i.e., the first trimester) or even prior to pregnancy.

Many researchers have the goal of reducing perinatal morbidity and mortality by developing preventions or treatments for perinatal complications. This is still true with today's emphasis on genetics, and more recently, "personalized health care." The new paradigm of personalized health care appears to have originally focused on genomics approaches (i.e., identification of each individual's genomic makeup and risk profile), followed by appropriate application of therapies to address them (i.e., pharmacogenomics; Li, 2011). However, nurse researchers could more fully develop the concept of "personalized" as meaning something much more than one's DNA sequence, structure, and function. Indeed, the growing body of evidence regarding gene–environment interaction makes it clear that we need to increase the number of studies that evaluate the interactions between physical, social, cultural, and psychological environment and the genome, or even the proteome, metabolome, or epigenome.

The field of epigenetics and developmental origins of adult health is a prime example of how important it is to consider how the environment (internal and external to the body) has far-reaching impact on health outcomes, including type 2 diabetes mellitus and cardiovascular disease later in life (Armitage et al., 2008). This challenges the health sciences (including nurse-researchers) to use a "big picture" vision. For example, rather than compartmentalizing personalized health care as solely a pharmacogenomics approach, nurse researchers can contribute to the new paradigm by ensuring the inclusion of relevant biobehavioral factors that influence health outcomes and contribute to health disparities.

CHAPTER SUMMARY

This chapter has attempted to provide the reader with an understanding of the role of genetics in two specific pregnancy complications: preterm birth and dysfunctional placental phenotype. Furthermore, some of the current genetic research findings in relation to preterm birth and placental phenotype have been discussed. There is much opportunity for nurse-researchers to contribute to perinatal genetics investigations, but this requires perseverance, increased levels of understanding and skills in genetics, and multidisciplinary mentorship. Furthermore, the conduct of genetics research, perinatal and otherwise, by nurse-investigators requires a substantial increase in knowledge base and exposure to interdisciplinary research teams during nursing education, in both undergraduate and graduate programs. This is essential for developing self-efficacy and confidence in perinatal genetics investigations.

REFERENCES

Aagaard-Tillery, K. M., Grove, K., Bishop, J., Ke, X., Fu, Q., McKnight, R., & Lane, R. H. (2008). Developmental origins of disease and determinants of chromatin structure: Maternal diet modifies the primate fetal epigenome. *Journal of Molecular Endocrinology*, *41*(2), 91–102.

Alfaidy, N., Li, W., MacIntosh, T., Yang, K., & Challis, J. (2003). Late gestation increase in 11beta-hydroxysteroid dehydrogenase 1 expression in human fetal membranes: A novel intrauterine source of cortisol. *Journal of Clinical Endocrinology and Metabolism*, *88*(10), 5033–5038.

Anderson, N. L., & Anderson, N. G. (1998). Proteome and proteomics: New technologies, new concepts, and new words. *Electrophoresis*, *19*(11), 1853–1861.

Annells, M. F., Hart, P. H., Mullighan, C. G., Heatley, S. L., Robinson, J. S., Bardy, P., & McDonald, H. M. (2004). Interleukins-1, -4, -6, -10, tumor necrosis factor, transforming growth factor-beta, FAS, and mannose-binding protein C gene polymorphisms in Australian women: Risk of preterm birth. *American Journal of Obstetrics and Gynecology*, *191*(6), 2056–2067.

Anum, E. A., Springel, E. H., Shriver, M. D., & Strauss, J. F., III. (2009). Genetic contributions to disparities in preterm birth. *Pediatric Research*, *65*(1), 1–9.

Armitage, J. A., Poston, L., & Taylor, P. D. (2008). Developmental origins of obesity and the metabolic syndrome: The role of maternal obesity. *Frontiers of Hormone Research*, *36*, 73–84.

Behrman, R. E., & Stith Butler, A. (2007). *Preterm birth: Causes, consequences, and prevention*. Washington, DC: National Academy Press.

Berger, S. L., Kouzarides, T., Shiekhattar, R., & Shilatifard, A. (2009). An operational definition of epigenetics. *Genes & Development*, *23*(7), 781–783.

Bodamer, O. A., Mitterer, G., Maurer, W., Pollak, A., Mueller, M. W., & Schmidt, W. M. (2006). Evidence for an association between mannose-binding lectin 2 (MBL2) gene polymorphisms and pre-term birth. *Genetics in Medicine*, *8*(8), 518–524.

Bruchova, H., Vasikova, A., Merkerova, M., Milcova, A., Topinka, J., Balascak, I., Pastorkova, A. . . . Brdicka, R. (2010). Effect of maternal tobacco smoke exposure on the placental transcriptome. *Placenta*, *31*(3), 186–191.

Buhimschi, C. S., Weiner, C. P., & Buhimschi, I. A. (2006). Clinical proteomics: A novel diagnostic tool for the new biology of preterm labor, part I: Proteomics tools. *Obstetric & Gynecological Survey*, *61*(7), 481–486.

Challis, J. R., Bloomfield, F. H., Bocking, A. D., Casciani, V., Chisaka, H., Connor, K., . . . Premyslova, M. (2005). Fetal signals and parturition. *Journal of Obstetrics and Gynaecology Research*, *31*(6), 492–499.

Chaudhari, B. P., Plunkett, J., Ratajczak, C. K., Shen, T. T., DeFranco, E. A., & Muglia, L. J. (2008). The genetics of birth timing: Insights into a fundamental component of human development. *Clinical Genetics*, *74*(6), 493–501.

Crider, K. S., Whitehead, N., & Buus, R. M. (2005). Genetic variation associated with preterm birth: A HuGE review. *Genetics in Medicine*, *7*(9), 593–604.

Dolan, S. M. (2010). Genetic and environmental contributions to racial disparities in preterm birth. *Mount Sinai Journal of Medicine*, *77*(2), 160–165.

Drake, A. J., Tang, J. I., & Nyirenda, M. J. (2007). Mechanisms underlying the role of glucocorticoids in the early life programming of adult disease. *Clinical Science*, *113*(5), 219–232.

Enquobahrie, D. A., Williams, M. A., Qiu, C., Muhie, S. Y., Slentz-Kesler, K., Ge, Z., & Sorenson, T. (2009). Early pregnancy peripheral blood gene expression and risk of preterm delivery: A nested case control study. *BMC Pregnancy and Childbirth*, *9*, 56.

Founds, S. A., Terhorst, L. A., Conrad, K. P., Hogge, W. A., Jeyabalan, A., & Conley, Y. P. (2011). Gene expression in first trimester preeclampsia placenta. *Biological Research for Nursing*, *13*(2), 134–139.

Gennari-Moser, C., Khankin, E. V., Schüller, S., Escher, G., Frey, B. M., Portmann, C. B., . . . Mohaupt, M. G. (2011). Regulation of placental growth by aldosterone and cortisol. *Endocrinology*, *152*(1), 263–271.

Giurgescu, C. (2009). Are maternal cortisol levels related to preterm birth? *Journal of Obstetric, Gynecologic, and Neonatal Nursing*, *38*(4), 377–390.

Goedhart, G., Vrijkotte, T. G., Roseboom, T. J., van der Wal, M. F., Cuijpers, P., & Bonsel, G. J. (2009). Maternal cortisol and offspring birthweight: Results from a large prospective cohort study. *Psychoneuroendocrinology*, *35*(5), 644–652.

Homan, A., Guan, H., Hardy, D. B., Gratton, R. J., & Yang, K. (2006). Hypoxia blocks 11beta-hydroxysteroid dehydrogenase type 2 induction in human trophoblast cells during differentiation by a time-dependent mechanism that involves both translation and transcription. *Placenta*, *27*(8), 832–840.

Honest, H., Bachmann, L. M., Gupta, J. K., Kleijnen, J., & Khan, K. S. (2002). Accuracy of cervicovaginal fetal fibronectin test in predicting risk of spontaneous preterm birth: Systematic review. *British Medical Journal*, *325*(7359), 301.

Jansson, T., & Powell, T. L. (2007). Role of the placenta in fetal programming: Underlying mechanisms and potential interventional approaches. *Clinical Science*, *113*(1), 1–13.

Jauniaux, E., & Burton, G. J. (2007). Morphological and biological effects of maternal exposure to tobacco smoke on the feto-placental unit. *Early Human Development*, *83*(11), 699–706.

Johnson, L. J., & Tricker, P. J. (2010). Epigenomic plasticity within populations: Its evolutionary significance and potential. *Heredity*, *105*(1), 113–121.

Johnstone, J. F., Bocking, A. D., Unlugedik, E., & Challis, J. R. (2005). The effects of chorioamnionitis and betamethasone on 11beta hydroxysteroid dehydrogenase types 1 and 2 and the glucocorticoid receptor in preterm human placenta. *Journal of the Society for Gynecologic Investigation*, *12*(4), 238–245.

Joss-Moore, L. A., Metcalfe, D. B., Albertine, K. H., McKnight, R. A., & Lane, R. H. (2010). Epigenetics and fetal adaptation to perinatal events: Diversity through fidelity. *Journal of Animal Science*, *88*(13, Suppl.), E216–E222.

Kanasaki, K., & Kalluri, R. (2009). The biology of preeclampsia. *Kidney International*, *76*(8), 831–837.

Khoury, M. J., Bertram, L., Boffetta, P., Butterworth, A. S., Chanock, S. J., Dolan, S. M., . . . Ioannidis, J. P. (2009). Genome-wide association studies, field synopses, and the development of the knowledge base on genetic variation and human diseases. *American Journal of Epidemiology, 170*(3), 269–279.

Kouzarides, T. (2007). Chromatin modifications and their function. *Cell, 128*, 693–705.

Latendresse, G., & Ruiz, R. J. (2011). Maternal corticotropin-releasing hormone and the use of selective serotonin reuptake inhibitors independently predict the occurrence of preterm birth. *Journal of Midwifery & Women's Health, 56*(2), 118–126.

Lee, K. J., Shim, S. H., Kang, K. M., Kang, J. H., Park, D. Y., Kim, S. H., . . . Cha, D. H. (2010). Global gene expression changes induced in the human placenta during labor. *Placenta, 31*(8), 698–704.

Li, C. (2011). Personalized medicine - the promised land: Are we there yet? *Clinical Genetics, 79*(5), 403–412.

Lieberman, E., Torday, J., Barbieri, R., Cohen, A., Van Vunakis, H., & Weiss, S. T. (1992). Association of intrauterine cigarette smoke exposure with indices of fetal lung maturation. *Obstetrics and Gynecology, 79*(4), 564–570.

MacLennan, N. K., James, S. J., Melnyk, S., Piroozi, A., Jernigan, S., Hsu, J. L., . . . Lane, R. H. (2004). Uteroplacental insufficiency alters DNA methylation, one-carbon metabolism, and histone acetylation in IUGR rats. *Physiolical Genomics, 18*(1), 43–50.

Macones, G., Parry, S., Elkousy, M., Clothier, B., Ural, S. H., & Strauss, J. F., III. (2004). A polymorphism in the promoter region of TNF and bacterial vaginosis: Preliminary evidence of gene-environment interaction in the etiology of spontaneous preterm birth. *American Journal of Obstetrics and Gynecology, 190*(6), 1504–1508.

Mark, P. J., Augustus, S., Lewis, J. L., Hewitt, D. P., & Waddell, B. J. (2009). Changes in the placental glucocorticoid barrier during rat pregnancy: Impact on placental corticosterone levels and regulation by progesterone. *Biology of Reprodroduction, 80*(6), 1209–1215.

Mark, P. J., & Waddell, B. J. (2006). P-glycoprotein restricts access of cortisol and dexamethasone to the glucocorticoid receptor in placental BeWo cells. *Endocrinology, 147*(11), 5147–5152.

Martin, J., Osterman, M., & Sutton, P. (2010). *Are preterm births on the decline in the United States? Recent data from the National Vital Statistics System.* Hyattsville, MD: National Center for Health Statistics.

Mayor-Lynn, K., Toloubeydokhti, T., Cruz, A. C., & Chegini, N. (2011). Expression profile of microRNAs and mRNAs in human placentas from pregnancies complicated by preeclampsia and preterm labor. *Reproductive Sciences, 18*(1), 46–56.

McGrath, S., & Smith, R. (2002). Prediction of preterm delivery using plasma corticotrophin-releasing hormone and other biochemical variables. *Annals of Medicine, 34*(1), 28–36.

Menon, R., Velez, D. R., Thorsen, P., Vogel, I., Jacobsson, B., Williams, S. M., & Fortunato, S. J. (2006). Ethnic differences in key candidate genes for spontaneous preterm birth: TNF-alpha and its receptors. *Human Heredity, 62*(2), 107–118.

Mericq, V., Medina, P., Kakarieka, E., Márquez, L., Johnson, M. C., & Iñiguez, G. (2009). Differences in expression and activity of 11beta-hydroxysteroid dehydrogenase type 1 and 2 in human placentas of term pregnancies according to birth weight and gender. *European Journal of Endocrinology, 161*(3), 419–425.

Moutquin, J. M. (2003). Socio-economic and psychosocial factors in the management and prevention of preterm labour. *British Journal of Obstetrics and Gynaecology, 110*(Suppl. 20), 56–60.

Murphy, V. E., & Clifton, V. L. (2003). Alterations in human placental 11beta-hydroxysteroid dehydrogenase type 1 and 2 with gestational age and labour. *Placenta, 24*(7), 739–744.

Myatt, L. (2006). Placental adaptive responses and fetal programming. *Journal of Physiology, 572* (Pt. 1), 25–30.

Ni, X. T., Duan, T., Yang, Z., Guo, C. M., Li, J. N., & Sun, K. (2009). Role of human chorionic gonadotropin in maintaining 11beta-hydroxysteroid dehydrogenase type 2 expression in human placental syncytiotrophoblasts. *Placenta*, *30*(12), 1023–1028.

Plunkett, J., & Muglia, L. J. (2008). Genetic contributions to preterm birth: Implications from epidemiological and genetic association studies. *Annals of Medicine*, *40*(3), 167–195.

Pregnancy Risk Assessment Monitoring System (PRAMS). Division of Reproductive Health, National Center for Chronic Disease Prevention and Health Promotion, Centers for Disease Control and Prevention. (2004). Retrieved from http://www.cdc.gov/nccdphp/publications/factsheets/Prevention/pdf/smoking.pdf

Romero, R., Kusanovic, J. P., Chaiworapongsa, T., & Hassan, S. S. (2011). Placental bed disorders in preterm labor, preterm PROM, spontaneous abortion and abruptio placentae. *Best Practice & Research. Clinical Obstetrics & Gynaecology*, *25*(3), 313–327.

Romero, R., Velez Edwards, D. R., Kusanovic, J. P., Hassan, S. S., Mazaki-Tovi, S., Vaisbuch, E., . . . Menon, R. (2010). Identification of fetal and maternal single nucleotide polymorphisms in candidate genes that predispose to spontaneous preterm labor with intact membranes. *American Journal of Obstetrics and Gynecology*, *202*(5), 431 e431–e434.

Rosario, J. F., Gomez, M. P., & Anbu, P. (2008). Does the maternal micronutrient deficiency (copper or zinc or vitamin E) modulate the expression of placental 11 beta hydroxysteroid dehydrogenase-2 per se predispose offspring to insulin resistance and hypertension in later life? *Indian Journal of Physiology and Pharmacology*, *52*(4), 355–365.

Sandman, C. A., Glynn, L., Schetter, C. D., Wadhwa, P., Garite, T., Chicz-DeMet, A., & Hobel, C. (2006). Elevated maternal cortisol early in pregnancy predicts third trimester levels of placental corticotropin releasing hormone (CRH): Priming the placental clock. *Peptides*, *27*(6), 1457–1463.

Sata, F., Toya, S., Yamada, H., Suzuki, K., Saijo, Y., Yamazaki, A., . . . Kishi, R. (2009). Proinflammatory cytokine polymorphisms and the risk of preterm birth and low birthweight in a Japanese population. *Molecular Human Reproduction*, *15*(2), 121–130.

Seckl, J. R., & Holmes, M. C. (2007). Mechanisms of disease: Glucocorticoids, their placental metabolism and fetal 'programming' of adult pathophysiology. *Nature Clinical Practice. Endocrinology & Metabolism*, *3*(6), 479–488.

Sherwood, L. (2004). *Human physiology: From cells to systems* (5th ed.). Belmont, CA: Brooks/Cole—Thomson Learning.

Sinclair, K. D., Lea, R. G., Rees, W. D., & Young, L. E. (2007). The developmental origins of health and disease: Current theories and epigenetic mechanisms. *Society of Reproduction and Fertility Supplement*, *64*, 425–443.

Smith, J. T., & Waddell, B. J. (2002). Leptin receptor expression in the rat placenta: Changes in ob-ra, ob-rb, and ob-re with gestational age and suppression by glucocorticoids. *Biology of Reproduction*, *67*(4), 1204–1210.

Staud, F., Mazancová, K., Miksík, I., Pávek, P., Fendrich, Z., & Pácha, J. (2006). Corticosterone transfer and metabolism in the dually perfused rat placenta: Effect of 11beta-hydroxysteroid dehydrogenase type 2. *Placenta*, *27*(2–3), 171–180.

Sun, M., Kingdom, J., Baczyk, D., Lye, S. J., Matthews, S. G., & Gibb, W. (2006). Expression of the multidrug resistance P-glycoprotein, (ABCB1 glycoprotein) in the human placenta decreases with advancing gestation. *Placenta*, *27*(6–7), 602–609.

Suter, M., Abramovici, A., Showalter, L., Hu, M., Shope, C. D., Varner, M., & Aagaard-Tillery, K. (2010). In utero tobacco exposure epigenetically modifies placental CYP1A1 expression. *Metabolism*, *59*(10), 1481–1490.

Suter, M., Ma, J., Harris, A. S., Patterson, L., Brown, K. A., Shope, C., . . . Aagaard-Tillery, K. M. (2011). Maternal tobacco use modestly alters correlated epigenome-wide placental DNA methylation and gene expression. *Epigenetics*, *6*(11).

Tamashiro, K. L., & Moran, T. H. (2010). Perinatal environment and its influences on metabolic programming of offspring. *Physiology & Behavior, 100*(5), 560–566.

Tattersall, M., Engineer, N., Khanjani, S., Sooranna, S. R., Roberts, V. H., Grigsby, P. L., . . . Johnson, M. R. (2008). Pro-labour myometrial gene expression: Are preterm labour and term labour the same? *Reproduction, 135*(4), 569–579.

U.S. Department of Health and Human Services. (2010). *How tobacco smoke causes disease: The biology and behavioral basis for smoking-attributable disease: A report of the surgeon general.* Atlanta, GA: U.S. Department of Health and Human Services, Centers for Disease Control and Prevention, National Center for Chronic Disease Prevention and Health Promotion, Office on Smoking and Health.

Varvarigou, A. A., Liatsis, S. G., Vassilakos, P., Decavalas, G., & Beratis, N. G. (2009). Effect of maternal smoking on cord blood estriol, placental lactogen, chorionic gonadotropin, FSH, LH, and cortisol. *Journal of Perinatal Medicine, 37*(4), 364–369.

Varvarigou, A. A., Petsali, M., Vassilakos, P., & Beratis, N. G. (2006). Increased cortisol concentrations in the cord blood of newborns whose mothers smoked during pregnancy. *Journal of Perinatal Medicine, 34*(6), 466–470.

Vedmedovska, N., Rezeberga, D., Teibe, U., Melderis, I., & Donders, G. G. (2011). Placental pathology in fetal growth restriction. *European Journal of Obstetrics, Gynecology, and Reproductive Biology, 155*(1), 36–40.

Velez, D. R., Fortunato, S. J., Thorsen, P., Lombardi, S. J., Williams, S. M., & Menon, R. (2008). Preterm birth in Caucasians is associated with coagulation and inflammation pathway gene variants. *PLoS ONE, 3*(9), e3283.

Wächter, R., Masarik, L., Bürzle, M., Mallik, A., & von Mandach, U. (2009). Differential expression and activity of 11beta-hydroxysteroid dehydrogenase in human placenta and fetal membranes from pregnancies with intrauterine growth restriction. *Fetal Diagnosis and Therapy, 25*(3), 328–335.

Wadhwa, P. D., Garite, T. J., Porto, M., Glynn, L., Chicz-DeMet, A., Dunkel-Schetter, C., & Sandman, C. A. (2004). Placental corticotropin-releasing hormone (CRH), spontaneous preterm birth, and fetal growth restriction: A prospective investigation. *American Journal of Obstetrics and Gynecology, 191*(4), 1063–1069.

Wilcox, A. J., Skjaerven, R., & Lie, R. T. (2008). Familial patterns of preterm delivery: Maternal and fetal contributions. *American Journal of Epidemiology, 167*(4), 474–479.

Winn, V. D., Haimov-Kochman, R., Paquet, A. C., Yang, Y. J., Madhusudhan, M. S., Gormley, M., . . . Fisher, S. J. (2007). Gene expression profiling of the human maternal-fetal interface reveals dramatic changes between midgestation and term. *Endocrinology, 148*(3), 1059–1079.

Woods, N. F., & Magyary, D. L. (2010). Translational research: Why nursing's interdisciplinary collaboration is essential. *Research and Theory for Nursing Practice, 24*(1), 9–24.

Yang, K., Julan, L., Rubio, F., Sharma, A., & Guan, H. (2006). Cadmium reduces 11 beta-hydroxysteroid dehydrogenase type 2 activity and expression in human placental trophoblast cells. *American Journal of Physiology. Endocrinology and Metabolism, 290*(1), E135–E142.

CHAPTER 17

Pediatric Hemoglobinopathies

From the Bench to the Bedside

Tonya A. Schneidereith

ABSTRACT

The β-chain hemoglobinopathies affect the β-globin gene on chromosome 11 and comprise some of the most prevalent genetic disorders in humans, including sickle cell disease (SCD) and β-thalassemia. The mutations associated with these diseases cause various symptoms and degrees of severity. Extensive research has sought to identify physiologic and genetic factors responsible for these variations, including the role of fetal hemoglobin (HbF) and its importance in the alleviation of symptoms. This chapter on the genomics of hemoglobinopathies addresses the interests of both the researcher and the caregiver. The pathophysiology of SCD and thalassemia are reviewed, as well as the state of the science on the regulation of HbF, including newly identified quantitative trait loci (QTLs), single nucleotide polymorphisms (SNPs), and suggested genetic mechanisms. Studies on the current therapies of hemoglobinopathies, both pharmacologic and non-pharmacologic, are also reviewed. Research reviews relevant to the care of children include physical and psychological sequelae, genetic counseling, and effects on learning. With a thorough understanding of the normal physiology of hemoglobin, the pathophysiology of SCD and the thalassemias, and the associated physical and psychological sequela, nurses can improve the quality of life for children and families living with these diseases.

http://dx.doi.org/10.1891/0739-6686.29.353

INTRODUCTION

The β-Chain hemoglobinopathies, including Sickle Cell Disease (SCD) and β-thalassemia (β-thal), are among the earliest described molecular diseases. Since the 1940s, scientists have appreciated the mutations associated with these diseases and have come to recognize the many variations in symptoms and severity. Extensive research has sought to identify physiologic and genetic factors responsible for these variations, including fetal hemoglobin (HbF) and its importance in the alleviation of symptoms. However, many of the molecular and genetic mechanisms responsible for increased HbF production, mainly in the form of F-cells, remain unclear. Increased understanding of these mechanisms may not only provide opportunity for direction of therapies for these diseases, but may also provide insight for mechanisms associated with other molecular diseases (Adegbola, 2009; Bhatnagar et al., 2011; Charache, 1990; Thein & Craig, 1998).

The β-chain hemoglobinopathies are conditions affecting the β-globin gene on chromosome 11 and comprise some of the most prevalent genetic disorders in humans (Pearson, 1996). β-Thal is the most common β-chain hemoglobinopathy worldwide, whereas SCD is more prevalent in the United States. These disorders are so prevalent that the World Health Organization (WHO) estimates that approximately 5% of the world's populations are carriers for hemoglobinopathies (WHO, 2011).

HEMOGLOBIN PHYSIOLOGY

In order to understand the pathophysiology of hemoglobinopathies and the basis for subsequent potential therapy, a brief review of normal hemoglobin production is warranted. Hemoglobin is a tetramer, formed from four globin chains. Normally, two chains are from the alpha (α)-like globin genes located on chromosome 16 and two are from the beta (β)-like globin genes located on chromosome 11 (see Figure 17.1). The α-like genes include one zeta (ζ) gene and two α genes, whereas the β-like genes include one epsilon (ε) gene, two gamma (γ) genes, one delta (δ) gene, and one β gene (Abboud & Musallam, 2009; Stamatoyannopoulos & Nienhuis, 1994).

During embryogenesis, the hemoglobin tetramer is assembled from two chains of zeta (ζ) globin or two chains of alpha (α) globin from chromosome 16 and two chains of epsilon (ε) globin then two chains of gamma (γ) globin from chromosome 11 (Stamatoyannopoulos & Nienhuis, 1994).

Thus, embryonic hemoglobin is found as $\zeta_2\varepsilon_2$, $\zeta_2\gamma_2$, or $\alpha_2\varepsilon_2$, with the predominant forms of embryonic hemoglobin assembled from ζ and ε chains for the first 6 weeks following conception. As embryonic development advances and

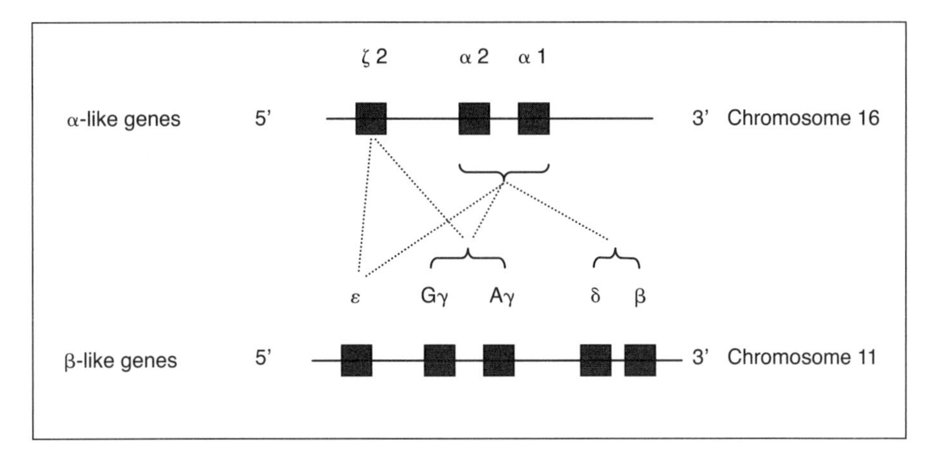

FIGURE 17.1 α-Globin and β-Globin Gene Clusters: Hemoglobin is synthesized from two α-globin genes and two β-globin genes. The figure shows how embryonic, fetal, and adult hemoglobins are paired.

hematopoiesis moves from the embryonic yolk sac to the fetal liver and spleen, the globin chains change from an embryonic, $\zeta_2\varepsilon_2$, to a fetal configuration, $\alpha_2\gamma_2$ (Stamatoyannopoulos & Nienhuis, 1994).

Finally, around the time of birth and concurrent with the shift of hematopoiesis to the bone marrow, the predominant form of hemoglobin changes from HbF ($\alpha_2\gamma_2$) to the major adult hemoglobin, HbA ($\alpha_2\beta_2$). The minor adult hemoglobin HbA_2 ($\alpha_2\delta_2$) is also assembled at a smaller concentration (Weatherall & Clegg, 1969).

What is imperative to note is that the switch from fetal to adult hemoglobin is not complete. There is continued synthesis of fetal hemoglobin throughout adulthood at levels usually 1%–2% of the total hemoglobin in red blood cells (see Table 17.1; Dover & Boyer, 1987; Keefer et al., 2006; Pearson, 1996). This topic will be addressed in depth in the following section on HbF.

PATHOPHYSIOLOGY OF THALASSEMIA

Thalassemia is a hemoglobinopathy because of the unbalanced synthesis of α- and β-globin and can cause various forms of α-thal and β-thal (see Table 17.2; Nienhuis, Ley, Humphries, Young, & Dover, 1990; Sankaran & Nathan, 2010). In the α-thalassemias, there is decreased production of α-globin chains, increased amounts of free β-globin, and subsequent assembly of a moderately unstable hemoglobin, HbH ($\alpha_1\beta_3$; β_4). HbH can cause a mild-moderate hemolytic and hypochromic anemia (Sankaran & Nathan, 2010).

TABLE 17.1
Hemoglobin Synthesis

Type of Hemoglobin	Site of Synthesis	Time of Synthesis	Globin Chains
Embryonic	Yolk sac	Up to week 6 post-conception	$\zeta_2\varepsilon_2$, $\zeta_2\gamma_2$, or $\alpha_2\varepsilon_2$
Fetal (HbF)	Fetal liver and spleen	Approx 6 weeks post-conception to 48 postnatal weeks; continues at low levels throughout adulthood	$\alpha_2\gamma_2$
Adult (HbA)	Bone marrow	Approx 6 weeks post-conception and throughout adulthood	$\alpha_2\beta_2$

The β-thalassemias lead to decreased or absent production of β-globin chains and unpaired α-globin chains. The unpaired α-globins can precipitate and cause injury to the membranes of erythroid precursors, thereby leading to cell death, ineffective erythropoiesis, and sequela associated with this disease (Nienhuis et al., 1990; Sankaran & Nathan, 2010).

TABLE 17.2
The Thalassemias

	α—Globin chains			
α-Thalassemia	α_2	α_1	β-Thalassemia	β-Globin
Normal	■	■	Normal	■
	■	■		■
Silent Carrier	■	■		
	■	□		
Trait	■	■	Trait/ minor	■
	□	□		□
HbH	■	□		
	□	□		
Major	□	□	Major	□
(Hydrops fetalis)	□	□	(Cooley's anemia)	□

Note. ■ Normal production. □ Absent production.

Individuals can be homozygous for β-thal (each of the two inherited chromosomes contain the β-thal mutation) or heterozygous for β-thal (only one of the two chromosomes has the β-thal mutation). Combinations of β-thal with other diseases, such as HbS, is often denoted as β^0(no formation of β-globin) or β^+(some production of β-globin). β^0-Thal, also known as Cooley's anemia, can lead to severe hemolytic anemia requiring repeated blood transfusions for survival. The severity of compound heterozygotes (β^+/HbS; β^+/HbC) depends on the presence of mutations on the non-affected chromosome 11 (Ashley-Koch, Yang, & Olney, 2000; Dover, 1998; Pearson, 1996; Sankaran & Nathan, 2010).

PATHOPHYSIOLOGY OF SICKLE CELL DISEASE

SCD is an autosomal recessive disorder and refers to a group of hematologic diseases distinguished by the presence of sickle hemoglobin (Abboud & Musallam, 2009; Zimmerman, Ware, & Kinney, 1997). This sickle hemoglobin (HbS) results from an altered codon that leads to substitution of the amino acid valine for glutamic acid in the sixth position of the β-globin chain (Pauling, Itano, Singer, & Wells, 1949). This can lead to polymerization and elongation of circulating red blood cells. Cycles of polymerization and depolymerization can lead to damaged red cell membranes, irreversible sickled cells, and subsequent vascular occlusions (Steinberg, 2008). Four SCD genotypes account for most SCD cases in the Unites States. These include Sickle Cell Anemia, Sickle-Hemoglobin C disease (HbSC), Sickle β^+-thalassemia (HbS/β^+-thal), and Sickle β^0- thalassemia (HbS/ β^0-thal; Kaye, 2006; Kenner, Gallo, & Bryant, 2005). The severity of these illnesses varies depending on the combination of abnormal hemoglobins, although those with heterozygous diseases (HbSC, HbS/β-thal) usually have less anemia (Vichinsky, MacKlin, Wayne, Lorey, & Olivieri, 2005; Zimmerman et al., 1997). HbSS results in the clinical condition known as Sickle Cell Anemia (SCA).

In the United States, SCD occurs in approximately 1 in 500 African American newborns (National Heart Lung and Blood Institute, 2009). It is most predominant in populations of African descent, although it is also present in other ethnic groups, including those from the Caribbean, Mediterranean, the Arabian Peninsula, and India (Ashley-Koch et al., 2000; Gribbons, Zahr, & Opas, 1995).

Antonarakis et al. (1984) determined three predominant sickle β-globin (β^S) haplotypes (patterns of nucleotide polymorphisms that can distinguish one group from another) among American and Jamaican individuals with SCD. Pagnier et al. (1984) found three common haplotypes in West Africans, thus named for the countries in which they were most dominant: Benin, Bantu, and Senegal.

The study of β^S haplotypes helped to establish the West African origin and subsequent migration of the majority of individuals with SCD. This includes the haplotypes identified in American and Jamaican populations (Stine, Dover, Zhu, & Smith, 1992).

It appears that HbS confers some resistance to *Plasmodium falciparum* malaria. Based on the early work by Allison (1954), individuals without a β^S mutation (AA) showed a similar rate of infection with the *P. falciparum* parasite as those with at least one β^S mutation (AS), however, there were fewer deaths in those with AS than those with AA. Nagel and colleagues performed in vitro experiments and showed that red blood cells from individuals with AS containing *P. falciparum* sickled eight times faster than those that were uninfected, thereby making them more readily detectable for removal by the spleen (Roth et al., 1978). In other words, red blood cells containing hemoglobin with the β^S mutation are protected against infection by the malaria parasite because they are removed from the circulation faster, secondary to increased sickling. In addition, the life span of the β^S red blood cells is shorter than unaffected cells, thereby decreasing the time available for maturation of the malaria parasite (Nagel, 1990).

SCA can lead to vaso-occlusive complications including extreme pain, stroke, retinopathy, pulmonary vaso-occlusion, infarction of the mesenteric, splenic, or hepatic vessels, leg ulcers, infection, and can also lead to death in severe cases. Vaso-occlusive pain is the most common problem encountered by patients with SCA. The pain, which is intermittent, often severe, and unpredictable, can start as early as 6–9 months of age, and continues throughout adulthood (Shapiro, Dinges, Orne, & Bauer, 1995). This pain leads to consequences beyond the physical effects, including implications on school attendance, learning, and psychological difficulties (Moran, Siegel, & DeBaun, 1998; Noll et al., 1996; Shapiro et al., 1995).

THE ROLE OF FETAL HEMOGLOBIN

In 1948, Watson, Stahman, and Bilello (1948) reported that red cells of newborns with SCA were not as quick to sickle as those of older children. Watson hypothesized that increased levels of HbF, present in the newborn, were responsible for this phenomena. Further laboratory studies confirmed that HbF has an inhibitory effect on polymerization of deoxygenated HbS (Bunn, 1994). Gamma chains, either as HbF ($\alpha_2\gamma_2$) or as a hybrid tetramer ($\alpha_2\gamma\beta S$), are not incorporated into the sickled polymer (Steinberg, 2008). Higher levels of HbF are also beneficial to those with β-thal by improving balance of α- and non-α-globin chains, thereby reducing the number of circulating inclusion bodies (Sankaran & Nathan, 2010).

Additional proof of the beneficial effect of HbF is found in a subgroup of adult individuals with sickle cell anemia who have increased levels of HbF. They have a milder disease or do not exhibit anemia or any other sequelae associated with SCA. This phenomenon is known as Hereditary Persistence of Fetal Hemoglobin (HPFH) and is usually a result of deletions of different sizes involving the β-globin gene cluster or point mutations in the γ-globin gene promoters, although only one point mutation (-187 in the promoter of the γ gene) has been associated with increased HbF in SCD (Forget, 1998; Pearson, 1996; Zago et al., 1979).

HPFH is classified into two forms, deletion and non-deletion, although they sometimes have overlapping phenotypes (Forget, 1998). δβ-thalassemia results from a deletion of the β-globin and δ-globin genes (Charache, 1990). This phenotype results in a modest increase of HbF heterogeneously distributed among HbF-containing cells (F-cells; Forget, 1998). The second form, HbS associated non-deletion HPFH, includes an intact β-globin gene cluster (Rochette, Craig, & Thein, 1994). This non-deletion phenotype has a significantly higher level of HbF with a heterocellular distribution (Forget, 1998; Zago et al., 1979).

Two groups included in the heterogeneously distributed form are individuals with HPFH who have inheritance patterns of higher levels of HbF and a hematologically normal British population who were noted to have an inherited occurrence of increased HbF (Zago et al., 1979). This evidence further suggests that there is an inherited component associated with higher levels of HbF other than chromosome 11. These other potential loci are discussed in more detail subsequently.

Among the three β^S globin haplotypes mentioned earlier, individuals with SCA who are homozygous for the Senegal β^S haplotypes have, on average, higher levels of HbF than do the those homozygous for the Benin and Bantu β^S haplotypes (Nagel et al., 1985). Labie and colleagues (1985) identified a mutation within the promoter region of the γ^G gene at -158bp in the Senegal haplotype that has been associated with increased HbF levels (Labie, 1985). Those with the Senegalese haplotype mutation are noted to have better health than those with the Benin or Bantu haplotypes, including decreased frequency of sickle cell crises, bone infarction, and infection (Powars, 1990).

F-CELLS

As stated earlier, all humans continue to synthesize fetal hemoglobin throughout adulthood. The small amount of HbF produced (HPFH excluded) is usually less than 2% of the total amount of hemoglobin produced and can be found in red blood cells known as F-cells (Boyer, Belding, Margolet, & Noyes, 1975; Dover & Boyer, 1987). Boyer found in normal adults a wide variation in HbF/F-cell of approximately 14%–28% of the total hemoglobin and an approximately 20-fold variation in F-cell levels (Boyer et al., 1975). In other words, there is a wide range in the amount of fetal hemoglobin per F-cell and a wide variation in the number of

F-cells produced. Because approximately 15%–20% of HbF in whole blood concentrations inhibits HbS polymerization, and there is a decrease in vaso-occlusive crises by 50% for patients with 4%–15% HbF, increasing HbF creates a potential therapy for ameliorating SCD (Bertles, 1974; Boyer et al., 1975; Platt et al., 1994). To review, there is a small amount of HbF produced in adulthood and that HbF is confined to a subset of red blood cells, known as F-cells. There is a wide variation in the amount of HbF per F-cell and a wide variation in the number of F-cells produced. Increasing the total amount of HbF could lead to decreased sequelae associated with SCD.

The levels of HbF in individuals with SCA are the conglomeration of three independent processes: F-cell production, HbF biosynthesis within F-cells, and preferential F-cell survival (Dover, Boyer, Charache, & Heintzelman, 1978). Each of these factors appeared to be separately regulated within an individual as well as between individuals (Boyer et al., 1984; Dover, Boyer, & Pembrey, 1981). In addition, the amount of HbF per F-cell appears to be regulated separately from the production of F-cells (Dover et al., 1978). Again, there is a wide variation in the amount of HbF per individual and there is a wide range of F-cells produced, both of which do not appear to be regulated by the same genes.

In 1981, Dover et al. showed that either F-cell or HbF levels in nonanemic sickle trait (AS) parents predicted HbF production in their SS children (Dover et al., 1981). Further research suggested that genetic factors that control the production of F-cells were heterogeneous. Moreover, Dover, Boyer, and colleagues determined that the amount of F-cells (as measured by their precursors, F-reticulocytes), showed continuous distribution between unrelated individuals from 2% to 50% (Dover et al., 1978; Dover et al., 1981). In other words, the data of Dover and colleagues supported the previous findings regarding the wide variation in F-cell production.

However, between sibling pairs, there was a significant correlation in F-reticulocyte levels (Dover et al., 1981). That is, the relative abundance of F-cells varies widely among unrelated individuals, but is less varied in sib pairs. It appears that the levels of HbF/F-cell are regulated by genes within the β-globin locus on Chromosome 11, whereas the amount of F-cells produced is regulated by genes not linked to the β-globin cluster (Boyer et al., 1984; Dover et al., 1981).

Miyoshi et al. (1988) studied 300 healthy Japanese men and women and found that distribution of F-cell levels was consistent with X-linked inheritance. Furthermore, there appeared to be a relationship between X-linked high F-cell levels and color-blindness in two of the families studied. The gene for color-blindness resides on the terminus of the long arm of Chromosome X (Xq28). Based on these findings, Dover et al. began a linkage analysis using DNA from Jamaican individuals with SCA as well as individuals without SCA. They could not show linkage to the region that Miyoshi determined, but did show linkage to a region on the short arm of Chromosome X. This region was designated the F-cell production (FCP) locus and was localized to Xp22.2 (Dover et al., 1992).

The relative importance of the X-linked FCP locus and β-globin haplotypes in determining HbF levels was determined by Chang et al. (1997) in different populations, including Jamaican, French, and an American kindred with HPFH but without SCA (Chang et al., 1997; Chang, Smith, Moore, Serjeant, & Dover, 1995). They analyzed the relative importance of five factors reported to be important in determining HbF levels (Chang et al., 1997). These factors are age, sex, β-globin haplotypes, α-globin gene number, and the FCP phenotype. They had four conclusions: (a) when each factor was analyzed alone, all five factors were significantly associated with HbF variation; (b) when all five factors were analyzed in a multiple regression analysis, the only significant association was the FCP locus, which was associated with 35%–41% of the overall variation in HbF levels; (c) variation within each FCP phenotype is modulated by genetic factors associated with the three common β^S haplotypes; and (d) approximately 50% of the variation in HbF levels could not be attributed to these five factors.

In addition to the X chromosome loci that contribute to F-cell variation in the Japanese (Xq) and SCD individuals of African descent (Xp), Thein and colleagues identified another locus that appears to influence F-cell levels mapped to chromosome 6q23 in a large kindred of Asian Indian origin (Garner et al., 1998; Thein & Craig, 1998). They further identified an intergenic fragment, HMIP-2, on chromosome 8q that appears to influence HbF-related traits in both individuals of African origin with SCD and in healthy individuals (Creary et al., 2009; Garner, Tatu, Best, Creary, & Thein, 2002).

It is important to note that the loci identified on Xq28 in a Japanese population and 6q23 in an Asian Indian kindred have no discernable influence in the Jamaican population (Chang et al., 1997). Furthermore, Thein and colleagues did not show linkage in their kindred to the FCP locus.

Recently, human genome-wide association studies (GWAS) identified another quantitative trait locus that influences the production of HbF (Menzel et al., 2007). This locus maps to 2p15, close to the oncogene BCL11A. Uda et al. (2008) studied over 4,000 Sardinians and identified nine single-nucleotide polymorphisms (SNPs) associated with HbF levels in adults, six of which fell within the β-globin gene cluster. The strongest association within the nine SNPs was in intron 2 of the BCL11A gene (Uda et al., 2008). Further studies genotyped additional BCL11A SNPs in two large SCD cohorts, the Cooperative Study of Sickle Cell Disease (CSSCD; $n = 1{,}275$) and an SCD cohort from Brazil ($n = 350$; Lettre et al., 2008). Lettre and colleagues reported common SNPs at three loci, including BCL11A, HBS1L-MYB, and β-globin (HBB), which accounted for > 20% of the variation in HbF levels. In addition, this group showed that high HbF-alleles at five common SNPs from three loci were associated with reduced pain crises rates in patients with SCD (Lettre et al., 2008). They suggest the possibility that genotyping the variants in patients with SCD may predict the risk of disease severity, allowing for the delivery of more directed therapies.

Further evidence for the role of BCL11A included knockdown experiments in erythroid progenitor cells ($CD34^+$). The results showed 28.9% HbF in $CD34^+$ BCL11A knockdown cells vs. 0.2% HbF in control cells (Xu et al., 2010). Experiments with SOX6, a gene shown to be coexpressed with BCL11A during erythroid development in both human and mice, also had an effect on HbF production. In the same set of $CD34^+$ experiments mentioned earlier, SOX6 knockdown experiments induced HbF to 4.7%. In combined BCL11A/SOX6 knockdown experiments, induction of HbF exceeded 45%, suggesting that these two genes work together to induce HbF production (Sankaran & Nathan, 2010; Xu et al., 2010)

To understand the mechanisms through which BCL11A acts on erythroid cells, Sankaran et al., investigated the role of this gene on *cis*-regulatory elements within the β-globin cluster (Sankaran et al., 2008; Sankaran & Nathan, 2010). They showed that BCL11A interacts directly with the β-globin cluster in $CD34^+$ cells (Sankaran et al., 2008; Sankaran & Nathan, 2010).

Bhatnagar and colleagues (2011) tested the role of BCL11A in an African ancestry cohort with SCD ($n = 440$) and confirmed the previous association of BCL11A. In addition, they also reported an independent effect of a BCL11A SNP in F-cell regulation (Bhatnagar et al., 2011). Furthermore, they reported associations for SNP's from PHEX and MAGEB18, both close to the Xp22.2 locus, mentioned previously (Bhatnagar et al., 2011; Dover et al., 1992). It is important to note that this group used F-cell number and not total HbF as the HbF phenotype, suggesting that F-cell number, not HbF/F-cell, determines HbF variation (Bhatnagar et al., 2011).

Further study of this promising gene and its effects on HbF variation can lead to targeted therapies for patients with SCD and β-thalassemia (Galarneau et al., 2010; Lettre et al., 2008; Sankaran et al., 2011; Sankaran et al., 2008; Sankaran & Nathan, 2010). These recent discoveries may also have a future role in genetic counseling, prenatal diagnosis, and newborn screening.

PHARMACOLOGIC MANIPULATION

In the early 1980s, researchers began investigating pharmacologic agents to increase production of HbF. An experiment by DeSimone and colleagues (1978) on anemic baboons treated with 5-Azacytidine (Aza), a nucleoside analogue, found that this agent increased production of HbF (DeSimone, Biel, & Heller, 1978). Dover and colleagues (1984) treated four SCA patients with Aza who had become refractory to chronic transfusion therapy. They showed that in the two individuals who were compliant and continued therapy, F-reticulocyte levels increased significantly. Because this agent was potentially carcinogenic and because potentially safer agents known to increase HbF were available (including hydroxyurea), protocols using Aza were discontinued (Charache, 1993; Dover, 1990).

It was noted that individuals with urea-cycle disorders who were treated with fatty acid analogues coincidentally had an increased level of HbF (Dover, Brusilow, & Charache, 1994). Sodium Phenylbutyrate (SPB), an orally administered agent developed to promote waste nitrogen excretion, was found to increase HbF during clinical trials of patients with urea-cycle disorders. Another analogue, phenylbutyrate, has also been found to increase the levels of γ-globin mRNA in treated cells (Fibach, Prasanna, Rodgers, & Samid, 1993). However, the mechanisms by which butyrate, phenylbutyrate, and other butyrate analogues activate the γ-globin gene are not known (Papayannopoulou, Torrealba de Ron, Veith, Knitter, & Stamatoyannopoulos, 1984). Clinical trials of individuals with β-thalassemia included an extraordinary daily dosage requirement of 40 tablets. In addition, these individuals noted adverse side effects including edema, epigastric discomfort, and an extremely offensive body odor (Collins et al., 1995). Although Collins et al. found an increase in HbF production, the difficulties with compliance was correlated with an inconsistent response to therapy.

Treatment of baboons with hydroxyurea, a chemotherapeutic agent cytotoxic to the cells in the S phase, was found to be a successful inducer of HbF (Lavelle, DeSimone, Heller, Zwiers, & Hall, 1986). Clinical trials began in 1987 to study the effects of hydroxyurea on individuals with SCD. The Multicenter Study of Hydroxyurea (MSH) established that 44% of crises were decreased in patients with a history of three or more crises per year, hydroxyurea decreased the number of transfusions required for patients with SCD, and hydroxyurea decreased the number of patients with vaso-occlusive chest crisis (Brawley et al., 2008; Inati, Chabtini, Mounayar, & Taher, 2009; Lanzkron, Haywood, Segal, & Dover, 2006; Ma et al., 2007). The risks of hydroxyurea were acceptable compared to untreated SCD (Brawley et al., 2008). Subsequently, hydroxyurea was approved for use in adults by the FDA (Brawley et al., 2008; Inati et al., 2009; Ma et al., 2007).

To establish hydroxyurea safety and dosing in children, the multicenter study (BABY HUG) followed the MSH study and also showed increases in HbF (approaching 20%), F-cell percentage, and reversible lab toxicities (Inati et al., 2009; Lanzkron et al., 2006; Steinberg, 2008). Although the results show beneficial effects, in children, hydroxyurea is underutilized (Brandow & Panepinto, 2011). Common reasons for underutilization include provider discontinuation because of the absence of clinical benefit and poor patient compliance (Brandow & Panepinto, 2011; Steinberg, 2008). Interestingly, hydroxyurea was prescribed for children with vaso-occlusive pain crises and chest syndrome, not as indicated for long-term therapy to preserve end-organ function (Brandow, Jirovec, & Panepinto, 2010; Brandow & Panepinto, 2011). Therefore, Brandow suggests that if patients did not have benefits during the vaso-occlusive crises, providers perceived a lack of response to hydroxyurea and discontinued the therapy (Brandow & Panepinto, 2011).

As stated earlier, a problem well-established with hydroxyurea is the variable response in vivo (Inati et al., 2009; Lanzkron et al., 2006; Ma et al., 2007). Although not well understood, some explanations have included poor patient compliance, decreased provider prescriptions, possible drug–drug interactions, and unidentified genetic elements (Boosalis et al., 2011; Brandow & Panepinto, 2011; Inati et al., 2009; Lanzkron et al., 2006; Ma et al., 2007). In addition, the likelihood for incomplete care is increased because of limited resources and lack of seasoned clinicians competent in caring for the ethnic minorities represented by these diseases (Brawley et al., 2008).

Ikuta and colleagues (2001) published data indicating a relationship between HbF expression and the soluble guanylate cyclase–cyclic GMP-dependent protein kinase pathway in K562 cells. Soluble guanylate cyclase converts GTP to cyclic GMP (cGMP). cGMP functions in second messenger pathways activated by signal transduction to alter expression of specific target proteins, including protein kinase G (Denninger & Marletta, 1999). Inhibition of soluble guanylate cyclase (reducing cGMP levels) or protein kinase G activity in K562 cells reduced the induction of γ-globin gene expression by arginine butyrate, hemin, and protoporphyrin IX (Ikuta et al., 2001). Haynes and colleagues (2004) further implicated L-arginine/nitric oxide/ cGMP in HbF induction using Zileuton, an analogue similar in structure to hydroxyurea. These studies offer a new mechanism for induction of HbF that does not necessarily involve histone deacetylation.

Keefer and Schneidereith further analyzed the role of cyclic nucleotides in both K562 cells and $CD34^+$ human erythroid cultures (Keefer et al., 2006). Using the HbF inducing agents hydroxyurea, sodium butyrate (SB), and 5-azacytidine (Aza) we found that cAMP is required for HbF production by all three agents (Keefer et al., 2006). Furthermore, adenylate cyclase inhibition appears to blunt the production of HbF, arguing that some amount of cAMP formation is required for induction activity by hydroxyurea, SB, and Aza. Interestingly, we found that cGMP did not have a role in HbF production in the $CD34^+$ system, contrasting with mechanisms suggested in the K562 system (Ikuta, Ausenda, & Cappellini, 2001; Keefer et al., 2006). This argues the role of different signaling programs in cells that terminally differentiate versus immortalized cell lines. This may also offer support for the role of the oncogene BCL11A and its role on the cell cycle as a mechanism for increased HbF production.

Most recently, Boosalis and colleagues (2011) identified four candidate compounds, from a library of 13,000 chemicals, which were able to induce γ-globin production 4- to 8-fold. Two of these candidates were able to induce increases without toxicity in vivo and may prove favorable for clinical application (Boosalis et al., 2011).

OTHER THERAPIES

Aside from pharmacologic therapies, other options available for patients with hemoglobinopathies include red blood cell (RBC) transfusions with chelation therapy and stem cell transplant (SCT).

RBC transfusions are given to increase oxygen-carrying capacity, dilute the number of sickled cells with normal RBCs, and to change flow characteristics of the circulating volume (Inati, 2009; Inati et al., 2009). This therapy is used in patients at high risk for stroke, for those who do not respond to hydroxyurea, and for those who require extra oxygen-carrying capacity with crises (Inati, 2009; Inati et al., 2009).

Individuals with β-thalassemia major require transfusions of RBC on a regular basis to survive and are necessary to reduce the effects that occur as a result of ineffective RBC production, including bone marrow expansion, bone destruction, and organ infiltration (Sankaran & Nathan, 2010). A problem associated with frequent transfusions includes iron overload, requiring chelation therapy. Iron can impact vital organs, including the heart and liver, as well as other tissues. Treatment with iron chelators, such as deferoxamine, binds to the iron and allow for physiologic removal (Inati, 2009; Inati et al., 2009; Sankaran & Nathan, 2010). Recent studies on Hepcidin, a primary regulator of iron homeostasis, may serve as a target for therapy aimed at reducing iron overload in patients who are chronically transfused (Sankaran & Nathan, 2010).

Unlike hydroxyurea and transfusion therapy that is aimed toward prevention of complications, stem cell transplant is the only potential cure. The indications for this extreme measure include acute chest syndrome, stroke, priapism, and the inability to receive RBC transfusions because of alloimmunization (Inati et al., 2009; Sankaran & Nathan, 2010). Only approximately 33% of children will meet the match requirements necessary for SCT, but for those who do, the mean overall survival is 95% (Inati et al., 2009; Walters et al., 2000). Additional considerations for SCT include graft vs. host disease, graft failure, or sepsis related to immunosuppressive therapy (Sankaran & Nathan, 2010).

NEWBORN SCREENING

As mentioned earlier, newborn screening programs have led to a decrease in mortality associated with hemoglobinopathies (Cunningham et al., 1998; Kladny, Williams, Gupta, Gettig, & Krishnamurti, 2011; Yanni, Grosse, Yang, & Olney, 2009). It is important to remember that infants have more fetal hemoglobin than adult hemoglobin; therefore, they do not manifest symptoms of disease until late infancy or early childhood.

In most states, newborn screening for hemoglobinopathies includes nonsickle hemoglobinopathies (including β-thal), hemoglobinopathy carriers (including sickle cell trait, HbC carrier, and HbE carrier), and α-thalassemia syndromes (α-thal minor, HbH; Kaye, 2006). Early identification allows for a number of benefits. First is the administration of prophylactic penicillin, a therapy known to decrease the incidence of pneumococcal sepsis by 84%. It is important to start this prophylaxis within the first 2 months of age in order to receive the most benefit (Kaye, 2006). Second, early identification allows for parent education regarding signs and symptoms of vaso-occlusive crises, including splenic sequestration and chest crises (Kaye, 2006; Steinberg, 2008). These, combined with pneumococcal conjugate and polysaccharide vaccines, have led to a decrease in mortality for children under the age of 4 years old (Kaye, 2006; Steinberg, 2008).

GENETIC COUNSELING

Although newborn screening has been available since the 1970s, it has only been mandated in all 50 states since 2005 (Kladny et al., 2011). Kladny et al. (2011) found that many parents are unaware of newborn screening, including the diseases that are included in the screen. It therefore can come as a surprise and shock when parents learn of their child's hemoglobinopathy.

Genetic counseling is often performed following diagnosis at the recommendation of or from their primary care provider/pediatrician (Kladny et al., 2011; Rowley, Loader, Sutera, & Kozyra, 1995). It is vital that the counselor understands the genetics of hemoglobinopathies, including inheritance patterns, diagnoses, and manifestations. Rowley and colleagues (1995) found that health care personnel trained in medical genetics are not sufficient to meet the needs of our growing population. They noted several advantages to having nurses provide counseling, including nurses inclusion of genetic counseling as part of their normal duties, nurse accessibility, and decreased cost of nurses' time when compared with physicians (Rowley et al., 1995).

Studies have shown that, following genetic counseling, parents are more open to discussing sickle cell disease, have decreased anxiety, and felt the overall experience was a positive one (Acharya, Lang, & Ross, 2009; Kladny et al., 2011). This research helps to recognize that genetic counseling provides more for families than simply providing information.

Genetic counseling is an important component to the overall wellness of people with hemoglobinopathies. It is important that those in both primary and tertiary settings have a general knowledge of the genetics of hemoglobinopathies and how to best direct these families toward comprehensive health care.

PHYSICAL COMPLICATIONS

Physical complications associated with SCD can consist of vaso-occlusive disorders including extreme pain; stroke; retinopathy; pulmonary vaso-occlusion; infarction of the mesenteric, splenic, or hepatic vessels; leg ulcers; infection; and death in severe cases. Primary hemorrhagic stroke is one of the most devastating neurological complications of SCD (Strouse, Hulbert, DeBaun, Jordan, & Casella, 2006). Vaso-occlusive pain is the most common problem encountered by patients with SCA and is related to formation of the HbS polymer and subsequent red cell sickling (Niscola, Sorrentino, Scaramucci, de Fabritis, & Cianciulli, 2009). The pain can be acute or chronic, severe, and unpredictable. It can start as early as 6–9 months of age and continues throughout adulthood (Niscola et al., 2009; Shapiro et al., 1995). The chronic pain associated with SCD may be because of complications, including leg ulcers and avascular necrosis (Niscola et al., 2009).

This pain leads to consequences beyond the physical effects, including implications on school attendance, learning, and psychological difficulties (Moran et al., 1998; Noll et al., 1996; Shapiro et al., 1995). Jacob and colleagues (2006) examined the effects of acute pain crises on sleep, activity level, and eating in 27 hospitalized children. Although they saw no significant changes in eating or activity levels, they did show data that suggested changes in nighttime sleep (Jacob et al., 2006).

MORTALITY

Since the late 1960s, improvements in health care, including comprehensive care for children with SCD, penicillin prophylaxis, newborn screening programs, and pneumococcal and *Haemophilus influenzae* vaccines have contributed to a decrease in mortality (Cunningham et al., 1998; Davis, Schoendorf, Gergen, & Moore, 1997; Yanni et al., 2009). In addition, careful follow-up and extensive parental education have also been shown to contribute to the lowered mortality (Davis, Gergen, & Moore, 1997).

Yanni and colleagues (2009) evaluated death certificates of children classified as Black with SCD and under the age of 15 years old. When they compared the 1999–2002 statistics to those from 1983–1986, there was a decrease in SCD-related mortality rates of 68% in Black children 0–3 years old, 39% for children 4–9 years old, and 24% for children 10–14 years old (Yanni et al., 2009).

Davis and colleagues analyzed where deaths occurred, including inpatient settings, outpatient/ED, or if the child was dead on arrival (DOA). Their analysis showed that children with SCD who died before hospital admission accounted for 41% of deaths among 1- to 4-year-olds with SCD, 27% among 5- to 9-year-olds with SCD, and 12% among 10- to 14-year-olds with SCD. Overall, deaths prior to hospital admission were 30% when age groups were combined (Davis, Moore, & Gergen, 1997).

These data make it apparent that further research of effective outpatient therapies for SCD should remain a priority for research, especially in the older pediatric age groups. This includes investigation into mechanisms of hemoglobin production because these could, in principle, ameliorate all aspects of SCD.

PSYCHOLOGICAL DIFFICULTIES

There is growing recognition of possible psychological difficulties that can occur as a result of the biological challenges associated with SCD. Because of the many stressful experiences that commonly occur, children with SCD are described as having the potential for excessive anxiety, depressed mood, poor self-concept, and difficulties with social acceptance (Noll et al., 1996). The combination of these psychological problems and physical symptoms such as learning disabilities, small stature, and chronic fatigue, places children with SCD at high risk for problematic peer relationships (Graff et al., 2010; Lutz, Barakat, Smith-Whitley, & Ohene-Frempong, 2004; Noll et al., 1996). The peers of adolescents with SCD viewed them as sick, less athletic, and frequently absent. Interestingly, these adolescents with SCD did not have fewer friendships, less likability, or a poor social reputation (Noll et al., 1996).

In addition, the relationship of family and siblings can also have an impact on the psychological state of the child with SCD. Many caregivers are unemployed or have jobs that earn less than $20,000/year, adding to the stress of the household (Gold, Mahrer, Treadwell, Weissman, & Vichinsky, 2008; Mitchell et al., 2007). Consideration should be given to how SCD has an impact on the caregiver's life, including family activities, friendships, and time spent with siblings (Gold et al., 2008; Graff et al., 2010; Mitchell et al., 2007).

EFFECTS ON LEARNING

Children with SCA are at risk for difficulties in learning for a number of reasons. Approximately 7% of children with SCA have a stroke before the age of 14 (Moran et al., 1998). Sadly, for those who have had a hemorrhagic stroke, 25%–50% will die within 2 weeks of the event (Strouse et al., 2006). The sequelae associated with strokes can include generalized motor impairment, decreased cognitive performance, and shortened attention span (Hoppe, Styles, & Vichinsky, 1998; Moran et al., 1998; Winrow, 1998).

Patients may experience acute and/or "silent strokes" caused by occluding cerebral and cerebellar vessels (Moran et al., 1998; Strouse et al., 2006). "Silent strokes" as defined by MRI may lead to specific functional deficits including memory, language, visual spatial skill, and response inhibition (Moran et al., 1998).

Also affecting learning are the child's emotional state, including distress and low self-esteem, as well as multiple days of school absences (Peterson, Palermo, Swift, Beebe, & Drotar, 2005; Shapiro et al., 1995). Shapiro et al. (1995) studied the impact of SCD-related pain on school attendance and the average consecutive number of school days missed. They found that patients missed about one-fifth of their school days and only one-half of those absences were associated with SCD-related pain. The absences on days without vaso-occlusive pain had a number of causes, including minor infections, clinic visits, or other medical problems associated with SCD (Shapiro et al., 1995). In addition, there is little information about how these students function when they *are* able to attend classes (Peterson et al., 2005). Furthermore, effects on learning can occur in early childhood, presenting with learning difficulties and functional academic needs as young as kindergarten age (Peterson et al., 2005).

During the school day, themes of wanting peer support and needing educated staff who are able to identify untoward behaviors and symptoms were identified in the qualitative study by Graff et al. (2010). Students with SCD reported being teased, isolated, and treated differently than other students. However, teachers and coaches who were aware of the special needs of these students made the parents feel more at ease and helped integrate the students into the classroom (Graff et al., 2010).

COST

SCD is a major health concern, causing frequent emergency department (ED) visits as well as inpatient hospitalizations. The costs of hospitalizations associated with SCD in the United States for the years 1989 through 1993 were estimated at an average of 75,000 hospitalizations, an average length of stay of 6.1 days, and a yearly total of 456,000 days of care. The estimated total direct cost of hospitalization per year for these individuals was $475.2 million (Ashley-Koch et al., 2000; Davis, Moore et al., 1997).

Lanzkron et al. (2006) found that there were no significant changes in the rates of hospitalization of children or in their average length of stay (LOS) from FY 1995 through FY 2003. However, their data showed a 40% increase in costs for pediatric SCA discharges in FY 2003 compared to adjusted FY 1995 costs. In 2003, they showed a total hospitalization cost for children with SCA of $2,720,957 that was 0.6% of the total hospital expenditures for children in Maryland (Lanzkron et al., 2006).

The most common diagnosis of sickle cell disease with crisis accounted for ~77% of overall hospitalizations and 64% of ED visits in the study by Brousseau and colleagues (2010). Rates of health care use and rehospitalization of those with SCD showed a 30-day rehospitalization rate of 33.4%, a 14-day rehospitalization rate of 22.1%, and approximately 37% of patients were under the age of 18. The 30-day rehospitalization numbers were higher than other chronic diseases, including asthma (3.4%), heart failure (16%), and diabetes mellitus (20%; Brousseau et al., 2010).

In the majority of cases, the primary expected payer is a public source. Research has shown that most of the hospitalizations are paid by government programs, including (but not limited to) both Medicare and Medicaid (Brousseau et al., 2010; Davis, Moore et al., 1997). Studies by Mvundura et al. (2009) showed the mean number of ED visits was 49% higher for Medicaid-enrolled children compared to those with private insurance and that the mean total expenditures for children with SCD were 25% higher for privately insured children than those that were enrolled in Medicaid ($14, 722 vs. $11, 075).

Hydroxyurea has proven to be a cost-effective therapy (Nietert, Silverstein, & Abboud, 2002; Moore et al., 2000). The Multicenter Study of Hydroxyurea in Sickle Cell Disease showed that admissions for acute pain crises were reduced by approximately 50% in those patients with SCA who were treated with hydroxyurea (Nietert et al., 2002). Furthermore, the annual mean for hospitalization for those treated with hydroxyurea was $12,160 versus $17,290 for those treated with placebo (Moore et al., 2000). However, the data obtained by Lanzkron's group suggests that hydroxyurea not had its intended impact. Patient concerns over costs, side effects, or lack of prescriptions may have lead to a decrease in the number of patients taking the drug (Adams-Graves et al., 2008; Lanzkron et al., 2006)

Finally, cost not only refers to direct costs, but also includes indirect and intangible costs. Patients with SCD and their family members incur indirect costs (time lost from school, lost wages, transportation costs, and income lost from early mortality) and intangible costs (pain, added stress, and altered family life;

Davis, Moore et al., 1997; Graff et al., 2010). Graff's qualitative study (2010) captured the feelings of parental distress with ED visits, balancing work and childcare, and the need for frequent clinic visits.

SUMMARY

Sickle cell disease and thalassemias are costly genetic disorders that affect approximately 5% of the world's population. The physical effects of SCD can cause vaso-occlusive crises, leading to acute and chronic pain, strokes, respiratory difficulties, and organ damage. Silent strokes can lead to decreased memory, learning disabilities, and poorer fine motor skills. The psychological impacts can lead to poor school performance, difficulty with peer relationships, anxiety, and family stress.

The sequela associated with hemoglobinopathies is no longer just about the single-gene mutation, but about the complex interactions between genes, genotype, the environment, and the individual's genomic pattern (Higgs & Wood, 2008; Kenner et al., 2005). Recent genome-wide association studies in multiple ethnic groups have established three major quantitative trait loci on chromosomes 2, 6, and 11, as well as an F-cell production locus on chromosome X, that are responsible for 20%–50% of the phenotypic variation in HbF levels (Dover et al., 1992; Garner et al., 1998; Higgs & Wood, 2008; Sankaran et al., 2011; Sankaran et al., 2008; Sankaran & Nathan, 2010). Two of the quantitative trait loci are oncogenes, reflecting the significance of cell differentiation on HbF production (Bhatnagar et al., 2011).

The severity of hemoglobinopathies because of abnormal β-chains is reduced in those who have higher HbF levels. Therefore, it is important to understand the genetic mechanisms involved in the normal variation of HbF as well as the pharmacologic mechanisms known to increase HbF.

This chapter has summarized current research relevant to the care of patients with hemoglobinopathies. With a thorough understanding of the normal physiology of hemoglobin, the pathophysiology of SCD and the thalassemias, and the associated physical and psychological sequela, nurses can improve the quality of life for children and families living with these diseases. Whether wearing the hat of researcher, practitioner, or educator, it is vital that nurses stay current with the state of nursing research and use that to continue to improve lives in both health and illness.

ACKNOWLEDGMENTS

Thank you to Dr. Christine Kasper and Dr. Kirby Smith for thoughtful review of this chapter and to Virginia Polley for helpful manuscript preparation.

REFERENCES

Abboud, M. R., & Musallam, K. M. (2009). Sickle cell disease at the dawn of the molecular era. *Hemoglobin*, *33*(Suppl. 1), S93–S106.

Acharya, K., Lang, C. W., & Ross, L. F. (2009). A pilot study to explore knowledge, attitudes, and beliefs about sickle cell trait and disease. *Journal of the National Medical Association*, *101*(11), 1163–1172.

Adams-Graves, P., Ostric, E. J., Martin, M., Richardson, P., & Lewis, J. B., Jr. (2008). Sickle cell hospital unit: A disease-specific model. *Journal of Healthcare Management*, *53*(5), 305–315.

Adegbola, M. A. (2009). Can heterogeneity of chronic sickle-cell disease pain be explained by genomics? A literature review. *Biological Research for Nursing*, *11*(1), 81–97.

Allison, A. C. (1954). Protection afforded by sickle-cell trait against subtertian malareal infection. *British Medical Journal*, *1*(4857), 290–294.

Antonarakis, S. E., Beohm, C. D., Serjeant, G. R., Theisen, C. E., Dover, G. J., & Kazazian, H. H. (1984). Origin of the beta S-globin gene in blacks: The contribution of recurrent mutation or gene conversion or both. *Proceedings of the National Academy of Sciences of the United States of America*, *81*(3), 853–856.

Ashley-Koch, A., Yang, Q., & Olney, R. S. (2000). Sickle hemoglobin (HbS) allele and sickle cell disease: A HuGE review. *American Journal of Epidemiology*, *151*(9), 839–845.

Bertles, J. F. (1974). Human fetal hemoglobin: Significance in disease. *Annals of the New York Academy of Sciences*, *241*(0), 638–652.

Bhatnagar, P., Purvis, S., Barron-Casella, E., DeBaun, M. R., Casella, J. F., Arking, D. E., & Keefer, J. R. (2011). Genome-wide association study identifies genetic variants influencing F-cell levels in sickle-cell patients. *Journal of Human Genetics*, *56*(4), 316–323.

Boosalis, M. S., Castaneda, S. A., Trudel, M., Mabera, R., White, G. L., Lowrey, C. H., . . . Perrine, S. P. (2011). Novel therapeutic candidates, identified by molecular modeling, induce γ-globin gene expression in vivo. *Blood Cells, Molecules, & Diseases*, *47*(2), 107–116.

Boyer, S. H., Belding, T. K., Margolet, L., & Noyes, A. N. (1975). Fetal hemoglobin restriction to a few erythrocytes (F cells) in normal human adults. *Science*, *188*(4186), 361–363.

Boyer, S. H., Dover, G. J., Serjeant, G. R., Smith, K. D., Antonarkis, S. E., Embury, S. H., . . . Bias, W. B. (1984). Production of F cells in sickle cell anemia: Regulation by a genetic locus or loci separate from the beta-globin gene cluster. *Blood*, *64*(5), 1053–1058.

Brandow, A. M., Jirovec, D. L., & Panepinto, J. A. (2010). Hydroxyurea in children with sickle cell disease: Practice patterns and barriers to utilization. *American Journal of Hematology*, *85*(8), 611–613.

Brandow, A. M., & Panepinto, J. A. (2011). Monitoring toxicity, impact, and adherence of hydroxyurea in children with sickle cell disease. *American Journal of Hematology*, *86*(9), 804–806.

Brawley, O. W., Cornelius, L. J., Edwards, L. R., Gamble, V. N., Green, B. L., Inturrisi, C., . . . Schori, M. (2008). National Institutes of Health Consensus Development Conference statement: Hydroxyurea treatment for sickle cell disease. *Annals of Internal Medicine*, *148*(12), 932–938.

Brousseau, D. C., Owens, P. L., Mosso, A. L., Panepinto, J. A., & Steiner, C. A. (2010). Acute care utilization and rehospitalizations for sickle cell disease. *Journal of the American Medical Association*, *303*(13), 1288–1294.

Bunn, H. F. (1994). Sickle hemoglobin and other hemoglobin mutants. In G. Stamatoyannopoulos, A. Nienhuis, P. Majerus, & H. Varmus (Eds.), *The molecular basis of blood disease* (2nd ed., pp. 207–256). Philadelphia, PA: W. B. Saunders Co.

Chang, Y. C., Maier-Redelsperger, M., Smith, K., Contu, L., Ducrocq, R., de Montalembert, M., . . . Girot, R. (1997). The relative importance of the X-linked FCP locus and beta-globin haplotypes in determining haemoglobin F levels: A study of SS patients homozygous for beta S haplotypes. *British Journal of Haematology*, *96*(4), 806–814.

Chang, Y. C., Smith, K. D., Moore, R. D., Serjeant, G. R., & Dover, G. J. (1995). An analysis of fetal hemoglobin variation in sickle cell disease: The relative contributions of the X-linked factor, beta-globin haplotypes, alpha-globin gene number, gender, and age. *Blood*, *85*(4), 1111–1117.

Charache, S. (1990). Fetal hemoglobin, sickling, and sickle cell disease. *Advances in Pediatrics*, *37*, 1–31.

Charache, S. (1993). Pharmacological modification of hemoglobin F expression in sickle cell anemia: An update on hydroxyurea studies. *Experientia*, *49*(2), 126–132.

Collins, A. F., Pearson, H. A., Giardina, P., McDonagh, K. T., Brusilow, S. W., & Dover, G. J. (1995). Oral sodium phenylbutyrate therapy in homozygous beta thalassemia: A clinical trial. *Blood*, *85*(1), 43–49.

Creary, L. E., Ulug, P., Menzel, S., McKenzie, C. A., Hanchard, N. A., Taylor, V., . . . Thein, S. L. (2009). Genetic variation on chromosome 6 influences F cell levels in healthy individuals of African descent and HbF levels in sickle cell patients. *PLoS One*, *4*(1), e4218.

Cunningham, G., Lorey, F., Kling, S., Soper, K., Urso, F., Harris, K., et al. (1998). Mortality among children with sickle cell disease identified by newborn screening during 1990–1994—California, Illinois, and New York. *Journal of the American Medical Association*, *279*(14), 1059–1060.

Davis, H., Gergen, P. J., & Moore, R. M., Jr. (1997). Geographic differences in mortality of young children with sickle cell disease in the United States. *Public Health Reports*, *112*(1), 52–58.

Davis, H., Moore, R. M., Jr., & Gergen, P. J. (1997). Cost of hospitalizations associated with sickle cell disease in the United States. *Public Health Reports*, *112*(1), 40–43.

Davis, H., Schoendorf, K. C., Gergen, P. J., & Moore, R. M., Jr. (1997). National trends in the mortality of children with sickle cell disease, 1968 through 1992. *American Journal of Public Health*, *87*(8), 1317–1322.

Denninger, J. W., & Marletta, M. A. (1999). Guanylate cyclase and the NO/cGMP signaling pathway. *Biochimica et Biophysica Acta, 1411*(2–3), 334–350.

DeSimone, J., Biel, S. I., & Heller, P. (1978). Stimulation of fetal hemoglobin synthesis in baboons by hemolysis and hypoxia. *Proceedings of the National Academy of Sciences of the United States of America, 75*(6), 2937–2940.

Dover, G. J. (1990). Pharmacologic manipulation of fetal hemoglobin: Update on clinical trials with hydroxyurea. *Annals of the New York Academy of Sciences*, *612*, 184–190.

Dover, G. J. (1998). Hemoglobin switching protocols in thalassemia. Experience with sodium phenylbutyrate and hydroxyurea. *Annals of the New York Academy of Sciences*, *850*, 80–86.

Dover, G. J., Boyer, S., Charache, S., & Heintzelman, K. (1978). Individual variation in the production and survival of F cells in sickle-cell disease. *New England Journal of Medicine*, *299*(26), 1428–1435.

Dover, G. J., & Boyer, S. H. (1987). Fetal hemoglobin-containing cells have the same mean corpuscular hemoglobin as cells without fetal hemoglobin: A reciprocal relationship between gamma- and beta-globin gene expression in normal subjects and in those with high fetal hemoglobin production. *Blood*, *69*(4), 1109–1113.

Dover, G. J., Boyer, S. H., & Pembrey, M. E. (1981). F-cell production in sickle cell anemia: Regulation by genes linked to beta-hemoglobin locus. *Science*, *211*, 1441.

Dover, G. J., Brusilow, S., & Charache, S. (1994). Induction of fetal hemoglobin production in subjects with sickle cell anemia by oral sodium phenylbutyrate. *Blood*, *84*(1), 339–343.

Dover, G. J., Charache, S., & Boyer, S. H. (1984). Increasing fetal hemoglobin in sickle cell disease: Comparisons of 5-azacytidine (subcutaneous or oral) with hydroxyurea. *Transactions of the Association of American Physicians*, *97*, 140–145.

Dover, G. J., Smith, K. D., Chang, Y. C., Purvis, S., Mays, A., Meyers, D. A., . . . Serjeant, G. (1992). Fetal hemoglobin levels in sickle cell disease and normal individuals are partially controlled by an X-linked gene located at Xp22.2. *Blood*, *80*(3), 816–824.

Fibach, E., Prasanna, P., Rodgers, G. P., & Samid, D. (1993). Enhanced fetal hemoglobin production by phenylacetate and 4-phenylbutyrate in erythroid precursors derived from normal donors and patients with sickle cell anemia and beta-thalassemia. *Blood*, *82*(7), 2203–2209.

Forget, B. G. (1998). Molecular basis of hereditary persistence of fetal hemoglobin. *Annals of the New York Academy of Sciences*, *850*, 38–44.

Galarneau, G., Palmer, C. D., Sankaran, V. G., Orkin, S. H., Hirschhorn, J. N., & Lettre, G. (2010). Fine-mapping at three loci known to affect fetal hemoglobin levels explains additional genetic variation. *Nature Genetics*, *42*(12), 1049–1051.

Garner, C. P., Mitchell, J., Hatzis, T., Reittie, J., Farrall, M., & Thein, S. L. (1998). Haplotype mapping of a major quantitative-trait locus for fetal hemoglobin production, on chromosome 6q23. *American Journal of Human Genetics*, *62*(6), 1468–1474.

Garner, C. P., Tatu, T., Best, S., Creary, L., & Thein, S. L. (2002). Evidence of genetic interaction between the beta-globin complex and chromosome 8q in the expression of fetal hemoglobin. *American Journal of Human Genetics*, *70*(3), 793–799.

Gold, J. I., Mahrer, N. E., Treadwell, M., Weissman, L., & Vichinsky, E. (2008). Psychosocial and behavioral outcomes in children with sickle cell disease and their healthy siblings. *Journal of Behavioral Medicine*, *31*(6), 506–516.

Graff, J. C., Hankins, J. S., Hardy, B. T., Hall, H. R., Roberts, R. J., & Neely-Barnes, S. L. (2010). Exploring parent-sibling communication in families of children with sickle cell disease. *Issues in Comprehensive Pediatric Nursing*, *33*(2), 101–123.

Gribbons, D., Zahr, L. K., & Opas, S. R. (1995). Nursing management of children with sickle cell disease: An update. *Journal of Pediatric Nursing*, *10*(4), 232–242.

Haynes, J., Jr., Baliga, B. S., Obiako, B., Ofori-Acquah, S., & Pace, B. (2004). Zileuton induces hemoglobin F synthesis in erythroid progenitors: Role of the L-arginine-nitric oxide signaling pathway. *Blood*, *103*(10), 3945–3950.

Higgs, D. R., & Wood, W. G. (2008). Genetic complexity in sickle cell disease. *Proceedings of the National Academy of Sciences of the United States of America*, *105*(13), 11595–11596.

Hoppe, C., Styles, L., & Vichinsky, E. (1998). The natural history of sickle cell disease. *Current Opinion in Pediatrics*, *10*(1), 49–52.

Ikuta, T., Ausenda, S., & Cappellini, M. D. (2001). Mechanism for fetal globin gene expression: Role of the soluble guanylate cyclase-cGMP-dependent protein kinase pathway. *Proceedings of the National Academy of Sciences of the United States of America*, *98*(4), 1847–1852.

Inati, A. (2009). Recent advances in improving the management of sickle cell disease. *Blood Reviews*, *23*(Suppl. 1), S9–S13.

Inati, A., Chabtini, L., Mounayar, M., & Taher, A. (2009). Current understanding in the management of sickle cell disease. *Hemoglobin*, *33*(S1), S107–S115.

Jacob, E., Miaskowski, C., Savedra, M., Beyer, J. E., Treadwell, M., & Styles, L. (2006). Changes in sleep, food intake, and activity levels during acute painful episodes in children with sickle cell disease. *Journal of Pediatric Nursing*, *21*(1), 23–34.

Kaye, C. I. (2006). Newborn screening fact sheets. *Pediatrics*, *118*(3), e957–e963.

Keefer, J. R., Schneidereith, T. A., Mays, A., Purvis, S. H., Dover, G. J., & Smith, K. D. (2006).Role of cyclic nucleotides in fetal hemoglobin induction in cultured $CD34^+$ Cells. *Experimental Hematology*, *34*(9), 1150–1160.

Kenner, C., Gallo, A. M., & Bryant, K. D. (2005). Promoting children's health through understanding of genetics and genomics. *Journal of Nursing Scholarship*, *37*(4), 308–314.

Kladny, B., Williams, A., Gupta, A., Gettig, E. A., & Krishnamurti, L. (2011). Genetic counseling following the detection of hemoglobinopathy trait on the newborn screen is well received, improves knowledge, and relieves anxiety. *Genetics in Medicine*, *13*(7), 658–666.

Labie, D., Dunda-Belkhodja, O., Rouabhi, F., Pagnier, J., Ragusa, A., & Nagel, R. L. (1985). The -158 site 5′ to the G gamma gene and G gamma expression. *Blood*, *66*(6), 1463–1465.

Lanzkron, S., Haywood, C., Jr., Segal, J. B., & Dover, G. J. (2006). Hospitalization rates and costs of care of patients with sickle-cell anemia in the state of Maryland in the era of hydroxyurea. *American Journal of Hematology*, *81*(12), 927–932.

Lavelle, D., DeSimone, J., Heller, P., Zwiers, D., & Hall, L. (1986). On the mechanism of Hb F elevations in the baboon by erythropoietic stress and pharmacologic manipulation. *Blood*, *67*(4), 1083–1089.

Lettre, G., Sankaran, V. G., Bezerra, M. A. C., Araújo, A. S., Uda, M., Sanna, S., . . . Orkin, S. H. (2008). DNA polymorphisms at the BCL11A, HBS1L-MYB, and beta-globin loci associate with fetal hemoglobin levels and pain crises in sickle cell disease. *Proceedings of the National Academy of Sciences of the United States of America*, *105*(33), 11869–11874.

Lutz, M. J., Barakat, L. P., Smith-Whitley, K., & Ohene-Frempong, K. (2004). Psychological adjustment of children with sickle cell disease: Family functioning and coping. *Rehabilitation Psychology*, *49*(3), 224–232.

Ma, Q., Wyszynski, D. F., Farrell, J. J., Kutlar, A., Farrer, L. A., Baldwin, C. T., & Steinberg, M. H. (2007). Fetal hemoglobin in sickle cell anemia: Genetic determinants of response to hydroxyurea. *The Pharmacogenomics Journal*, *7*(6), 386–394.

Menzel, S., Garner, C., Gut, I., Matsuda, F., Yamaguchi, M., Heath, S., . . . Thein, S. L. (2007). A QTL influencing F cell production maps to a gene encoding a zinc-finger protein on chromosome 2p15. *Nature Genetics*, *39*(10), 1197–1199.

Mitchell, M. J., Lemanek, K., Palermo, T. M., Crosby, L. E., Nichols, A., & Powers, S. W. (2007). Parent Perspectives on pain management, coping, and family functioning in pediatric sickle cell disease. *Clinical Pediatrics*, *46*(4), 311–319.

Miyoshi, K., Kaneto, Y., Kawai, H., Ohchi, H., Niki, S., Hasegawa, K., . . . Yamano, T. (1988). X-linked dominant control of F-cells in normal adult life: Characterization of the Swiss type as hereditary persistence of fetal hemoglobin regulated dominantly by gene(s) on X chromosome. *Blood*, *72*(6), 1854–1860.

Moore, R. D., Charache, S., Terrin, M. L., Barton, F. B., & Ballas, S. K. (2000). Cost-effectiveness of hydroxyurea in Sickle Cell Anemia. Investigators of the Multicenter Study of Hydroxyurea in Sickle Cell Anemia. *American Journal of Hematology*, *64*(1), 26–31.

Moran, C. J., Siegel, M. J., & DeBaun, M. R. (1998). Sickle cell disease: Imaging of cerebrovascular complications. *Radiology*, *206*(2), 311–321.

Mvundura, M., Amendah, D., Kavanagh, P. L., Sprinz, P. G., & Grosse, S. D. (2009). Health care utilization and expenditures for privately and publicly insured children with sickle cell disease in the United States. *Pediatric Blood & Cancer*, *53*(4), 642–646.

Nagel, R. L. (1990). Innate resistance to malaria: The intraerythrocytic cycle. *Blood Cells*, *16*(2–3), 321–329.

Nagel, R. L., Fabry, M. E., Pagnier, J., Zohoun, I., Wajcman, H., Baudin, V., & Labie, D. (1985). Hematologically and genetically distinct forms of sickle cell anemia in Africa: The Senegal type and the Benin type. *New England Journal of Medicine*, *312*(14), 880–884.

National Heart Lung and Blood Institute. (2009). *Disease and conditions index. Sickle cell anemia: Who is at risk?* Bethesda, MD: U.S. Department of Health and Human Services.

Nienhuis, A. W., Ley, T. J., Humphries, R. K., Young, N. S., & Dover, G. J. (1990). Pharmacological manipulation of fetal hemoglobin synthesis in patients with severe β-thalassemia. *Annals of the New York Academy of Sciences*, *612*, 198–211.

Nietert, P. J., Silverstein, M. D., & Abboud, M. R. (2002). Sickle cell anemia: Epidemiology and cost of illness. *Pharmacoeconomics*, *20*(6), 357–366.

Niscola, P., Sorrentino, F., Scaramucci, L., de Fabritiis, P., & Cianciulli, P. (2009). Pain syndromes in sickle cell disease: An update. *Pain Medicine*, *10*(3), 470–480.

Noll, R. B., Vannatta, K., Koontz, K., Kalinyak, K., Bulowski, W. M., & Davies, W. H. (1996). Peer relationships and emotional well-being of youngsters with sickle cell disease. *Child Development*, *67*(2), 423–436.

Pagnier, J., Mears, J. G., Dunda-Belkhodja, O., Schaefer-Rego, K. E., Beldjord, C., Nagel, R. L., & Labie, D. (1984). Evidence for the multicentric origin of the sickle cell hemoglobin gene in Africa. *Proceedings of the National Academy of Sciences of the United States of America*, *81*(6), 1771–1773.

Papayannopoulou, T., Torrealba de Ron, A., Veith, R., Knitter, G., & Stamatoyannopoulos, G. (1984). Arabinosylcytosine induces fetal hemoglobin in baboons by perturbing erythroid cell differentiation kinetics. *Science*, *224*(4649), 617–618.

Pauling, L., Itano, H., Singer, S. J., & Wells, I. C. (1949). Sickle cell anaemia: A molecular disease. *Science*, *110*(2865), 543–548.

Pearson, H. A. (1996). Pharmacologic manipulation of fetal hemoglobin levels in sickle cell diseases and thalassemia: Promise and reality. *Advances in Pediatrics*, *43*, 309–334.

Peterson, C. C., Palermo, T. M., Swift, E., Beebe, A., & Drotar, D. (2005). Assessment of psycho-educational needs in a clinical sample of children with sickle cell disease. *Children's Health Care*, *34*(2), 133–148.

Platt, O. S., Brambilla, D. J., Rosse, W. F., Milner, P. F., Castro, O., Steinberg, M. H., & Klug, P. P. (1994). Mortality in sickle cell disease. Life expectancy and risk factors for early death. *The New England Journal of Medicine*, *330*(23), 1639–1644.

Powars, D. R. (1990). Sickle cell anemia and major organ failure. *Hemoglobin*, *14*(6), 573–598.

Rochette, J., Craig, J. E., & Thein, S. L. (1994). Fetal hemoglobin levels in adults. *Blood Reviews*, *8*(4), 213–224.

Roth, E. F., Friedman, M., Ueda, Y., Tellez, I., Trager, W., & Nagel, R. (1978). Sickling rates of human AS red cells infected in vitro with Plasmodium falciparum malaria. *Science*, *202*(4368), 650–652.

Rowley, P. T., Loader, S., Sutera, C. J., & Kozyra, A. (1995). Prenatal genetic counseling for hemoglobinopathy carriers: A comparison of primary providers of prenatal care and professional genetic counselors. *American Journal of Human Genetics*, *56*(3), 769–776.

Sankaran, V. G., Menne, T. F., Šćepanović, D., Vergilio, J. A., Ji, P., Kim, J., . . . Losdish, H. F. (2011). MicroRNA-15a and -16-1 act via MYB to elevate fetal hemoglobin expression in human trisomy 13. *Proceedings of the National Academy of Sciences of the United States of America*, *108*(4), 1519–1524.

Sankaran, V. G., Menne, T. F., Xu, J., Akie, T. E., Lettre, G., Van Handel, B., . . . Orkin, S. H. (2008). Human fetal hemoglobin expression is regulated by the developmental stage-specific repressor BCL11A. *Science*, *322*(5909), 1839–1842.

Sankaran, V. G., & Nathan, D. G. (2010). Thalassemia: An overview of 50 years of clinical research. *Hematology/Oncology Clinics of North America*, *24*(6), 1005–1020.

Shapiro, B. S., Dinges, D. F., Orne, E. C., & Bauer, N. (1995). Home management of sickle cell-related pain in children and adolescents: Natural history and impact on school attendance. *Pain*, *61*(1), 139–144.

Stamatoyannopoulos, G., & Nienhuis, A. W. (1994). Hemoglobin switching. In G. Stamatoyannopoulos, A. Nienhuis, P. Majerus, & H. Varmus (Eds.), *The molecular basis of blood disease* (2nd ed., pp. 107–156). Philadelphia, PA: W. B. Saunders Co.

Steinberg, M. H. (2008). Sickle cell anemia, the first molecular disease: Overview of the molecular etiology, pathophysiology, and therapeutic approaches. *The Scientific World Journal*, *8*, 1295–1324.

Stine, O. C., Dover, G. J., Zhu, D., & Smith, K. D. (1992). The evolution of two west African populations. *Journal of Molecular Evolution*, *34*(4), 336–344.

Strouse, J. J., Hulbert, M. L., DeBaun, M. R., Jordan, L. C., & Casella, J. F. (2006). Primary hemorrhagic stroke in children with sickle cell disease is associated with recent transfusion and use of corticosteroids. *Pediatrics*, *118*(5), 1916–1924.

Thein, S. L., & Craig, J. E. (1998). Genetics of Hb F/F cell variants in adults and heterocellular hereditary persistence of fetal hemoglobin. *Hemoglobin*, 22(5–6), 401–414.

Uda, M., Galanello, R., Sanna, S., Lettre, G., Sankaran, V. G., Chen, W., . . . Cao, A. (2008). Genome-wide association study shows BCL11A associated with persistent fetal hemoglobin and amelioration of the phenotype of beta-thalassemia. *Proceedings of the National Academy of Sciences of the United States of America*, *105*(5), 1620–1625.

Vichinsky, E. P., MacKlin, E. A., Wayne, J. S., Lorey, F., & Olivieri, N. F. (2005). Changes in the epidemiology of thalassemia in North America: A new minority disease. *Pediatrics*, *116*(6), e818–e825.

Walters, M. C., Storb, R., Patience, M., Leisenring, W., Taylor, T., Sanders, J. E., . . . Sullivan, K. M. (2000). Impact of bone marrow transplantation for symptomatic sickle cell disease: An interim report. Multicenter investigation of the bone marrow transplantation for sickle cell disease. *Blood*, *95*(6), 1918–1924.

Watson, J., Stahman, A. W., & Bilello, F. P. (1948). The significance of the paucity of sickle cells in newborn Negro infants. *American Journal of Medical Science*, *215*(4), 419–423.

Weatherall, D. J., & Clegg, J. B. (1969). The control of human hemoglobin synthesis and function in health and disease. *Progress in Hematology*, *6*, 231–304.

Winrow, N. (1998). *The relationship between cerebral blood flow velocity and neuropsychology.* Paper presented at the Twenty-fourth Annual Alan Coopersmith Professorship, Johns Hopkins University.

World Health Organization. (2011). Sickle-cell disease and other haemoglobin disorders. *World Health Organization Fact Sheet, 2011*. Retrieved from http://www.who.int/mediacentre/factsheets/fs308/en/index.html

Xu, J., Sankaran, V. G., Ni, M., Menne, T. F., Puram, R. V., Kim, W., & Orkin, S. H. (2010). Transcriptional silencing of γ-globin by BCL11A involves long-range interactions and cooperation with SOX6. *Genes & Development*, *24*(8), 783–798.

Yanni, E., Grosse, S. D., Yang, Q. H., & Olney, R. S. (2009). Trends in pediatric sickle cell disease-related mortality in the United States, 1983–2002. *Journal of Pediatrics*, *154*(4), 541–545.

Zago, M. A., Wood, W. G., Clegg, J. B., Weatherall, D. J., O'Sullivan, M., & Gunson, H. (1979). Genetic control of F cells in human adults. *Blood*, *53*(5), 977–986.

Zimmerman, S. A., Ware, R. E., & Kinney, T. R. (1997). Gaining ground in the fight against sickle cell disease. *Contemporary Pediatrics*, *14*(10), 154–177.

Index

Note: Page references followed by "f" and "t" denote figures and tables, respectively.